Also in the Variorum Collected Studies Series

PAUL KUNITZSCH
The Arabs and the Stars: Texts and Traditions on the Fixed Stars and Their Influence in Medieval Europe

MENSO FOLKERTS
Essays on Early Medieval Mathematics: The Latin Tradition

RICHARD LORCH
Arabic Mathematical Sciences: Instruments, Texts and Transmission

GEORGE MOLLAND
Mathematics and the Medieval Ancestry of Physics

WESLEY M. STEVENS
Cycles of Time and Scientific Learning in Medieval Europe

BRUCE S. EASTWOOD
The Revival of Planetary Astronomy in Carolingian and Post-Carolingian Europe

BRUCE S. EASTWOOD
Astronomy and Optics from Pliny to Descartes:
Texts, Diagrams and Conceptual Studies

EDWARD S. KENNEDY
Astronomy and Astrology in the Medieval Islamic World

DANIEL MARTIN VARISCO
Medieval Folk Astronomy and Agriculture in Arabia and the Yemen

DAVID A. KING
Astronomy in the Service of Islam

A. I. SABRA
Optics, Astronomy and Logic: Studies in Arabic Science and Philosophy

JULIO SAMSÓ
Islamic Astronomy and Medieval Spain

CURTIS WILSON
Astronomy from Kepler to Newton

RAMZI BAALBAKI
Grammarians and Grammatical Theory in the Medieval Arabic Tradition

VARIORUM COLLECTED STUDIES SERIES

Stars and Numbers

Paul Kunitzsch

Stars and Numbers

Astronomy and Mathematics in the Medieval Arab and Western Worlds

Routledge
Taylor & Francis Group
LONDON AND NEW YORK

First published 2004 by Ashgate Publishing

2 Park Square, Milton Park, Abingdon, Oxfordshire OX14 4RN
711 Third Avenue, New York, NY 10017

Routledge is an imprint of the Taylor & Francis Group, an informa business

First issued in paperback 2018

ISBN 978-0-86078-968-0 (hbk)
ISBN 978-1-138-37568-0 (pbk)

British Library Cataloguing-in-Publication Data
Kunitzsch, Paul
Stars and Numbers: Astronomy and Mathematics in the Medieval Arab and Western Worlds. – (Variorum Collected Studies Series: CS791)
1. Ptolemy, 2nd Cent. 2. Astronomy – History – To 1500. 3. Astronomy, Medieval. 4. Astronomy, Arab. 5. Mathematics, Medieval. 6. Astrolabes – History – To 1500.
I. Title
520. 9'02

US Library of Congress Cataloging-in-Publication Data
Kunitzsch, Paul.
Stars and Numbers: Astronomy and Mathematics in the Medieval Arab and Western Worlds / Paul Kunitzsch.
p. cm. – (Variorum Collected Studies Series: CS791).
Includes bibliographical references and indexes.
1. Astronomy – Mathematics. 2. Astronomy, Medieval. 3. Astronomy, Arab – History. 4. Astronomy – Europe – History. I. Title. II. Collected Studies: CS791.

QB47.K86 2004
520'.9'02–dc22 2003063852

VARIORUM COLLECTED STUDIES SERIES CS791

Contents

This volume contains xiv + 340 pages

Publisher's Note

The articles in this volume, as in all others in the Variorum Collected Studies Series, have not been given a new, continuous pagination. In order to avoid confusion, and to facilitate their use where these same studies have been referred to elsewhere, the original pagination has been maintained wherever possible.

Each article has been given a Roman numeral in order of appearance, as listed in the Contents. This number is repeated on each page and quoted in the index entries.

Preface

This is a second collection of my articles on Arabic-Islamic astronomy and its reception in medieval Europe – the first being *The Arabs and the Stars* (CS 307, 1989) – this time concluded by four articles of mathematical interest.

General remarks on the importance of Arabic-Islamic astronomy, its growth based on translations mostly of Greek, but also of some Middle-Persian and Indian works, and its outstanding role for the development of this science in medieval Europe have been made in the Preface to the first volume and do not need to be repeated here. The articles presented here follow the same line as those in the first volume and bring forward more pieces of evidence for several items in this historical process. The Arabic material discussed in these articles mostly stems from the early period of the development of Arabic-Islamic astronomy, i.e. up to ca. 1000 AD, whereas the Latin material belongs to the first stage of Western contact with Arabic science at the end of the tenth century and to the peak of the Arabic-Latin translations in Spain in the twelfth century.

For practical reasons the articles in this volume have been arranged in four subgroups. The first subgroup, "Ptolemy in the Arabic-Latin Tradition", contains contributions on the Arabic reception of Ptolemy's *Almagest* (III, IV, V), Gerard of Cremona's translation from Arabic into Latin (I, II), the transmission of Ptolemy's *Planisphaerium* (VI, VII, VIII) and general observations on Gerard's and other Latin translations from Arabic (IX, X). Here it may be added that my edition of Ptolemy's star catalogue mentioned in the Preface of the 1989 volume, p. xiv, has been completed in the meantime: Claudius Ptolemäus, *Der Sternkatalog des Almagest. Die arabisch-mittelalterliche Tradition*, vol. I: *Die arabischen Übersetzungen*, 1986; vol. II: *Die lateinische Übersetzung Gerhards von Cremona*, 1990; vol. III: *Gesamtkonkordanz der Sternkoordinaten*, 1991, published by Otto Harrassowitz Verlag, Wiesbaden, Germany. As for the *Planisphaerium*, there may be further mentioned P. Kunitzsch – R. Lorch, *Maslama's Notes on Ptolemy's Planisphaerium and Related Texts*, München: C. H. Beck, 1994 (Bayerische Akademie der Wissenschaften, Philos.-hist. Klasse, Sitzungsberichte 1994:2), where Maslama's notes are edited in Arabic and in three Latin translations.

The second subgroup, "Arabic Astronomy", presents the edition of a chapter on the fixed stars in an early Arabic translation from Middle-Persian (XI), two articles centered on the Persian astronomer al-Ṣūfī, 903–986 (XII, XIII), and the description of an Islamic celestial globe in the Schmidt collection in Vienna, which is a replica of the famous globe in Dresden of Muḥammad ibn Muʾayyad al-ʿUrḍī and of which the date, as here demonstrated, can be fixed at ca. 1305 AD (XIV; only the English version of the article is reproduced, translated by C. Embleton; the original German version preceding the English text in the journal is not given here). It had also been planned to include here (as in the 1989 volume, items XVIII–XX) some entries from the *Encyclopaedia of Islam*, new edition, *viz.* the major articles *Minṭaḳat al-Burūdj* ("the Zodiac") and *al-Nudjūm* ("the Stars") and the shorter *al-Shiʿrā* ("Sirius"), for which the publisher, Brill in Leiden, had kindly provided the permission for reproduction. But in the end, for lack of space, they could not be included. In any case, it may be much easier for interested readers to find copies of the *Encyclopaedia* in scholarly libraries than to locate the various sources of the other items contained in the present volume.

The third subgroup, "Arabic Astronomy in the West", comprises eleven articles, most of which treat details from the first period of Western contact with Arabic astronomy, particularly the astrolabe. XV is a general survey of the situation at the time of Gerbert of Aurillac. XVI describes the drawings, in MS Paris, BnF lat. 7412, of an Andalusian (Arabic) astrolabe, which may be regarded as the earliest Andalusian astrolabe so far known. XVII discusses the origin of the table of the seven *climata* in the "old corpus" of Latin treatises on the astrolabe. XVIII analyses the stars found on a Latin astrolabe perhaps dating from around 1000 AD, whose authenticity, however, is still debated. (More information on this instrument, formerly in the possession of the late Marcel Destombes, Paris, and not quite aptly called by him "the Carolingian Astrolabe", is to be found in *Physis*, N.S. 32, Fasc. 2–3, 1995, containing the Proceedings of a special session on the instrument at the XIXth International Congress of History of Science, Saragossa, 22–29 August, 1993, ed. by W. Stevens.) XIX and XX discuss details on the stars mentioned in the oldest Latin astrolabe texts. XXI brings the explanation for two misleading terms in the treatise on the astrolabe by Ascelinus of Augsburg (early eleventh century). XXII presents an anonymous Latin text on six kinds of astrolabe, certainly based on Arabic material. XXIII shows a table of astrolabe stars of mixed character (13th century), oddly added in a manuscript to the treatise on the construction of the astrolabe of

Hermann the Lame (d. 1054). XXIV describes a document of more recent times (1681–83), a celestial globe of V. Coronelli carrying the constellation names in Arabic, in Arabic script; the names appear to be derived from the Book on the Constellations of al-Ṣūfī (on him, cf. above, items XII–XIII, and item XI in the 1989 volume). XXV discusses a set of star names of unknown origin and meaning appearing for the first time in a modern star atlas (A. Bečvář, 1951), which look as if intended to imitate true Arabic star names.

The last four items treat details in the transmission of Euclid's *Elements* (XXVI, XXVII), the correspondence of the letters used in geometrical diagrams between Greek, Arabic and Latin texts (XXVIII) and the transmission of our "Arabic" numerals from India to the Arabs and from them further on to Europe (XXIX; this article is based on a lecture given in a conference in the MIT, Cambridge, Mass., in 1998). With regard to the last item it is worth mentioning a recent article by C. Burnett, Indian Numerals in the Mediterranean Basin in the Twelfth Century, with Special Reference to the "Eastern Forms", in: *From China to Paris: 2000 Years Transmission of Mathematical Ideas*, ed. Y. Dold-Samplonius, J. W. Dauben, M. Folkerts and B. van Dalen, Stuttgart: Franz Steiner Verlag, 2002, pp. 237–288, where especially the occurrence of the Eastern Arabic forms of the numerals in Latin and Byzantine manuscripts is discussed.

At the end it may be mentioned that the *Liber de stellis beibeniis* discussed in items XII–XIV of the 1989 volume has recently been edited, in two Arabic versions and the Latin translation together with the text of a second Latin version as edited by C. Burnett and D. Pingree (1997): *Liber de stellis beibeniis*, cura et studio Paul Kunitzsch, in: *Hermetis Trismegisti Astrologica et Divinatoria*, Turnhout: Brepols, 2001, pp. 9–107, followed by its Hebrew version, *Sefer Hermes*, cura et studio Fabrizio Lelli, *ibid.*, pp. 109–137 (Corpus Christianorum, Continuatio Mediaevalis CXLIV C = Hermes Latinus t. IV, pars IV).

It is a pleasure to express my sincere thanks to Prof. M. Folkerts for help in typesetting the Preface and the Indexes, to the publishers of the original papers for giving permission to reproduce these papers and to Dr. J. Smedley of Ashgate Publishing Ltd. for including these texts in another volume of the Variorum Reprint Series.

Munich, PAUL KUNITZSCH
July 2003

Acknowledgements

Grateful acknowledgement is made to the following for kindly permitting the reproduction of articles originally published by them: Biblioteca Statale di Cremona (I); Peter Lang Verlag, Frankfurt am Main (II); Franz Steiner Verlag Wiesbaden GmbH, Stuttgart (III, XXVI); Brepols Publishers, Turnhout (IV); Prof. F. Sezgin, Institut für Geschichte der Arabisch-Islamischen Wissenschaften, Frankfurt am Main (V–VII, XI, XIII, XVI, XIX, XXIV, XXVIII); Blackwell Publishing Ltd., Oxford (VIII, XXII); WILEY-VCH Verlagsgesellschaft, Weinheim (IX); Harrassowitz Verlag, Wiesbaden (X); Institut Français de Recherche en Iran (IFRI), Téhéran (XII); Internationale Coronelli-Gesellschaft, Vienna (XIV); Société "Lettres, Sciences et Arts" La Haute-Auvergne, Aurillac (XV); Georg Olms Verlag AG, Hildesheim (XVII); The Dean, Facultat de Filologia, Universitat de Barcelona (XVIII, XX); Dr. Elly Dekker, Linschoten, The Netherlands (XVIII); Taylor & Francis Group plc, Milton Park, Abingdon [http://www.tandf.co.uk] (XXI); Prof. R. Halleux, editor, *Archives internationales d'histoire des sciences*, Liège (XXIII); Prof. M. Folkerts, editor, *Algorismus*, Munich (XXV); Rodopi Publishers, Amsterdam (XXVII); The MIT Press, Cambridge, Massachusetts (XXIX).

I

GERARD'S TRANSLATIONS OF ASTRONOMICAL TEXTS, ESPECIALLY THE *ALMAGEST*

To speak about Gerard of Cremona and his works is a risky undertaking still in our days. As with most of the ancient and medieval authors, we know very little of his life and his professional career. For Gerard, at least there exist a *Vita*[1] and a list of 71 titles of works said to have been translated by him from Arabic into Latin.[2] This list is said to have been appended by his *socii*, i.e. by students or collaborators of his, to one of the last works translated by him, Galen's *Tegni* with the commentary of ʿAlī ibn Riḍwān, where as the *Vita* appears to have been composed separately, after the bibliography, under unknown circumstances. Mostly, but not always, the *Vita* and the bibliography are transmitted together in the manuscripts. On the other hand, normally his works were not signed by him, the manuscripts – with only a few exceptions – do not mention his name as the translator, or his name is given to texts which most certainly were not translated by him.

Modern research is still far behind in the systematic exploration of Gerard's works according to the standards of historical and philological methods of our times. For these shortcomings a number of reasons can be named, such as the terribly great number of manuscript sources many of which are insufficiently catalogued so that every researcher meets with enormous difficulties in locating the texts and obtaining copies of them. Further, in order to edit and analyse Gerard's works appropriately, three prerequisits must be assembled in scholars tackling this task: good knowledge of Latin, especially the current medieval usage of it; secondly, knowledge of the sciences treated in the respective works; and thirdly, and no less important, good knowledge of Arabic in order to determine what is « correct » or not, and what is really intended in Gerard's text.

Of Gerard's astronomical and astrological translations (astrological will here be included together with the astronomical works in the same group, since in antiquity and in medieval times the two together formed one unit), perhaps all exist in old printed editions from the 16th century or so. But these old editions are very unreliable according to our standards of editing works

1. I follow the edition by K. SUDHOFF, *Die kurze « Vita » und das Verzeichnis der Arbeiten Gerhards von Cremona*, in « Archiv für Geschichte der Medizin », 8 (1914), 73-82; other editions were made by B. BONCOMPAGNI, *Della vita e delle opere di Gherardo Cremonese...*, Rome, 1851, and by F. WÜSTENFELD, *Die Übersetzungen Arabischer Werke in das Lateinische seit dem XI. Jahrhundert*, Göttingen, 1877, pp. 57f.

2. Cf. the discussions of the list in G. SARTON, *Introduction to the History of Science*, vol. II, part. I (Washington, 1931; repr. 1950), pp. 338-44; R. LEMAY, article *Gerard of Cremona*, in *Dictionary of Scientific Biography*, XV (New York, 1978), pp. 173-92, esp. 176-89.

because they have been modified and « corrected » according to contemporary stages of knowledge. Therefore, in order to establish Gerard's true texts, for all of them modern editions are needed. There exist already, of some of those works, modern editions since the middle of, or late 19th century. But also these cannot all be regarded as sufficient according to our present standards because they often were made from one, or a few manuscripts just being within the reach of the editor while many more exist and the use of all of them might lead to important modifications in the edition.

The *Vita* tells us that Gerard died in 1187, in Toledo as can be inferred from a poetical eulogy transmitted together with the *Vita*, at an age of 73 years. From this it can be deduced that he was born in 1114, in Cremona as we know. The *Vita* further tells us that Gerard, inspired by love for the *Almagest – amore almagesti –* the text of which did not exist in the Latin world in his time, moved to Spain in order to find Arabic copies of it and to communicate it to the West in Latin. If we make the assumption that he went to Spain at an age of about thirty years, that would mean that he came to Toledo roughly in 1145 and that he, then, spent there roughly fourty years dedicated to the work of translation from Arabic into Latin.

This relatively long period makes it completely possible that he really translated the 71 works listed in the old bibliography by his companions, the *socii*, and perhaps a number of further texts which are suspected to be due to him and which are not mentioned in the *socii* list.

As is clearly indicated in the *Vita*, and as is corroborated by the state of transmission, Gerard's main objective in Spain was the translation of the *Almagest*, that is the great astronomical handbook of Claudius Ptolemaeus of Alexandria written around A.D. 150.

Through happy circumstances, there has survived the report of an English student of the sciences, Daniel of Morley, who was in Toledo some time before 1175 and who gives, in his *Philosophia*, a vivid description of a lecture of Gerard on the *Introductorium maius*, an astrological work by the ninth century Arabic astrologer Abū Maʿshar, and in the course of which Daniel began a rather frank dispute with Gerard; when naming Gerard, Daniel explicitly refers to his translation of the *Almagest*.[3]

From this report we learn that Gerard's work in Toledo, apart from the translations, also included teaching and – as in the present case – lecturing on scientific Arabic texts that are not among the titles translated by himself.

Further, the remark on Gerard's translation of the *Almagest* seems to underline that the translation of this outstanding work of astronomy conferred upon Gerard a special reputation already during his lifetime. For the succeeding centuries, and until the beginning of the « new astronomy » with Copernicus, Tycho Brahe, Galilei, and Kepler, Gerard's Latin version of the

3. See the recent edition by G. MAURACH, *Daniel von Morley, »Philosophia«*, in « Mittellateinisches Jahrbuch », 14 (1979), 204-55, esp. § 192-95.

Almagest remained the most renowned and most celebrate astronomical textbook in Europe. There has been made, around 1150, a direct Latin translation of the *Almagest* from the Greek, in Sicily, and probably by Hermannus de Carinthia.[4] But this never gained any importance or influence in Europe; it seems that works derived from Arabic sources were regarded as superior in authority to anything else in the middle ages and until Renaissance times when the Antiarabist movement came into being.

The astronomical and astrological works translated by Gerard from the Arabic include both texts of classical Greek authors transmitted in Arabic translations, and original writings of Arabic-Islamic authors. Also, among Gerard's translations in this field there are works which have been translated – perhaps before Gerard – by other Latin translators, too. The question of the selection of Arabic texts for translation has not been analysed so far. There are two elements that could have been of influence in this matter: first, the availability of texts in Muslim Spain; it can be observed, for example, that outstanding names of Eastern Arabic scholarship, such as the mathematician Abu l-Wafā', or the universal al-Bīrūnī, are not contained among the translated texts; it seems, therefore, that their works had not reached Muslim Spain in sufficient time to be included in the work of the 12th century translators. Second, the translators were dependent on what they could reach, in the reconquered Christian area of Spain, from the scientific manuscripts existing in the Muslim part of the land; R. Lemay and J. Vernet have quoted an interesting detail from a Spanish-Arabic handbook for market supervisors where a supervisor is formally advised to forbid the selling of scientific books to Christians.[5] That means, the translators were not completely free in the selection of works for translation; for a good part, they depended on the chance of obtaining, or not, the manuscripts they wanted.

Another point of importance, in our context, is: which was the standard of knowledge of the Arabic language on the side of the translators? Regarding Gerard specifically, we know that he came to Spain from Italy, that means that he could not have had any knowledge of Arabic in advance. He will have learnt the language in Toledo. But what sort of Arabic is it that he could have learnt there? The areas dominated by the Arabic language are known for their « diglossia », that is that there always existed – and still exist today – two languages side by side: the spoken colloquial Arabic generally used in

4. Cf. R. LEMAY, *Hermann de Carinthie. Auteur de la traduction « sicilienne » de l'Almageste à partir du grec (ca. 1150 A.D.)*, in *La diffusione delle scienze islamiche nel medio evo europeo*, Convegno internazionale (Roma, 2-4 ottobre 1984), Rome, 1987, pp. 428-84.

5. R. LEMAY, *Abu Ma'shar and Latin Aristotelianism in the Twelfth Century*, Beirut, 1962, p. 15, note 1; idem, *Gerard of Cremona* (*op. cit.* in note 2, above), p. 175; J. VERNET, *Les traductions scientifiques dans l'Espagne du Xe siècle*, in « Les Cahiers de Tunisie », 18 (1970), 47-59, esp. p. 58; cf. also J. A. GARCÍA-JUNCEDA y ALVAREZ-QUIÑONES, *El problema de las fuentes...*, in *Actas de las Jornadas de Cultura Arabe y Islamica* (1978), Madrid, 1981, pp. 319-25, esp. 324f and notes 10-1.

oral speach, and the language of writing which is strictly dominated by the rules of the *fuṣḥā*, the classical literary Arabic. All the current contacts that Gerard had with the local native speakers of Arabic will necessarily have been in the local colloquial, that is the Spanish-Arabic dialect.[6] The situation in his time, in this concern, may not have been different from our times which means that the average native speakers were not generally equally well versed with the classical, written Arabic as with the colloquial dialect, and much of what they could advise him did not correspond to the standards of literary Arabic – quite apart from the problems posed by the technical language of the various disciplines on which he was working. Clear evidence for the influence of the local dialect can be found in some of the transliterations of Arabic technical terms, or names, which he occasionally – though not often – uses in his translations. Another hint in this direction is again given by the above mentioned Daniel of Morley: in his notice on Gerard's translation of the *Almagest* he writes that Gerard here had the help of a mozarab (i.e., one of the Christians in Spain who had become arabicized under Arabic dominion) called *Galippus* (= Ġālib)[7] of whom it can be naturally expected that he was fully versant with the current local dialect but whose capacity in understanding or explaining written works of astronomy or other sciences must be somehow doubted.

There exist, until now, almost no detailed studies of Gerard's method of translation and of the style of his language. In 1959, a German scholar, Ilona Opelt, published a paper on this topic which, unfortunately, must be regarded as a complete failure.[8] She departed from the still unedited text of Gerard's translation of Aristotle's *De celo et mundo*[9] and, because of insufficient knowledge of Arabic, she compared the Latin directly to the Greek; in the few instances where she dared to include Arabic material (from manuscripts) she has committed gross errors. In sum, her paper is no real contribution to the problem. Partial, and reliable, analyses have been included by some editors in their editions of some of Gerard's works; of these could be mentioned P. L. Schoonheim's edition of the first book of Aristotle's *Meteorology*,[10] or my own study of the transmission of the *Almagest*[11] followed presently by the edition of the star catalogue.[12]

6. On this, see F. CORRIENTE, *A grammatical sketch of the Spanish Arabic dialect bundle*, Madrid, 1977.
7. *Op. cit.* above in note 3, § 192.
8. ILONA OPELT, *Zur Übersetzungstechnik des Gerhard von Cremona*, in « Glotta », 38 (1959), 135-70.
9. An edition, by the same I. Opelt, was published in 1971; see note 25, below.
10. P. L. SCHOONHEIM, *Aristoteles' Meteorologie in arabischer und lateinischer Uebersetzung*, Leiden 1978.
11. P. KUNITZSCH, *Der Almagest. Die Syntaxis Mathematica des Claudius Ptolemäus in arabisch-lateinischer Überlieferung*, Wiesbaden 1974.
12. CLAUDIUS PTOLEMÄUS, *Der Sternkatalog des Almagest. Die arabisch-mittelalterliche Tradition*, vol. I: *Die arabischen Übersetzungen*, ed. P. Kunitzsch, Wiesbaden 1986; vol. II: *Die lateinische Übersetzung Gerhards von Cremona*, ed. P. Kunitzsch, in the press; vol. III: *Gesamtkonkordanz der Sternkoordinaten*, by P. Kunitzsch, in preparation.

The outstanding feature of Gerard's translations, and perhaps the characteristic mark by which genuine works of his can be recognized, is the extreme literalness, to the extent that one even could say it is some sort of Arabic in Latin words. This also explains my earlier statement that everyone wishing to edit or analyse Gerard's works must at the same time be in good command of Arabic. The best way of editing his translations will always be to present both the Arabic source and the Latin version together. This has successfully been done by Schoonheim for the *Meteorology* (Book I),[13] and the same procedure has been followed in my present edition of Ptolemy's star catalogue.[14]

Let me add here another aspect of Gerard's ability as a translator, notably of classical texts transmitted by the Arabs: it can be nicely observed that he must have possessed a rather good education in classical Western traditions, for very often – though not always – he inserts the correct names of Greek authorities or other Greek names, and sometimes technical terms, which appeared in the Arabic sources in more or less corrupt transliterations. One of the mischiefs, in this field, is his famous *Abrachis*, for Hipparchus.[15] Another point which needs further study is how far he was versant with, and applied, astronomical and other scientific technical terms introduced and transmitted in pertaining texts of classical and late Roman authors.[16]

Speaking about Gerard's ability as a translator, also his technical competence in all those branches of science from which he translated texts must be considered. I have already mentioned the wide horizon of his education which no doubt has enabled him to master the difficulties of several disciplines. Another strong support against major slips may be seen in the literalness of his translations: as long as he closely stuck to the Arabic wording he at least could avoid committing new errors beyond those possibly contained in the Arabic texts themselves. But examples in the star catalogue show that this did not prevent him from producing some of the most ridiculous absurdities.[17] The problem here lay in wrong readings of Arabic words which lead to those absurd renderings from which neither his own critical mind nor the assistance of the native helper seem to have protected him. A more serious case will be mentioned soon when speaking about the *Almagest* in greater detail. It remains, however, uncertain whether such oddities must be attributed to a lack of competence on Gerard's side, or rather to the fully conscious method of translating which did not allow the translator to alter anything in his source.

13. See above, note 10.
14. *Op. cit.* in note 12, above, vols. I and II.
15. Cf. KUNITZSCH (*op. cit.* above in note 11), p. 160.
16. For the terminology of astronomy in classical Latin cf. the Paris dissertation by A. LE BOEUFFLE, *Le vocabulaire latin de l'astronomie* (thèse de Doctorat d'État ès Lettres, Sorbonne, le 11 décembre 1970; typoscript).
17. Cf. KUNITZSCH (*op. cit.* above in note 11), p. 107.

Let us now see which works of astronomy and astrology Gerard has translated. In this connexion we shall quickly pass over the treatises on the sphere by Autolycus, by Menelaus (whom Gerard calls *mileus*), and by Theodosius (on whose treatise dr. R. Lorch in Munich has presently started work), as well as Theodosius's *De habitationibus* which all are rather mathematical than astronomical works, and also the *Liber de crepusculis*, falling under physics rather than astronomy, which earlier had been regarded a work by the famous Arabic scientist Ibn al-Haytham but which A. I. Sabra has shown[18] to be by the much less known Spanish-Arabic astronomer Abū ʿAbdallāh Ibn Muʿādh (of whom Gerard, according to the findings of H. Hermelink,[19] has also translated the so-called *Tabule iahen*, a work of astronomical tables).

Astronomical works more properly are the following (I first name classical, then Arabic authors, finally I shall deal with the *Almagest* in greater detail).[20]

Greek texts

Liber introductorius ptolemei ad artem spericam, a text which shows much resemblance to the *Eisagoge* of Geminus; the Latin version has been edited (from two manuscripts of the 14th and 15th centuries, respectively) by K. Manitius as an appendix to his edition of the Greek text, in the Teubneriana, 1898.[21]

Liber esculegii, that is Hypsikles's *Anaphorikos*. Of this, the Greek and Arabic texts have been edited in 1966, by V. De Falco and M. Krause respectively (the last one posthumously), accompanied by German translations of both of them.[22] In the apparat of the Arabic text, and in the notes to its German translation, Krause has added several readings from the Latin version using the Paris Ms. B.N. lat. 9335 and an insufficient edition by K. Manitius (1888).[23]

For their astronomical relevance, I shall mention here also Aristotle's *Meteorology* (Book I edited recently, in Arabic and Latin together, by P. L. Schoonheim, 1978),[24] and Aristotle's *De celo et mundo* (edited by I. Opelt, in vol. V, part I of the new edition of the works of Albertus Magnus, 1971).[25]

18. In « Isis », 58 (1967), 77ff.
19. H. Hermelink, *Tabule Jahen*, in « Archive for History of Exact Sciences », 2 (1964), 108-12.
20. Reference is made in the following notes to F. J. Carmody, *Arabic Astronomical and Astrological Sciences in Latin Translation, a Critical Bibliography*, Berkeley-Los Angeles 1956.
21. Not mentioned by Carmody.
22. Hypsikles, *Die Aufgangszeiten der Gestirne*, Herausgegeben und übersetzt von V. De Falco und M. Krause (†) mit einer Einführung von O. Neugebauer, Göttingen 1966.
23. K. Manitius, *Des Hypsikles Schrift Anaphorikos...*, in « Programm des Gymnasium zum heiligen Kreuz in Dresden », 1. Abt., Dresden, 1888.
24. *Op. cit.* above in note 10.
25. Alberti Magni, *Opera omnia*, t. V, pars 1: *De caelo et mundo* (ed. P. Hossfeld), Münster, 1971; Gerard's translation, edited by I. Opelt, is here printed in the lower part of the pages of the volume.

Arabic texts

The oldest author of the Islamic Orient to be named here is Māshā'allāh, a Jew with a non-Islamic name, one of the early astronomers and astrologers of Baghdad, at the court of the caliphs (2nd half of the 8th into early 9th century). Gerard calls him *Messahala*, the *socii* list gives the title of the work as *De orbe*. There exist two old editions of 1504 and 1546, but no modern edition. It is also cited in Albertus Magnus's *Speculum astronomie* (chapter 2). The corresponding Arabic text has not yet been found.[26]

Chronologically the next author is al-Farghānī, *Alfraganus* in Latin (Baghdad, 9th century), whose *Astronomy*[27] has been translated twice, first in 1135 by Johannes Hispalensis, and some time later by Gerard.[28] It is a simplified survey of the teachings of the *Almagest*, in thirty chapters. Both translations exist in modern editions; Johannes Hispalensis was edited in 1943 by F. J. Carmody,[29] Gerard in 1910 by R. Campani.[30] The text with its simplified presentation of the Ptolemaic astronomy was very popular throughout the middle ages, and it could be imagined that Gerard translated it before tackling the *Almagest* itself, as sort of an exercise and preparation for the major work. But this is not certain, the chronology of Gerard's translations has not yet been established, and it will hardly be possible to do that because the manuscripts do not mention dates for the translations (which is done in some works of other translators). Of Gerard's version it can be added that it had reached Germany already shortly after 1200; at that time the famous poet Wolfram of Eschenbach has used in the second of his two great epics, the *Willehalm*, a group of Oriental geographical names extracted from Gerard's version (of course, without naming the source).[31]

Then follows Thābit ibn Qurra (in Latin, *Thebit*), one of the Baghdad scholars (836-901), active and productive in many fields of the sciences, and also translating from the Greek. Of his astronomical works, the *socii* list names two as being translated by Gerard: *Liber de expositione nominum almagesti* which seems to be the text otherwise called *De his que indigent expositione antequam legatur almagesti*[32] and might correspond to the Arabic *Kitāb tahyi'at qirā'at al-majasṭī* (« Book of the preparation to reading the *Almagest* », still preserved in two manuscripts);[33] there exists a modern edi-

26. Cf. F. Sezgin, *Geschichte des arabischen Schrifttums*, vi: *Astronomie*, Leiden, 1978, pp. 127 ff., work no. 2; Carmody, p. 32, work no. 8.
27. Sezgin, *op. cit*, p. 149ff., work no. 1.
28. Carmody, p. 113 f., work no. 1.
29. Al-Farghānī, *Differentie Scientie Astrorum*, ed. F. J. Carmody, Berkeley 1943.
30. *Alfragano: Il « libro dell'aggregazione delle stelle »*, ed. R. Campani, Città di Castello, 1910.
31. Cf. P. Kunitzsch, *Die orientalischen Ländernamen bei Wolfram (Wh. 74,3ff.)*, in *Wolfram-Studien*, ed. W. Schröder, ii (Berlin, 1974), 152-73.
32. Carmody, p. 118, work no. 2.
33. Sezgin, *op. cit.* pp. 163ff., work no. 19 (see also p. 90, work no. 5,a).

tion, but very confused and unsatisfactory, by F. J. Carmody.[34] The second is *Liber de motu accessionis et recessionis*, on the theory of trepidation.[35] The ascription of this work to Thābit is highly doubtful;[36] on the other hand, F. Sezgin quotes from the *zīj* of the Egyptian astronomer Ibn Yūnus (died 1009) the mentioning of the title of a treatise by Thābit which points to the very topic treated in the Latin version.[37] Of the Latin, there exist several modern editions, also again a bad one by Carmody (in the same book of 1960),[38] and a useful English translation with commentary by O. Neugebauer.[39]

Next comes Ǧābir ibn Aflaḥ (in Latin, *Geber*, but *ieber* in the *socii* list), an astronomer from Muslim Spain, 1st half of the 12th century. He wrote a description of the *Almagest*, including some fundamental criticism, in nine books (still existing in some manuscripts); the Latin version is called *Elementa astronomica*. It was printed in 1534, but there is no modern edition.[40] The work has been investigated by R. Lorch.[41]

The translation of the astronomical tables of Abū ʿAbdallāh Ibn Muʿādh, *Tabule iahen*, has already been mentioned earlier.

Apart from these texts mentioned in the *socii* list, there are some more astronomical texts which might also have been translated by Gerard (I refer to the list in R. Lemay's exhaustive article on Gerard in *Dictionary of Scientific Biography*).[42] But these, like most of the aforementioned, still need good modern editions and detailed research, always together with the corresponding Arabic sources, in order to arrive at convincing conclusions.

Further, I shall mention two other works also registered in the *socii* list which are not directly astronomical works but which contain astronomical material.

Liber anoe (which is the Arabic « Book of *al-anwā'* », i.e. seasons indicated by certain asterisms). This is a calendar put together from a calendar in the Arabic *anwā'* – style by ʿArīb ibn Saʿd (died 980) and a Christian liturgical calendar by the bishop Rabīʿ ibn Zayd, also called Recemundo, both in 10th century Spain.[43] The Latin text had first been edited by G. Libri;[44] later,

34. F. J. Carmody, *The Astronomical Works of Thabit b. Qurra*, Berkeley-Los Angeles, 1960, pp. 131-39; see also pp. 117f., 120ff.

35. Carmody, p. 117, work no. 1.

36. Cf. recently the dissertation by F. J. Ragep, *Cosmography in the « Tadhkira » of Nasir al-Din al-Tusi*, Harvard University, 1982.

37. Sezgin, *op. cit.*, p. 168, work no. 18.

38. *Op. cit.* in note 34, above, pp. 102-8; see also pp. 84ff.

39. O. Neugebauer, *Thâbit ben Qurra « On the Solar Year » and « On the Motion of the Eighth Sphere »*, in « Proceedings of the American Philosophical Society », 106 (1962), 264ff., esp. 290-9.

40. Carmody, p. 163, work no. 1; Sezgin, *op. cit.*, p. 93, work no. 26.

41. R. Lorch, in an unpublished dissertation of 1970; idem, article « Jābir b. Aflaḥ », in *Dictionary of Scientific Biography*, vii (New York, 1973), pp. 37-9; idem, *The Astronomy of Jābir ibn Aflaḥ*, in « Centaurus », 19 (1975), 85-107.

42. Cf. above, note 2.

43. Cf. F. Sezgin, *Geschichte des arabischen Schrifttums*, vii: *Astrologie - Meteorologie und Verwandtes*, Leiden, 1979, pp. 355f.

44. G. Libri, *Histoire des sciences mathématiques en Italie*, i (Paris, 1838), pp. 461ff.

the Latin was again edited, together with the Arabic, by R. Dozy;[45] finally, these two were re-edited, augmented by a French translation of the Arabic and an ample commentary, by Ch. Pellat.[46] It shall be added here that in the last years a second Latin translation of the calendar, from the 1st half of the 13th century, has been found; it was edited by J. Martínez Gázquez and J. Samsó.[47]

As no. 69, the *socii* list has *Liber alfadol id est arab de bachi*. This work has an exciting story. In 1967, B. F. Lutz has edited in a Ph. D. thesis in Heidelberg, from a manuscript of the late 15th century in Vienna, the German translation of a lot-book (sortilegium) ascribed to an author *alphadol von merenga*.[48] This lot-book is constructed according to the twelve signs of the zodiac and the twelve astrological « houses », it contains a system of 144 questions with twelve answers assigned to each question. The names of the « judges » of the 144 questions are mostly Arabic star names, among which the names of the 28 lunar mansions. Because of the possible Arabic implications I was asked for help in the analysis of the text. In the course of time, I found a number of Arabic manuscripts which contain a lot-book with the same system of questions and answers, but naming various different authors, and giving no names for the 144 judges, but only numbers.[49] At last, however, after the publication of Sezgin's seventh volume of his *Geschichte des arabischen Schrifttums*,[50] still another manuscript became known, in Teheran, of which I succeeded to obtain photocopies. It contains only the introduction and the table of the 144 questions. While the questions here also are only marked by numbers and not by star names, the manuscript at least has the author's name al-Faḍl ibn Sahl which also appears in the Western tradition of the book.[51] Further, several Latin manuscripts of the work also exist and were investigated in this connexion (the German version was of course derived from the Latin). Of these, the manuscript in the University Library of Cambridge (Kk.4.2) has the lot-book in a version appearing to be a direct translation from the Arabic. Three other manuscripts (in Florence, Paris and Berlin) contain the same work in a somewhat vulgarized form. It seems quite probable that this is the work translated by Gerard. The Arabic author's name

45. R. Dozy (ed.), *Le Calendrier de Cordoue*, Leiden, 1873.
46. Ch. Pellat, *Le Calendrier de Cordoue publié par R. Dozy*, Nouvelle édition, Leiden, 1961.
47. J. Martínez Gázques y Julio Samsó, *Una nueva traducción latina del calendario de Córdoba (siglo XIII)*, Barcellona, 1981 (Facultad de Filosofía y Letras, Universidad Autónoma de Barcelona).
48. B. F. Lutz, *Das Buch « Alfadol ». Untersuchung und Ausgabe nach der Wiener Handschrift 2804*, mit einem Nachtrag von P. Kunitzsch, *Die arabischen Losrichternamen*, Phil. Diss., Heidelberg 1967.
49. Cf. P. Kunitzsch, *Zum « Liber Alfadhol », eine Nachlese*, in « Zeitschrift der Deutschen Morgenländischen Gesellschaft », 118 (1968), 297-314.
50. *Op. cit.* in note 43, above, p. 115f.
51. Cf. P. Kunitzsch, *Eine neue Alfadhol-Handschrift*, in « Zeitschrift der Deutschen Morgenländischen Gesellschaft », 134 (1984), 280-5.

given in the Cambridge manuscript as *Alfadhol filius sehel* has been recently identified in the Teheran manuscript as al-Faḍl ibn Sahl (of Persian origin, vizier of the caliph al-Ma'mūn, and renowned for his interest in astrology etc.; the ascription of the book to him might be spurious like all the others in the other manuscripts). On the other hand, the 144 judges are given Arabic names, mostly star names, only in the Western versions while all the Arabic texts found and analysed so far just mark them by numbers. The work in its Arabic version shows elements of earlier Persian origin and might even be related to some older, Hellenistic tradition.

The Almagest

Now, eventually, we come to Gerard's astronomical *opus magnum*, the *Almagest*. What I have to say about this is the result of many years' research in the course of which I have already published an analysis of the transmission[52] and which presently ends up with the edition of the Arabic and Latin texts of the star catalogue. The first volume containing the Arabic translations has already appeared, in 1986;[53] the second volume containing Gerards version is in the hands of the publisher and can be expected in about two years.

Of this fundamental book, in the Orient were made one Syriac and four Arabic translations, the Syriac one at an unknown date, perhaps in the 7th or 8th century, and the Arabic versions from the early to the late 9th century.[54] As is testified by the Spanish-Arabic astronomer Ǧābir ibn Aflaḥ (whom I have mentioned earlier), two of these were known and in use in Muslim Spain, namely the translation of al-Ḥajjāj (dated 827/28), and the translation of Isḥāq ibn Ḥunayn (datable around 879-890) with the emendations of Thābit ibn Qurra (died 901).[55] Manuscripts of both these translations have survived into our times so that it was possible to edit the star catalogue in both versions together.

Of Gerard's translation I have located a total of 42 manuscripts.[56] One of these, at Salamanca, remained inaccessible to me; seven others were defective in one way or the other. For the edition, therefore, 34 manuscripts could be used.

In addition to the manuscripts, there also exists the printed edition of 1515, by Petrus Liechtenstein, in Venice, which of course was also used for the current edition.

A superficial inspection of the manuscripts shows, for example, a cha-

52. See note 11, above.
53. See note 12, above.
54. Cf. Kunitzsch (*op. cit.* in note 11, above), pp. 15ff.
55. Cf. Kunitzsch, *op. cit.*, p. 36 note 87; see also Lorch (1975, *op. cit.* above in note 41), p. 97.
56. The details will be found in the introduction to vol. II of my edition (cf. above, note 12).

racteristic difference in the star catalogue: here in the coordinates the latitudes of certain stars are given in one group of manuscripts as 300 degrees (and compositions with 300), and in another group as 60 degrees (and compositions with 60). The reason for this gross mistake has already been explained by E. B. Knobel:[57] in the Arabic texts the numerical values of the coordinates were expressed by letters, according to the so-called *abjad*-system where the units, the tens, and the hundreds are each represented by one letter. Now, in this system with some of the letters a difference had developped between the Eastern and the Western Arabic notation; in the Eastern system the number 60 was represented by the letter *sīn*, and the same letter got the value of 300 in the Western, the so-called maghrebi system. It is then clear that Gerard who had learnt his Arabic and who was working in Spain, that is in the area of the maghrebi style, translated from an Arabic manuscript where those numbers were written, according to the Eastern system, with the letter *sīn* and that he took this letter in the current maghrebi meaning of 300. It is remarkable that he ever wrote down this absolute nonsense; it is one of the basic elements of astronomical knowledge that the latitudes only reach up, or down, until 90 degrees north or south of the ecliptic (or, accordingly, the declinations from the equator). But I dare not decide whether this has to be explained as a sign of Gerard's incompetence in astronomy, or rather as another testimony for his absolute fidelity vis-à-vis the source which he simply had to render and not to modify. Whatever that was, in the second group of manuscripts the mistake has been corrected and the correct values of 60 (and compositions with 60) were inserted.

This mistake of « 300 » instead of « 60 » can serve as a mark for a rough distinction of two different groups of manuscripts. With the consideration of this and a number of additional characteristic differences it was possible for me to establish in the transmission of Gerard's text two classes which I have called the 𝔄- and 𝔅-classes.[58]

Both classes clearly contain the same text, i.e. Gerard's version of the *Almagest*, but the 𝔄-class has it in an incomplete and incorrect form whereas the 𝔅-class shows the corrected definite form of the work.

In comparing Gerard's text to the two Arabic versions known to have existed in Spain in the 12th century, i.e. the versions of al-Ḥajjāj and of Isḥāq, it is found that he began the translation with a copy of al-Ḥajjāj's version and followed it until the end of Book IX. From Book X to the end (Book XIII), however, he followed Isḥāq's version. The star catalogue, though contained

57. C. H. F. Peters and E. B. Knobel, *Ptolemy's Catalogue of Stars, A Revision of the Almagest*, Washington, 1915, introduction (by Knobel), p. 14.

58. Cf. Kunitzsch (*op. cit.* in note 11, above), pp. 102-4 (here called « p-Klasse » and « fmw-Klasse », respectively). See also the introduction to vol. II of my edition (mentioned above in note 12).

in Books VII-VIII (i.e., in the Ḥajjāj section), is translated from an Isḥāq version, but this was a special copy of it which was interspersed with a great number of verbal and numerical elements borrowed from the Ḥajjāj version.[59]

By some chance, the surviving manuscript of Isḥāq in Tunis shows almost exactly the same construction of the star catalogue as Gerard's text, including the numerous forms and numbers borrowed from al-Ḥajjāj. Even the graphical arrangement of the star tables, on 35 pages, has been imitated by Gerard in exactly the same form as they appear in the Tunisian manuscript. For several reasons the Tunisian manuscript (which is dated A.D. 1085) cannot itself have been the copy from which Gerard worked; but Gerard's source (for the star catalogue) must have been a very close relative of the Tunisian manuscript.[60]

To describe the two classes of transmission, one could say that the 𝔄-class is marked by the above mentioned mistake in the star coordinates (i.e., « 300 » for « 60 » in certain places), and further by defective or unsatisfactory translations and by mistakes originating within the Latin transmission. In the 𝔅-class all these mistakes are corrected, and the deviating phrases in the translations are recast so as to better correspond to the Arabic and following its order of words more closely; further, the manuscripts of the 𝔅-class contain a number of parallels translated from the Isḥāq version and added to the text already translated from al-Ḥajjāj (or vice versa), among which the entire first chapter of Book I.

Of the 34 manuscripts inspected, 17 belong to the 𝔄-class purely, 11 to the 𝔅-class purely, and 6 show mixtures between the two classes; the edition of 1515 also has a mixed version.

To fix a date for Gerard's translation of the *Almagest*, a clue is given by the Florence manuscript Laurentianus 89, supra 45 (from the beginning of the 13th century). At the end it contains the colophon of one Thadeus Ungarus who states to have finished a copy of the text in 1175. Now this manuscript contains the text with all the corrections made in the 𝔅-class, but not yet all the parallel translations to al-Ḥajjāj from Isḥāq that are found in the other manuscripts of class 𝔅. That means that the Florence manuscript shows the text in a state before all the elements of class 𝔅 had been completed, but after the addition of all the corrections to 𝔄 that were made in 𝔅. This leads us to the conclusion that Gerard must have begun the translation of the *Almagest* long time before 1175, and that he has finished it some time after 1175. In other words, we can say the translation of the *Almagest* can roughly be attributed to the years from 1150 to 1180. This would not badly suit what we generally know about Gerard's intentions in Toledo: his main objective there, the conversion of Ptolemy's *Almagest* from Arabic into Latin,

59. Cf. KUNITZSCH, *op. cit* above in note 11, pp. 38-40, 69, etc., and vols. I and II of the edition mentioned above in note 12.

60. See the introduction to vol. II of the edition mentioned above in note 12.

seems to have extended over most of the period of his working life in Toledo. There will have been interruptions because he also must have had the time to translate all the other works mentioned in the *socii* list and perhaps more which are not registered there, among them such voluminous titles as Avicenna's *Canon* – but the *Almagest* seems to have accompanied him through most of the years spent in Toledo.

Here we must also take into consideration that it will not have been easy for him to obtain the Arabic text copies he needed, or wanted. It would rather seem that he got such copies with some difficulty, in greater intervals, and perhaps not all of them were complete (like most of the surviving manuscripts), so that he had to search again and wait for better copies.

Let me add, finally, a few words on the transmission of another fundamental work of medieval science, Euclid's *Elements*, which can very well be compared to the transmission of the *Almagest*.

Of the *Elements*, too, there existed several Arabic translations among which notably two of al-Ḥajjāj and one of Isḥāq, the latter again emended by Thābit. In the West, there is a greater number of Latin versions made from the Arabic among which one by Gerard (it is also mentioned in the *socii* list under no. 4). A few years ago I have made an investigation based on the printed texts available (the results were published in the *Festschrift* for Helmuth Gericke, 1985).[61] I found that the situation in the *Elements* – apart from being much more complicated and less perspicuous than in the *Almagest* – was different also with regard to Gerard. In the *Elements* he seems to have based his translation on a copy of Isḥāq's version to which, however, were added several borrowings from al-Ḥajjāj. Further, the transmission of Gerard's version is much poorer than that of the *Almagest*; the number of existing manuscripts is remarkably less than for the *Almagest*, and all of them date from the 14th century only or later (compare the *Almagest*: here we have one manuscript of the late 12th, and many of the 13th century). Only rarely the style of the version ascribed to Gerard (which I used in the recent edition of H.L.L. Busard)[62] shows the extremely close imitation of the Arabic which is so characteristic for his translations; mostly the language in the *Elements* has been polished and transformed to a standardized mathematical style and to a more conventional Latin expression. Thus I concluded that what we now have as Gerard's translation of the *Elements* is but a later redaction and not the original text of his version. This could also be one reason for the lack of early manuscripts; the redaction may have been made sometime in the 13th century and may, then, have superseded the original text of Gerard.

61. P. KUNITZSCH, *Findings in some texts of Euclid's « Elements » (Mediaeval transmission, Arabo-Latin)*, in *Mathemata, Festschrift für Helmuth Gericke*, Stuttgart 1985 (Boethius, vol. 12), pp. 115-28.

62. *The Latin Translation of the Arabic Version of Euclid's « Elements » Commonly Ascribed to Gerard of Cremona*, ed. by H.L.L. Busard, Leiden, 1984.

This was a survey over a small sector of Gerard's extensive work. In conclusion, it remains to be admitted that our knowledge of Gerard and his works is still rather limited and much must be done to work out his profile in more detail. In order to arrive at conclusive results it will, however, always be necessary to include the Arabic element also in all our efforts.

II

Gerhard von Cremona als Übersetzer des Almagest

Zu den historisch wirkungsvollsten Übersetzungsleistungen gehört die Periode der Übersetzungen arabischer fachwissenschaftlicher Texte ins Lateinische in Spanien im 12. Jahrhundert. Wie schwerfällig und unvollkommen auch immer diese Übersetzungen gewesen sein mögen und wie sehr sie auch später von den am klassischen Latein orientierten und wieder direkt mit griechischen Texten arbeitenden Humanisten verurteilt wurden, so bleibt doch unbestritten, daß diese Übersetzungen in Europa nach langem Niedergang und Stillstand wieder die Beschäftigung mit den Wissenschaften – abseits der eng begrenzten christlich-theologischen Studien – anregten und damit eine Entwicklung in Gang setzten, die letztlich zur Herausbildung unserer modernen technologischen Zivilisation führte.

Unter den zahlreichen Übersetzern, die damals in den von den Christen zurückeroberten Städten Spaniens tätig waren,[1] ragt besonders Gerhard von Cremona hervor. Ein Verzeichnis der von ihm übersetzten Werke, das von seinen Schülern und Mitarbeitern (*socii*)[2] einer der spätesten Übersetzungen angehängt wurde,[3] zählt 71 Titel auf; wahrscheinlich lassen sich dem aber noch weitere Werke hinzufügen.[4] Eine nach seinem Tode entworfene Vita[5] gibt an, daß er, *amore almagesti* getrieben, aus seiner italienischen Heimat nach Toledo gegangen sei, um hier den ara-

[1] Vgl. die Übersichten bei Wüstenfeld und Sarton; speziell für Astronomie und Astrologie siehe auch F.J. Carmody, *Arabic Astronomical and Astrological Sciences in Latin Translation, A Critical Bibliography*, Berkeley/Los Angeles 1956.

[2] *Vita*, Sudhoff S. 75, Zeile 15.

[3] Meist zusammen überliefert mit der *Vita*.

[4] Vgl. die Aufzählungen bei Sarton und Lemay [1]. Da die meisten Texte noch nicht ediert und systematisch untersucht sind, bleiben alle Aussagen hierzu provisorisch.

[5] Lemay [1], S. 173f. nimmt aufgrund interner Evidenz an, daß zwar das Werkeverzeichnis von Gerhards *socii* stammt, daß aber die *Vita* unabhängig davon und später verfaßt sei.

bischen Text dieses wichtigen, bis dahin im Westen unzugänglichen Werkes aufzufinden und es dem Westen in lateinischer Sprache mitzuteilen. Im Alter von 73 Jahren sei er dann 1187 in Toledo[6] gestorben.

Hieraus könnte man auf ein Geburtsjahr um 1114 schließen. Nimmt man an, daß er mit ca. 30 Jahren, also um 1145, nach Spanien gegangen wäre, so ließe sich die Dauer seines Wirkens in Toledo auf rund 40 Jahre ansetzen. Diese verhältnismäßig lange Zeitspanne läßt es als durchaus möglich erscheinen, daß Gerhard tatsächlich die vielen ihm zugeschriebenen Werke aus den verschiedensten Wissensgebieten – Philosophie, Mathematik, Astronomie und Astrologie, Medizin, Geheimwissenschaften – übersetzt hat.

Ein zeitgenössisches Zeugnis zu Gerhards Wirken in Toledo findet sich in der „Philosophia" des Daniel von Morley (verfaßt nach 1175, bis 1200[7]); dieser hatte sich zu Studienzwecken zunächst in Paris und dann in Toledo aufgehalten und schildert einen recht kühnen Disput zwischen ihm und Gerhard während einer Vorlesung Gerhards über Abū Maᶜšars *Introductorium maius*[8]. Anläßlich der Erwähnung Gerhards (den er übrigens *Girardus Tholetanus* nennt) fügt Daniel von Morley die Bemerkung an: *qui Galippo mixtarabe interpretante almagesti latinavit.*

Daniels Bericht läßt keine chronologischen Rückschlüsse auf Gerhards Almagestarbeit zu, d.h. ob die Übersetzung zur Zeit seines dortigen Aufenthalts bereits abgeschlossen oder noch im Gange war. Andererseits erfahren wir eine wichtige Einzelheit über Gerhards Übersetzungstechnik bei der Arbeit am Almagest: er bediente sich dabei der Hilfe eines Mozarabers (*mixtarabs*) namens Ġālib (*Galippus*).

Hier kommen wir an eine der vielen grundsätzlichen Fragen, die sich um diese mittelalterlichen arabisch-lateinischen Übersetzungen ranken und die alle noch weit von einer Lösung entfernt sind, zumal ja die meisten Texte dieser Übersetzungsliteratur noch nicht ediert, geschweige denn genauer untersucht sind: hat Gerhard, haben überhaupt die lateinischen Übersetzer sprachkundige einheimische Helfer gehabt? Aufgrund

---

[6] Ich folge hier den überzeugenden Argumenten von Lemay [1] S. 173.
[7] Vgl. Maurach S. 209.
[8] „Philosophia" § 192-195 Maurach.

einer Angabe in der Einleitung der Übersetzung von Avicennas „De anima", von Ioannes Auendehut [= Johannes Hispalensis] und Dominicus Gundisalvi, könnte man annehmen, es sei üblich gewesen, daß jeweils ein (westlicher) Übersetzer, der die altspanische Umgangssprache und Latein (und vielleicht mehr oder weniger etwas Arabisch) beherrschte, mit einem Einheimischen (Mozaraber oder Juden), der ebenfalls die altspanische Umgangssprache und dazu Arabisch beherrschte, zusammenarbeitete, wobei dieser den arabischen Text in die altspanische Umgangssprache und jener ihn daraus weiter ins Lateinische übertrug.[9] Solche Mitarbeiter von Übersetzern werden sonst namentlich praktisch nirgends erwähnt,[10] und Lemay möchte derartige Zusammenarbeit für den Sonderfall halten, im allgemeinen hätten die Übersetzer allein gearbeitet.[11]

Für Gerhard ist *Galippus* nur durch Daniel von Morley, und nur für den Almagest, bezeugt. Aber selbst hierbei ergeben sich weitere Fragen. Denn wie wir gleich sehen werden, dürfte sich Gerhards Arbeit am Almagest über rund drei Jahrzehnte erstreckt haben, etwa von 1150-1180. Es ist kaum anzunehmen, daß *Galippus* über diese gesamte Zeitdauer hin als Mitarbeiter zur Verfügung gestanden haben wird; aber an welchen der sogleich zu erwähnenden Stadien der Almagestarbeit mag er dann mitgearbeitet haben?

Die *Mathematikḗ Sýntaxis* des Claudius Ptolemäus (geschr. um 140 n.Chr. in Alexandria) war für Spätantike und Mittelalter bis zum Heraufkommen der neuen, kopernikanischen Astronomie das wichtigste Lehrwerk der Himmelskunde. Durch mehrere arabische Übersetzungen wurde es auch im arabisch-islamischen Kulturraum zu einer Hauptautorität.[12]

---

9 *Cf.* Kunitzsch [1], S. 85 f. mit Anm. 224 (Zitat aus O. Bardenhewer, 1882).

10 Hier kann man noch den Juden *Abuteus* anfügen, der mit Michael Scotus zusammengearbeitet haben soll; *cf.* Sarton II,1 S. 581.

11 Lemay [1], S. 174.

12 Der Name *Almagest*, unter dem das Abendland dieses Werk seit dem Mittelalter vorzüglich kennt, beruht auf der Latinisierung der arab. Form *al-maǧasṭī*, die ihrerseits vermutlich auf einer älteren Pahlavi-

(Fortsetzung...)

Um 1150 (aber vor 1154) entstand in Sizilien davon eine lateinische Übersetzung aus dem griechischen Urtext, wahrscheinlich von Hermannus de Carinthia.[13] Aber diese griechisch-lateinische Übersetzung fand wenig Verbreitung und gewann in Europa keinen nachweisbaren Einfluß. Der für Europa ausschlaggebende Text sollte Gerhards Version werden.[14]

Gerhard von Cremona hat vom Ruhm des ptolemäischen Werkes schon in Italien erfahren, falls wir der *Vita* in dieser Hinsicht trauen können. Selbst den so großartig und verführerisch klingenden arabisierten Titel *Almagest* kann er bereits gekannt haben, da dieser in älteren arabisch-lateinischen Übersetzungen vielfach zitiert war.[15] Als er in der Mitte des 12. Jahrhunderts nach Spanien kam, zirkulierten dort nach dem Zeugnis des zeitgenössischen spanisch-arabischen Astronomen Ǧābir ibn Aflaḥ[16] zwei von den vier bekannten arabischen Textfassungen: diejenige von al-Ḥaǧǧāǧ ibn Yūsuf ibn Maṭar (entstanden 212 H/827-28) und diejenige von Isḥāq ibn Ḥunayn (entstanden um 879-890; hier zweifellos gemeint: in der Bearbeitung, *iṣlāḥ*, durch Ṯābit ibn Qurra, gest. 901). Von beiden haben sich Handschriften bis in unsere Zeit erhalten.[17]

Wie der Hinweis der *Vita*, aber auch Bedeutung und Umfang des Werkes nahelegen, scheint die Übersetzung des Almagest Gerhards Hauptanliegen in Toledo gewesen zu sein. Das wird durch die interne Evidenz seiner Version erhärtet, aus deren erhaltenen Textzeugen (vgl.

12(...Fortsetzung) Form beruht, die aus einem im griech. Original noch nicht nachgewiesenen Superlativ *Megístē* [sc. *Sýntaxis*] abgeleitet ist; *cf.* dazu Kunitzsch [1], S. 115-125. Zu den arabischen Übersetzungen s. ebda. S. 15 ff.

13 Siehe Lemay [2].

14 Hiervon konnten bisher 42 vollständige oder einen großen Teil des Werkes enthaltende Handschriften nachgewiesen werden; s. demnächst Kunitzsch [2], II, Einleitung.

15 *Cf.* Kunitzsch [1], S. 115 mit Anm. 2 und 4.

16 *Cf.* Kunitzsch [1], S. 36 f. Anm. 87; 86; 97, etc.

17 S. zuletzt Kunitzsch [2], I, S. 3 f.; hier auch die Edition des Sternkatalogs aus diesen beiden Versionen.

oben Anm. 14) sich ein Fortschreiten der Arbeit ablesen läßt, die sich über viele Jahre, grob von 1150-1180, erstreckt haben dürfte.

Das einzige konkrete Datum in diesem Zusammenhang liefert die Hs. Florenz, Laur. 89, sup. 45, welche zwar selbst vom Anfang des 13. Jahrhunderts stammt, worin aber am Ende ein älterer Kolophon mitüberliefert ist, in dem ein Thadeus Ungarus angibt, den Almagest 1175 in Toledo abgeschrieben zu haben. Das bedeutet zumindest, daß der Text in der Fassung dieser Handschrift 1175 in Toledo vorlag.

Nun werden aus der Textüberlieferung folgende Arbeitsetappen Gerhards am Almagest erkennbar: 1) Beginn der Übersetzung nach dem arabischen Text von al-Ḥaǧǧāǧ, und so durch bis Ende Buch IX; 2) Fortsetzung und Abschluß der Arbeit (Buch X-XIII) nach Isḥāq; 3) Überarbeitung der Übersetzung, Berichtigung von Fehlern, Vereinheitlichung der Terminologie, Herstellung engerer Übereinstimmung mit dem arabischen Wortlaut, Hinzufügung von Parallelen aus Isḥāq im Ḥaǧǧāǧ-Teil sowie aus al-Ḥaǧǧāǧ im Isḥāq-Teil. Hiernach gliedern wir die Überlieferung in die (ältere) 𝔄-Klasse (nur die Elemente von 1) und 2) enthaltend) und die (demgegenüber jüngere) 𝔅-Klasse (nun auch die Elemente von 3) enthaltend). Die erwähnte Handschrift aus Florenz gehört prinzipiell zur 𝔅-Klasse, d.h., sie weist schon alle Berichtigungen und Modifizierungen des Wortlauts auf, aber noch nicht alle Parallelzitate aus der jeweils anderen arabischen Version. Daraus läßt sich ableiten, daß die Grundübersetzung (𝔄-Klasse) sowie die meisten Abänderungen der 𝔅-Klasse um 1175 bereits vorlagen, aber nicht alle: viele Parallelübersetzungen aus Isḥāq (im Ḥaǧǧāǧ-Teil) sind offenbar erst noch nach 1175 hinzugekommen. Gemäß dieser Überlieferungslage setzen wir die Zeit für Gerhards Almagestübersetzung auf „ca. 1150-1180" an, d.h. weit vor 1175 (denn bis dahin war die Hauptarbeit einschließlich der Abänderungen in der 𝔅-Klasse erfolgt), und noch über 1175 hinaus (denn hernach kamen noch zahlreiche Parallelübersetzungen aus der jeweils anderen arabischen Version hinzu).

Zu den weiteren Fragen, die sich über das chronologische Problem hinaus aufdrängen, gehören vor allem die nach Gerhards fachlicher Kompetenz (bei der Vielzahl der Fachgebiete, aus denen er Texte übersetzte!), nach der Güte seiner arabischen Sprachkenntnisse und, *last not least*, nach seiner Übersetzungsmethode an sich.

Was die fachliche Kompetenz angeht, so deuten Beobachtungen im Text der 𝔄-Klasse darauf hin, daß Gerhard, zumindest in der Anfangsphase, mit dem Gebiet der Astronomie noch nicht genügend vertraut war. Als auffälligstes Merkmal nenne ich, aus dem Sternkatalog, jene Fälle (im ersten Drittel, bei den nördlichen Sternbildern), wo er mit 60 zusammengesetzte Breitenwerte fälschlich als mit 300 zusammengesetzte Werte übertrug (also z.B. statt 62° : 302° – die Breiten können aber nach Nord oder Süd nur je bis 90° ansteigen). Die Ursache hierfür dürfte darin gelegen haben, daß er in diesem Bereich des Textes nach einer Handschrift ostarabischer Herkunft gearbeitet hat, in der der Wert 60 nach ostarabischer Weise im *abǧad*-System mit dem Buchstaben *sīn* ausgedrückt war; er jedoch, im maghrebinisch-westarabischen Raum lebend und wirkend, faßte *sīn* mit dem abweichenden Zahlenwert 300 auf, den dieser Buchstabe im Westen hatte (60 ist dort *ṣād*). Freilich können wir kaum entscheiden, ob sich darin wirklich fachliche Inkompetenz äußerte, oder aber vielmehr das Ethos eines *fidus interpres*, der seine Vorlage getreulich nachzubilden hatte und daran keinerlei eigenmächtige Veränderungen vornehmen durfte. Immerhin hat Gerhard auf diese „getreuliche" Weise auch eine Reihe grotesker, komischer Fehlübersetzungen geliefert,[18] die auch in der 𝔅-Klasse nicht verbessert wurden. Andererseits hat er jene falschen Breitenwerte der 𝔄-Klasse später in der 𝔅-Klasse selbst berichtigt.

Die absolute Treue zum Original scheint bei ihm einen hohen Wert genossen zu haben. Wenn auch bisher nur wenige seiner Arbeiten ediert sind bzw. näher untersucht wurden,[19] so läßt sich doch – auch im Ver-

---

[18] *Cf.* Kunitzsch [1], S. 107.

[19] Zum Almagest s. Kunitzsch [1], S. 83 ff., besonders 104-112, 131-149, 171 f., 212-217; zu Aristoteles' „Meteorologie" (Buch I) s. Schoonheim, besonders S. 23-28; der Aufsatz von I. Opelt bezieht sich auf das noch nicht edierte „De caelo et mundo" (Aristoteles) und ist leider methodisch und in den Ergenissen verfehlt, da das Arabische gar nicht bzw. nur stellenweise und dann unzulänglich einbezogen wurde.

gleich zu den Übersetzungen der anderen zeitgenössischen Übersetzer – als hervorstechendstes Merkmal seiner Übersetzungen auf jeden Fall die äußerst enge, wörtliche Nachahmung des arabischen Wortlauts im Lateinischen hervorheben. Auch diese war in der 𝔄-Klasse noch nicht voll ausgebildet, wurde dann aber in der 𝔅-Klasse durchgehend hergestellt. Diese Wörtlichkeit der Übersetzung geht soweit, daß man starke Zweifel daran haben muß, wie weit die mittelalterlichen Leser seine Übersetzungen überhaupt verstehen konnten. Moderne Bearbeiter und Herausgeber können, wenn sie sachgerecht und erfolgreich arbeiten wollen, an seine Texte nur herangehen, wenn sie neben Latein auch Arabisch beherrschen und seine Übersetzung Wort für Wort in der arabischen Vorlage verfolgen und belegen können. Weiter ist zu beobachten, daß Gerhard um Gleichmäßigkeit und Genauigkeit bei der Wiedergabe der Terminologie bemüht ist: dieselben arabischen Termini werden stets durch dasselbe lateinische Wort wiedergegeben; umgekehrt werden abgeleitete oder anderweitig variierte Formen des Arabischen im Lateinischen gleichfalls durch ähnliche Ableitungen oder Variationen bezeichnet. Gerhards Wörtlichkeit und Konsistenz im Sprachgebrauch gehen soweit, daß man von Fall zu Fall korrupte arabische Lesarten anhand seiner Version restituieren kann.

Gerhard scheint auch – ähnlich wie im Arabischen der große Ḥunayn ibn Isḥāq[20] – eine Art Textkritik getrieben zu haben, zumindest bei einem so großen und wichtigen Text wie dem Almagest. So muß er sich – wie die Entwicklung der Arbeit in den 𝔄- und 𝔅-Klassen zeigt – nach und nach zusätzliche Handschriften des Almagest beschafft haben. Zunächst scheint ihm nur eine Handschrift der Ḥaǧǧāǧ-Version vorgelegen zu haben, die noch dazu Fehler aufwies, nicht immer gut lesbar und vielleicht unvollständig war. Im Laufe der Zeit muß ihm dann auch die Isḥāq-Version zugänglich geworden sein,[21] nach der er seine Übersetzung

---

[20] *Cf.*, der Kürze halber, F. Rosenthal, *Das Fortleben der Antike im Islam*. Zürich/Stuttgart 1965, S. 36-38.

[21] Die Beschaffung arabischer wissenschaftlicher Texte wird für Christen in Spanien nicht immer ganz leicht gewesen sein, vgl. den Hinweis auf (Fortsetzung...)

ab Buch X fortsetzte und abschloß. In den folgenden Jahren scheint er immer mehr Handschriften verglichen zu haben, woraus dann die Verbesserungen und Abänderungen der 𝔅-Klasse resultierten. Diese können in manchen Fällen auch ein Auswechseln von Wörtern, Formeln oder Zahlenwerten der einen Version gegen solche der anderen Version umfassen.

Arabische Sprachkenntnisse wird Gerhard zweifellos gehabt haben; er wird am Beginn seines Aufenthaltes in Toledo wohl als erstes arabischen Unterricht genommen haben und wird seine Kenntnisse im Verlaufe seines vierzigjährigen dortigen Wirkens ständig verbessert haben. Andererseits gehörte Toledo, das 1085 von den Christen zurückerobert worden war, zu Gerhards Zeit seit langem nicht mehr direkt zum arabischen Andalus. Auch die Verwendung des einheimischen Mozarabers *Galippus* läßt aufhorchen. Man wird wohl annehmen dürfen – wie auch ein Vergleich aus unserer Zeit, bei der Heranziehung von gerade am Ort vorhandenen „Muttersprachlern", bestätigen würde –, daß das Arabisch, das Gerhard kennenlernte, kein lupenreines „Hocharabisch" war, sondern eine mehr oder weniger stark vom spanisch-arabischen Dialekt beeinflußte Form des Arabischen. Diese Annahme wird bestätigt durch die Wiedergabe einiger arabischer Fremdwörter und sonstiger Formen, die Gerhard in der Almagestübersetzung (speziell im Bereich des Sternkatalogs) beibehielt.

Hierzu gehören folgende Formen, die spanisch-arabische umgangssprachliche Aussprache reflektieren: *mirac, mirach* (aus [*al-*]*mirāqq*, statt [*al-*]*marāqq*,[22] Sterne Nr. 25 und 977[23]); *meizer* (aus [*al-*]*mayzar*, statt [*al-*] *mi'zar*,[24] Stern Nr. 103; daneben aber auch „klassisch" *mizar* = [*al-*] *mi'zar*, Stern Nr. 346); *masim, mahasim* (aus [*al-*]*maʿṣim*, statt [*al-*] *miʿṣam*,[22] Sterne Nr. 126, 225, 227); *thauguebe* (aus [*aḏ-*]*ḏawēbe*, statt

---

[21](...Fortsetzung) ein entsprechendes Verbot im *ḥisba*-Werk von Ibn ʿAbdūn, bei Lemay [1], S. 175.

[22] *Cf.* Corriente 5.1.11.

[23] Nach der laufenden Numerierung der ptolemäischen Sterne von Baily, die auch in den Editionen bei Kunitzsch [2], I und II beibehalten ist.

[24] *Cf.* Corriente 2.28.4.

[*ad-*]*du'āba*, in der Schlußsumme am Ende des Sternkatalogs). Ferner gehört dazu auch die Wiedergabe der starken *imāla* im Spanisch-Arabischen, z.B. *thauguebe* (s.o.); *ascimech* = *as-simāk* (Sterne Nr. 110, 510, 526); *aliouze* und *ieuze* = (*al-*)*ğawzā'* (Überschriften zu Gemini und Orion); *echiguen* = *akīwēn*, *'kyw'n* (korrumpiert aus *'ktw's*, arabische Transkription von griechisch *Ichthýes*, Überschrift Pisces). Auch die häufige Wiedergabe von arabisch *ğ* mit lateinisch *i* könnte einer Eigenheit des spanisch-arabischen Dialekts entsprechen:[25] (*al-*)*maiarati* = *al-mağarra*; *aliouze, ieuze* (s.o.); *asuia* = *aš-šuğā'* (Hydra).

Die Assimilation des arabischen Artikels wird ursprünglich durchgehend ausgedrückt: *ascimech* (s.o.), *asuia* (s.o.), *azubenen* = *az-zubānān* (pro *az-zubānayān*, Libra), *aridf* (= *ar-ridf*, Stern Nr. 163). Spätere Abschreiber fügen gelegentlich sekundär ein *l* ein, da ihnen das *al* am Anfang arabischer Fremdwörter bereits geläufig war (z.B. Stern Nr. 390: *aldebaran*, neben vereinzeltem *adebaran*, = *ad-dabarān*).

Auch die Übersetzung einzelner arabischer Wörter kann durch lokale Sprachverhältnisse beeinflußt sein, wie z.B. das in der Astronomie seit langem unerklärte *lupus* „Wolf" für griechisch *Thēríon*, arabisch *as-sabu'*.[26]

Griechische Wörter und Eigennamen, die im Arabischen zum Teil in Transkription beibehalten und dabei im Laufe der Überlieferung oft stark entstellt wurden, hat Gerhard teilweise gut wiedererkannt, was für seinen weiten Bildungshintergrund spricht: *perseus, delphinus, andromade* (genit.), *cetus, orion, (h)ydra, centaurus*; stärker entstellte oder ihm unbekannte Namen konnte er dagegen nicht restituieren: *Abrachis* (für Hipparchus), *Arsatilis* (für Aristyllus), *Agrinus* (für Agrippa) usw., sowie *theguius* [*gu* = w!] für Bootes, *allore* für Lyra (mit beigefügtem arabischem Artikel), *afeichus* für Ophiuchus, *calurus* für griech. *kollórrhobos*, usw.

---

25 *Cf.* Corriente 2.19.5.

26 Vgl. *Glossarium Latino-Arabicum*, ed. C.F. Seybold, Berlin 1900, S. 299 *s.v. lupus*: *saba'* (umgangssprachlich so vokalisiert); Pedro de Alcalá, *Petri Hispani de lingua Arabica libri duo*, ed. P. de Lagarde, Göttingen 1883, S. 294b, *s.v. lobo*: *çábá* (dgl.).

Welche Hilfsmittel Gerhard und den anderen arabisch-lateinischen Übersetzern, über gelegentlich greifbare einheimische Sprachkenner hinaus, zur Verfügung gestanden haben, wissen wir nicht. Glossare von der Art des *Glossarium Latino-Arabicum* (vgl. Anm. 26) waren in ihrem Wortschatz viel zu begrenzt, ganz zu schweigen von der Fachterminologie der verschiedenen Wissensgebiete. Sicherlich konnte Gerhard auf eine ganz gute Ausbildung im Lateinischen zurückgreifen, die ihm bereits einen Grundstock an astronomisch-astrologischer Terminologie vermittelt haben dürfte.

Auf jeden Fall sind die damals unter so schwierigen Voraussetzungen zustande gekommenen arabisch-lateinischen Übersetzungen – bei aller Schwerfälligkeit und Unvollkommenheit – als bedeutende Sprachleistungen anzusehen. Ohne nennenswerte Vorstadien wurde hier Neuland betreten: unvermittelt mußten schwierige Fachtexte aus dem so ganz anders strukturierten semitischen Idiom in das seit vielen Jahrhunderten in sich selbst ruhende, gegen Fremdeinflüsse weitgehend abgeschirmte Latein umgesetzt werden. Bei Würdigung der historischen Voraussetzungen und Begleitumstände werden wir Heutigen daher nicht in die abfälligen Urteile der Antiarabisten der Renaissance einstimmen, sondern jenen Übersetzungen durchaus den Rang eigenwilliger schöpferischer Sprachgestaltung zuerkennen.

Wenn auch die ptolemäische Astronomie, und damit Gerhards Version des Almagest, nun schon seit Jahrhunderten überholt und verdrängt sind, so leben immerhin einige Namen und Wörter aus Gerhards Übersetzung doch in der modernen Astronomie noch weiter.[27]

---

[27] Für Beispiele s. P. Kunitzsch, *Arabische Sternnamen in Europa*, Wiesbaden 1959.

Bibliographie

Daniel von Morley: K. Sudhoff, „Daniels von Morley Liber de naturis inferiorum et superiorum", in: *Archiv für die Geschichte der Naturwissenschaften und der Technik* 8 (1917-18), 1-40 (dazu bessere Lesarten aus der Berliner Hs., von A. Birkenmajer, ebda. 9 [1920-22], 45-51); G. Maurach, Daniel von Morley, »Philosophia«, in: *Mittellateinisches Jahrbuch* 14 (1979), 204-255.

Corriente: F. Corriente, *A grammatical sketch of the Spanish Arabic dialect bundle*, Madrid 1977.

Kunitzsch [1]: P. Kunitzsch: *Der Almagest. Die Syntaxis Mathematica des Claudius Ptolemäus in arabisch-lateinischer Überlieferung*, Wiesbaden 1974.

Kunitzsch [2]: Claudius Ptolemäus, *Der Sternkatalog des Almagest. Die arabisch-mittelalterliche Tradition*, I: *Die arabischen Übersetzungen*, ed. P. Kunitzsch, Wiesbaden 1986; II: *Die lateinische Übersetzung*, ed. P. Kunitzsch, Wiesbaden (z.Zt. im Druck).

Lemay [1]: R. Lemay, Art. „Gerard of Cremona", in: *Dictionary of Scientific Biography*, XV (New York 1978), S. 173-192.

Lemay [2]: R. Lemay, „Hermann de Carinthie. Auteur de la traduction «sicilienne» de l'Almageste à partir du grec (ca. 1150 A.D.)", in: *La diffusione delle scienze islamiche nel medio evo europeo*, Convegno internazionale (Roma, 2-4 ottobre 1984), Rom: Acc. Naz. dei Lincei, 1987, S. 428-484.

Opelt: I. Opelt, „Zur Übersetzungstechnik des Gerhard von Cremona", in: *Glotta* 38 (1959), 135-170.

Sarton: G. Sarton, *Introduction to the History of Science*, vol. II, part 1 (Washington 1931; repr. 1950), S. 338-344.

Schoonheim: P.L. Schoonheim, *Aristoteles' Meteorologie*, Leiden 1978.

Vita: Ausgaben von B. Boncompagni (Rom, 1851); K. Sudhoff (in: *Archiv für Geschichte der Medizin* 8 [1914], 73-82); Wüstenfeld S. 57 f. (kommentiertes Werkeverzeichnis 58-77; Lobpreis 77). Engl. Übers. von M. McVaugh, in: E. Grant, *A Source Book in Medieval Science*, Cambridge, Mass., 1974, S. 35-38.

Wüstenfeld: F. Wüstenfeld, *Die Übersetzungen Arabischer Werke in das Lateinische seit dem XI. Jahrhundert*, Göttingen 1877 (Abh. Kgl. Ges. d. Wiss. zu Göttingen, 22), S. 55-81.

III

ÜBER EINIGE SPUREN
DER SYRISCHEN ALMAGESTÜBERSETZUNG

Wie bei der arabisch-islamischen Übernahme vieler anderer antiker Werke ging auch bei der 'Mathematike Syntaxis' des Claudius Ptolemäus ('Almagest') den Übersetzungen ins Arabische zeitlich eine syrische Übersetzung voraus. Dieser Tatbestand als solcher ist seit langem aus entsprechenden bibliographischen Notizen einschlägiger arabischer Autoren bekannt, ohne daß dort allerdings nähere Angaben darüber gemacht werden, wann, wo und durch wen die syrische Übersetzung angefertigt wurde [1]. Da die älteste arabische Version, die sogenannte 'alte' oder 'ma'mūnische' Übersetzung, etwa zu Beginn des 9. Jahrhunderts anzusetzen ist, müßte die syrische Übersetzung folglich im 8. Jahrhundert oder auch noch früher erfolgt sein.

In einschlägigen kosmologisch-kosmographischen Schriften syrischer Autoren wird keine Beziehung auf eine etwa benutzte vorliegende syrische Almagestübersetzung sichtbar: Hiob von Edessa, um 817 in Bagdad, folgt einer rein aristotelischen Tradition [2]. Sergius von Rēšᶜaynā (gest. 536) kannte immerhin die 'Mathematike Syntaxis' [3], aber näheres wissen wir bisher nicht. Barhebraeus (1225/26 - 1286) schließlich gehört der Spätzeit an und konnte, neben dem griechischen Urtext, auch auf mehrere vorhandene arabische Übersetzungen zurückgreifen, so daß sein Verhältnis zu den Quellen nur sehr schwer zu entwirren sein dürfte; an einigen Beispielen der Nomenklatur konnte der 'gemischte' Charakter seines 'Livre de l'ascension de l'esprit' (geschr. 1272-1279) immerhin nachgewiesen werden [4]. Für unsere Betrachtung am interessantesten wäre der um 660 schreibende Severus Sēḇōḵt (gest. 666/67), Bischof von Qinnasrīn, der rein zeitlich der syrischen Almagestübersetzung sehr nahegestanden haben könnte. Aber auch bei ihm findet sich kein direkter Hinweis auf eine solche Übersetzung. In seinem 'Traité sur les constellations' (geschr. 660) [5] behandelt er z.B. die 48 klassischen ptolemäischen Sternbilder (bei ihm zu 46 umgeordnet), aber er kann darin ebenso gut nach dem griechischen Urtext selbst wie nach einer existierenden syrischen Übersetzung oder auch syrischen Résumées oder Exzerpten vorgegangen sein [6].

Aus der einschlägigen arabischen Fachliteratur war bisher erst eine einzige Stelle bekannt, die eindeutig syrische Herkunft erkennen läßt und die der syrischen Almagestübersetzung entstammen könnte: der Windname 'zhfrs' (= ζέφυρος) bei al-Battānī (aus der Horizonttafel in 'Almag.' VI, 11), auf den C.A. NALLINO deswegen besonders hingewiesen hatte [7].

Alle die genannten Autoren und Stellen lassen freilich nur indirekte Schlüsse auf die Quellenlage und auf eine mögliche Benutzung der syrischen Almagestversion zu.

Daneben gibt es, in der arabischen Fachliteratur, aber auch ein paar wenige Stellen, welche die syrische Almagestübersetzung ganz explicit als Quelle namhaft machen und daraus zitieren. Diesen soll unser hiesiger Aufsatz zu Ehren des Jubilars, in dessen Arbeiten der 'Almagest' stets auch eine wesentliche Rolle spielt, gewidmet sein.

Die entsprechenden Testimonia fallen zeitlich ins 12. sowie ins 13. Jahrhundert oder später und beziehen sich auf Passagen im Bereich des ptolemäischen Sternkatalogs ('Almag.' VII,5 - VIII,1).

Die ältere dieser beiden Reihen ist namentlich genau greifbar: Der Arzt, Naturwissenschaftler und wichtige Kritiker zahlreicher älterer Überlieferungen Abū l-Futūḥ Aḥmad ibn Muḥammad ibn as-Sarī, bekannt als Ibn aṣ-Ṣalāḥ (gest. 1154), hat in einer besonderen Schrift die Ursachen der Fehler in der Koordinatenüberlieferung des Sternkatalogs aus dem Almagest untersucht [8]. Als Quellenmaterial bediente er sich dabei fünf verschiedener Textfassungen des 'Almagest': der syrischen sowie vier arabischer (dies sind: die 'alte, maʾmūnische'; diejenige von al-Ḥaǧǧāǧ ibn Yūsuf ibn Maṭar; diejenige von Isḥāq ibn Ḥunayn, und zwar sogar im Autograph; und Ṯābit ibn Qurra's Verbesserung [iṣlāḥ] von Isḥāqs Version). Zu 88 von den 1025 Sternen des ptolemäischen Katalogs sowie zu einigen weiteren Fällen breitet Ibn aṣ-Ṣalāḥ seine Kritik aus und zitiert dazu durchgehend aus diesen fünf Textfassungen des 'Almagest', von denen uns heute in direkter Überlieferung nur noch zwei handschriftlich erhalten sind (al-Ḥaǧǧāǧ und Isḥāq in Ṯābits Verbesserung). Dabei ergeben sich 60 aus der syrischen Version direkt zitierte Koordinatenwerte; in weiteren 11 Fällen dürfte die syrische Version mit eingeschlossen sein, wenn Ibn aṣ-Ṣalāḥ 'alle übrigen Versionen' zusammengefaßt einem bestimmten Sonderwert gegenüberstellt. Diese Koordinatenzitate und ihre Auswertung sind in der in Anm. 8 genannten Ibn aṣ-Ṣalāḥ-Ausgabe zu finden. Hier sei zusammenfassend nur noch soviel hinzugefügt, daß von den zitierten Werten 26 direkt und 8 indirekt zitierte mit dem edierten griechischen Text von HEIBERG übereinstimmen; ferner gibt es folgende Übereinstimmungen mit Varianten in HEIBERGs kritischem Apparat: 7 direkte Zitate sind identisch mit A^1, ein direktes Zitat mit A; 8 direkte und ein indirektes mit B; 6 direkte und ein indirektes mit C; 3 direkte mit D. Außerdem bieten 24 direkte und ein indirektes Zitat noch sonstige Werte, die bei HEIBERG gar nicht in Erscheinung treten.

Neben den Koordinatenwerten bringt Ibn aṣ-Ṣalāḥ auch zwei terminologische Zitate aus der syrischen Version, die im folgenden behandelt werden sollen.

Die Testimonia der zweiten, jüngeren Reihe gehören in die Nachfolge des Werkes von aṭ-Ṭūsī und umfassen drei terminologische Zitate aus der syrischen Version. Zu aṭ-Ṭūsīs Bearbeitung des 'Almagest' (Taḥrīr al-mǧsṭī) von 1247 hat um 1415 Qāḍīzāde ar-Rūmī, aus dem Astronomenkreis um Uluġ Bēg in Samarqand, einen Kommentar geschrieben [9], worin zwei dieser Zitate vorkommen. Sämtliche drei Zitate erscheinen daneben als Randglossen in der Pariser Ṭūsī-Handschrift B.N. ar. 2485 (15. Jh.). Die letztliche Herkunft dieser Zitate ist noch nicht ermittelt. Sie stimmen im Wortlaut untereinander völlig überein; eine der drei in der Pariser Ṭūsī-Handschrift vorhandenen Stellen fehlt jedoch bei Qāḍīzāde. Vermutlich haben also der Glossator und Qāḍīzāde beide unabhängig voneinander eine entsprechende einschlägige Schrift benutzt und ihre Zitate daraus entnommen. Welche Schrift dies gewesen sein könnte, bleibt noch aufzudecken [10].

Insgesamt liegen damit terminologische Zitate aus der syrischen Version für folgende vier Sterne vor [11]:

1. 23. Stern des Schützen (= $\beta^{1,2}$ Sagittarii) [12]:
ὁ ἐπὶ τοῦ ἐμπροσθίου καὶ ἀριστεροῦ σφυροῦ [13] 'der auf dem vorderen linken Knöchel' (d.h. Knöchel des linken Vorderfußes). Hierfür al-Battānī (vermutlich nach der 'alten, maʾmūnischen' Version des 'Almagest'): 'allaḏī ʿalā ʿurqūb ar-rāmī al-muqaddam al-aysar' 'derjenige auf der vorderen linken Achillessehne des Schützen'; Übersetzung al-Ḥaǧǧāǧ: 'al-kawkab

allaḏī ʿalā l-ʿurqūb al-aysar ʿalā muqaddamihī' 'der Stern auf der linken Achillessehne, und zwar auf deren vorderem [Teil]' (hiernach wiederum die lateinische Übersetzung des Gerhard von Cremona: 'Que est super cauillam sinistram super antecedens ipsius'); Version Isḥāq/Ṯābit: 'allaḏī ʿalā l-kaʿb al-muqaddam al-aysar' 'derjenige auf dem vorderen linken Knöchel'. Gemäß Ibn aṣ-Ṣalāḥ (Stern Nr. 58) in der s y r i s c h e n und in der 'alten, maʾmūnischen' Version: 'allaḏī ʿalā l-ʿurqūb al-aysar' 'derjenige auf der linken Achillessehne'.

2. 22. Stern des Schiffes Argo (= Lac. 3580) [14]:
τῶν ἐπὶ ταῖς ἀσπιδίσκαις ὡς ἐπὶ τῆς ἱστοδόκης $\overline{\gamma}$ ὁ βόρειος [Varianten laut Apparat: für ἱστοδόκης haben mss ACD ἱστο$^{\Delta}$, stark abgekürzt] [15] 'von den dreien auf den kleinen Schilden und gleichsam auf dem Mastbehälter der nördliche'. Dafür al-Ḥaǧǧāǧ: 'al-kawkab aš-šamālī min aṯ-ṯalāṯa allatī fī atrās ad-daqal' [var. atrās: raʾs, ad-daqal: ar-riǧl] 'der nördliche Stern von den dreien auf den Schilden des Mastes' [16]; Isḥāq/Ṯābit: 'aš-šamālī min aṯ-ṯalāṯa allatī fī t-turaysāt wa-ka-annahū ʿalā d-daqal' 'der nördliche von den dreien auf den kleinen Schilden und gleichsam auf dem Mast' [17] (hiernach Gerhard von Cremona: 'Septentrionalis trium que sunt in scutellis et quasi sint super costatum'). Die Pariser Ṭūsī-Handschrift hat auf fol. 62va zu at-turaysāt am Rande die Glosse: 'fī n-naql al-maʾmūni wa-fī s-suryānī turs aš-širāʿ' "In der maʾmūnischen und in der s y r i s c h e n Version: 'der Schild des Segels'" [18]; diese Notiz auch bei Qāḏīzāde fol. 100^{v},1 (wobei statt 'turs' in flüssigem 'nastaʿlīq' ein Wort dasteht, das 'bds' [unpunktiert] oder 'mds' gelesen werden könnte).

3. 33. Stern des Schiffes Argo (= σ Puppis) [19]:
ὁ μεταξὺ τῶν πηδαλίων ἐν τῇ τρόπει [var. B τροπῆι , C τροπη] [20] 'derjenige zwischen den [beiden] Steuerrudern, auf dem Kiel'. Dafür al-Ḥaǧǧāǧ: 'al-kawkab al-awsaṭ bayna [var. min] miǧḏāfay aṣ-ṣadr' 'der mittlere Stern zwischen [var. von] den beiden Rudern des Bugs'; Isḥāq/Ṯābit: 'allaḏī fī-mā bayna s-sukkānayn fī l-ḫašaba allatī ʿalayhā mabnā [var. baytā, bany] as-safīna' 'derjenige zwischen den beiden Steuerrudern, auf dem Holz(balken), auf dem sich der Aufbau des Schiffes befindet' (hiernach Gerhard von Cremona blind wörtlich: 'Que est in eo quod est inter duos temones in ligno super quod est fabricatio nauis'). Die Pariser Ṭūsī-Handschrift hat auf fol. 62va zu 'as-sukkānayn' am Rande die Glosse: 'fī n-naql al-qadīm wa-fī s-suryānī al-awsaṭ allaḏī fī miǧdāf aṣ-ṣadr' "In der alten [arabischen] und in der s y r i s c h e n Version: 'der mittlere, der auf dem Ruder des Bugs ist'" [21]; diese Notiz auch bei Qāḏīzāde fol. 100^{v},6 (wobei die Handschrift statt 'miǧdāf' schlecht 'mḥlʾb' schreibt, unpunktiert).

4. 8. Stern des Kentauren (= ψ Centauri) [22]:
τῶν ἐν τῷ θύρσῳ $\overline{\delta}$ τῶν ἡγουμένων $\overline{\beta}$ ὁ βορειότερος [var.: D οὐραίῳ statt θύρσῳ] [23] 'von den zwei vorangehenden der vier auf dem Thyrsosstab [var. D: Schwanz] der nördlichere'. Dafür al-Ḥaǧǧāǧ: 'al-kawkab aš-šamālī min al-iṯnayn al-mutaqaddimayn min al-arbaʿa allatī fī t-turs' [var. at-turs: ar-raʾs] 'der nördliche Stern von den beiden vorangehenden von den vier, die auf dem Schild sind'; Isḥāq/Ṯābit: 'aš-šamālī min al-iṯnayn al-muqaddamayn min al-arbaʿa allatī fī qaḍīb al-karm' 'der nördliche von den beiden vorderen von den vier, die auf dem Zweig des Weinstocks sind' (hiernach Gerhard von Cremona: 'Septentrionalis duarum antecedentium quattuor que sunt in | clipeo', mit Einsetzung des Terminus selbst nach der Ḥaǧǧāǧ-Form). Die Pariser Ṭūsī-Handschrift hat auf fol. 63ra zu 'qaḍīb al-karm' die Glosse: 'fī n-naql al-qadīm wa-s-suryānī fī t-turs' "In der alten [arabischen] und der s y r i s c h e n Version: 'auf dem Schild'" [24]. Diese Notiz fehlt bei Qāḏīzāde an

der entsprechenden Stelle. Zum Beschreibungstext dieses Sterns nimmt auch Ibn aṣ-Ṣalāḥ zweimal, in der Einleitung seines Traktats sowie unter seinem Stern Nr. 80 [25], Stellung und vermerkt, daß der betreffende griechische Terminus (scil. θύρσος) von Isḥāq mit 'quḍbān al-karm' 'die Weinreben, -zweige' [26] übersetzt worden sei, in der s y r i s c h e n Fassung und derjenigen von al-Ḥaǧǧāǧ (?) [27] mit at-turs 'der Schild' und in derjenigen von al-Ḥasan ibn Qurayš, d.h. also in der 'alten, maʾmūnischen', mit 'al-ḥarba' 'die Lanze'.

Diese leider nur sehr wenigen - vier [28] von insgesamt 1025 - quellengeschichtlich abgesicherten Zitate aus den Sternbeschreibungen der syrischen Almagestversion lassen es immerhin doch zu, daraus gewisse Schlüsse auf den Gehalt und das gegenseitige Verhältnis der syrischen und arabischen Almagestübersetzungen abzuleiten.

Hinsichtlich der 'alten, maʾmūnischen' Übersetzung ist zu bemerken, daß deren Wortlaut in Fall 1, 2 und 3 mit dem Wortlaut der syrischen Version übereinstimmt. In Fall 4 dagegen haben die beiden Versionen voneinander verschiedene Formulierungen (syr. 'at-turs' / 'maʾmūn.' 'al-ḥarba').

Die Übersetzung von al-Ḥaǧǧāǧ sodann zeigt eine größere Eigenständigkeit gegenüber der syrischen Version: glatte Übereinstimmung herrscht nur mehr bei Fall 1 und 4; Fall 3 weist eine Verfeinerung gegenüber der syrischen Version auf, und Fall 2 ist aus einer eigenen griechischen Lesart hervorgegangen, also 'eo ipso' von der syrischen Fassung unabhängig.

Völlige Unabhängigkeit von der syrischen Version und wesentlich engere, genauere Wiedergabe des griechischen Originals finden wir schließlich in Isḥāqs Übersetzung, die in sämtlichen Fällen eigene Formulierungen gebraucht und, anstatt zu vereinfachen oder zu vergröbern, in der Bemühung um größtmögliche Genauigkeit eher zu umständlichen weitschweifigen Umschreibungen greift.

Von keiner dieser drei arabischen Versionen läßt sich also aussagen, daß sie komplett und total aus der altersmäßig vorangehenden syrischen Übersetzung hervorgegangen wäre. Diese Meinung hatte ich aufgrund der bibliographischen Berichte und der internen Evidenz der Texte selbst auch schon an anderem Ort geäußert [29]. Wohl aber scheint es zulässig anzunehmen, daß die jeweils jüngeren arabischen Versionen, die ihrerseits auch stets aus dem Griechischen stattfanden, sich der voraufgegangenen älteren als Hilfsmittel und Korrektiv bedienten und mehr oder weniger häufig bei entsprechenden Situationen der Bequemlichkeit und Einfachheit halber bereits früher gefundene Übersetzungsformeln weiter übernahmen.

Das gilt, wie man sehen kann, in besonderem Maße für die älteste arabische Almagestübersetzung, die sogenannte 'maʾmūnische', die rein chronologisch der syrischen Version noch am nächsten stand. Ähnlich wie bei den oben vorgeführten textlichen Beispielen geschieht es übrigens auch bei den Sternkoordinaten mehrfach, daß die 'maʾmūnische' Version einen eigenen, von der syrischen Übersetzung unabhängigen Zahlenwert aufweist (neunmal unter den oben erwähnten 60 Direktzitaten [30]), während in den übrigen Fällen entweder beide Fassungen denselben Wert haben oder das Zitat nur für jeweils eine der beiden Fassungen vorliegt.

Weitere Aufschlüsse über die Textgestalt der 'alten, maʾmūnischen' Version lassen sich aus al-Battānīs Sternkatalog gewinnen, dessen Beschreibungsformeln von daher übernommen zu sein scheinen, wie gewisse Vergleiche mit Ibn aṣ-Ṣalāḥs Zitaten und den jüngeren Textfassungen zeigen [31]. Leider stehen dafür - über die oben dargestellten hinaus - keine parallelen Zitate aus der syrischen Version weiter zur Verfügung, so daß hier der terminologische Vergleich nicht weiter in die ältere Zeit hinein und auf die syrische Version ausgedehnt werden kann.

Bei al-Ḥaǧǧāǧ ist infolge der quellengeschichtlichen Situation meist nicht eindeutig festzulegen, ob die bei ihm auftretenden Bezüge zur syrischen Terminologie unmittelbar auf einer Mitbenutzung der syrischen Version selbst beruhen, oder ob er sie nicht vielmehr lediglich über das Medium der 'alten' arabischen Version aufnahm, die ihrerseits die syrische Fassung mitverwertet hatte.

Wenn man das Vokabular der Sternbeschreibungen durchsieht, so stößt man auf viele arabische Termini, die sich mit entsprechenden syrischen Wörtern berühren. Das muß aber nicht unbedingt in jedem Falle die Übernahme syrischer Formeln durch die arabischen Übersetzer bedeuten, sondern das kann weitgehend einfach auf der gemeinsamen semitischen Sprachverwandtschaft der beiden Idiome beruhen, die es eben mit sich bringt, daß beide für dieselben Gegenstände etymologisch verwandte Benennungen haben [32]. An echte Syriazismen sollte man nur dann denken, wenn der arabische Übersetzer ein dem Syrischen nahestehendes Wort wählt, obwohl es auch andere Möglichkeiten im Arabischen gegeben hätte, die dann später teilweise auch wirklich realisiert wurden [33].

An den vorstehenden Vergleichen läßt sich ablesen, wie die arabischen Almagesttexte von Stufe zu Stufe eine zunehmende Verselbständigung und Verbesserung durchmachen. Das ist auf der anderen Seite auch daran zu verfolgen, wie die Astronomen der arabisch-islamischen Welt, für die der 'Almagest' ein praktisches Handwerkszeug darstellte, sich immer der jeweils 'besten' brauchbaren Textfassung zuwandten. Mit der 'alten' Version scheinen nur frühe Astronomen, wie al-Battānī [34], gearbeitet zu haben. Die Ḥaǧǧāǧ-Fassung wurde u.a. benutzt von al-Farġānī, al-Kindī und dem Historiker al-Yaᶜqūbī, im 9. Jahrhundert [35]; im islamischen Westen war sie auch noch im 12. Jahrhundert in Gebrauch, wie einer Angabe des dortigen Astronomen Ǧābir ibn Aflaḥ (gest. um 1150) zu entnehmen ist [36], und nach ihr hat auch Gerhard von Cremona in Toledo seine lateinische Almagestübersetzung begonnen (Buch I-IX; der Rest nach Isḥāq/Ṯābit), die 1175 abgeschlossen war. Ibn aṣ-Ṣalāḥ zitiert die Ḥaǧǧāǧ-Fassung übrigens auffälligerweise sehr selten, wobei obendrein in den Handschriften noch merkwürdige Unsicherheiten und Lücken auftreten [37]. Die später meist - im islamischen Osten wohl sogar allein - benutzte Fassung scheint dann diejenige von Isḥāq in Ṯābits Bearbeitung gewesen zu sein: nach dieser arbeiten und zitieren z.B. aṣ-Ṣūfī, Ibn aṣ-Ṣalāḥ und aṭ-Ṭūsī (und über diese weiterhin Uluġ Bēg, Qāḍīzāde und andere) [38].

Wenn so auch die ältesten Almagestversionen, die syrische und die 'alte' arabische, allmählich ganz aus dem Blickfeld der praktizierenden Astronomen geraten zu sein scheinen, beweisen doch die oben vorgelegten Testimonia, daß sie nichtsdestoweniger noch viele Jahrhunderte lang weiter existierten und greifbar blieben. Historisch interessierte Kritiker und Kommentatoren konnten auch im 12. und 13. Jahrhundert und sogar noch später immer wieder darauf zurückgreifen. Um 1150 zitiert Ibn aṣ-Ṣalāḥ in Bagdad in aller Ausführlichkeit daraus; vereinzelte Anzeichen bei aṭ-Ṭūsī selbst machen es wahrscheinlich, daß auch er, um 1250, über die reine Isḥāq/Ṯābit-Tradition hinaus zusätzliches Material, darunter womöglich auch syrisches, benutzte [39]; und in späteren Ṭūsī-Kommentaren (Qāḍīzāde, um 1415; der Glossator der Pariser Ṭūsī-Handschrift) kommen erneut wörtliche Zitate aus der syrischen und der 'alten' arabischen Version zu Tage.

Die wissenschaftsgeschichtliche Tradition aus der Spätantike in die arabisch-islamische Welt hatte also nicht mit den klassischen Übersetzungen des 9. Jahrhunderts ihren Abschluß gefunden. Die Kontinuität blieb bis in die Spätzeit gewahrt, wenngleich das Material nun nicht mehr primär naturwissenschaftlichen Zwecken diente, sondern zu historisch-philologischen Analysen und Vergleichen verwandt wurde.

Anmerkungen

[1] Cf. P. KUNITZSCH, 'Der Almagest. Die Syntaxis Mathematica des Claudius Ptolemäus in arabisch-lateinischer Überlieferung', Wiesbaden 1974 (künftig zitiert als: KUNITZSCH [1]), S. 7 f., 17 ff., 23 f., 59 f. (Fihrist, Ibn al-Qifṭī, Ibn aṣ-Ṣalāḥ, Ḥāǧǧī Ḫalīfa).

[2] 'Encyclopaedia of Philosophical and Natural Sciences as Taught in Bagdad about A.D. 817 or Book of Treasures by Job of Edessa', ed. u. übers. A. MINGANA, Cambridge 1935. Cf. MINGANAs Einleitung S. XIX und XLVI f.

[3] Bei ihm genannt 'Das Rechenbuch des Qlaudios Ptolemaios', cf. E. HONIGMANN, 'Die sieben Klimata und die ΠΟΛΕΙΣ ΕΠΙΣΗΜΟΙ ', Heidelberg 1929, S. 117.

[4] Cf. P. KUNITZSCH, 'Arabische Sternnamen in Europa', Wiesbaden 1959, S. 32 f.

[5] Ed. u. übers. F. NAU, in: 'Revue de l'Orient Chrétien' 27 (1929/30), 327-410, und 28 (1931/32), 85-100.

[6] Cf. auch KUNITZSCH [1], S. 8 f., 66, 78[198], 80[205], 194[145].

[7] 'Al-Battānī sive Albatenii Opus astronomicum', ed., übers. u. komm. C.A. NALLINO; hier Text vol. 3 (Mailand 1899), S. 243; Übers. vol. 2 (ib. 1907), S. 92 (Kommentar dazu ib. S. 232-235); cf. auch KUNITZSCH [1], S. 7 f. Die Übersetzung von Isḥāq ibn Ḥunayn hat dafür übrigens 'zāfurus', mit ā für ε (ms 'rʾqws').

[8] Cf. die Ausgabe von P. KUNITZSCH, 'Ibn aṣ-Ṣalāḥ: Zur Kritik der Koordinatenüberlieferung im Sternkatalog des Almagest', Göttingen 1975 (Abh. d. Akad. d. Wiss. in Göttingen, Philol.-hist. Kl., 3. F., Nr. 94), künftig zitiert als: KUNITZSCH [2].

[9] Erhalten in ms Berlin, or. oct. 274 (= Ahlwardt 5657); cf. die ausführliche Darstellung in 'Anhang V' der oben in Anm. 8 erwähnten Ibn aṣ-Ṣalāḥ-Ausgabe.

[10] Der 1305 vollendete Kommentar von Niẓām ad-Dīn an-Nīsābūrī zu aṭ-Ṭūsīs 'Taḥrīr al-mǧsṭī' kann es nicht sein, wie an der in Anm. 9 genannten Stelle ermittelt wurde.

[11] Für die Handschriften und Fundstellen der verschiedenen im folgenden zitierten Almagestversionen kann grundsätzlich verwiesen werden auf KUNITZSCH [1], ebenso für den Text von Ibn aṣ-Ṣalāḥ auf KUNITZSCH [2].

[12] Cf. KUNITZSCH [1], S. 296 f., Nr. 409.

[13] Ptolemaeus, Opera, ed. J.L. HEIBERG, vol. I, pars I, Leipzig 1903, S. 114,14.

[14] Cf. KUNITZSCH [1], S. 329, Nr. 561/562. Bei al-Battānī ist hier ein Blatt des arabischen Textes verloren gegangen, so daß seine Formulierung nicht authentisch nachprüfbar ist; NALLINO zitiert dafür ersatzweise die altspanische Battānī-Übersetzung in ms Paris, Arsenal 8322, was für unsere Vergleiche jedoch nicht ausreicht.

[15] Ptolemaeus, Opera (wie Anm. 13), S. 148,19.

[16] D.h. der Übersetzer hatte den griechischen Text in folgender Form gelesen: τῶν ἐπὶ ταῖς ἀσπιδίσκαις [ὡς ἐπὶ] τοῦ ἱστοῦ γ̄ ὁ βόρειος

[17] Der Übersetzer gibt zwar den bei HEIBERG vorhandenen griechischen Text vollständig wieder, las jedoch statt des (vermutlich zu stark abgekürzten, cf. die Variante aus mss ACD) τῆς ἱστοδόκης ebenfalls abweichend: τοῦ ἱστοῦ .

[18] Hier liegt eine Lesung ... τῆς ἀσπιδίσκης [ὡς ἐπὶ] τοῦ ἱστίου ...

im Griechischen zugrunde. Diese Lesung (τῆς ἀσπιδίσκης , im Singular) wirkt umso plausibler, als Ptolemäus in den Beschreibungsformeln des Sternkatalogs die Präposition ἐπί überwiegend mit dem Genitiv (und nicht mit dem Dativ) gebraucht.

[19] Cf. KUNITZSCH [1], S. 331 f., Nr. 567/568. Betreffs al-Battānī cf. oben Anm. 14.

[20] Ptolemaeus, Opera (wie Anm. 13), S. 150,12.

[21] Hier (und ebenso bei al-Ḥaǧǧāǧ) muß statt ἐν τῇ τρόπει ein anderes griechisches Wort mit der Bedeutung 'Bug' gelesen worden sein, also vermutlich ... πηδαλίων ἐν τῇ πρῴρᾳ (oder ....' τῆς πρῴρας).

[22] Cf. KUNITZSCH [1], S. 330 f., Nr. 602. Der Battānī-Text ist hier zwar wieder vollständig, aber diesen Stern hat al-Battānī in seinen gegenüber dem 'Almagest' um rund die Hälfte gekürzten Sternkatalog nicht mit aufgenommen.

[23] Ptolemaeus, Opera (wie Anm. 13), S. 158,14.

[24] Die Übersetzung 'Schild' beruht auf einer Verlesung ἐν τῷ θυρεῷ für ἐν τῷ θύρσῳ im griechischen Text.

[25] Cf. KUNITZSCH [2], S. 45 und S. 70 f., Nr. 80.

[26] Sic im Plural; die gesamte sonstige Überlieferung dagegen hat einheitlich immer den Singular.

[27] Die Handschriften haben hier (in der Einleitung) versehentlich 'der maʾmūnischen' statt 'derjenigen von al-Ḥaǧǧāǧ'. Die 'maʾmūnische' wird jedoch sogleich anschließend noch extra mit der Übersetzung 'al-ḥarba' zitiert, und aus der Überlieferung ist at-turs von al-Ḥaǧǧāǧ her bekannt; also sind die Ibn aṣ-Ṣalāḥ-Handschriften an dieser Stelle entsprechend zu emendieren. Derselbe Fehler bei der Zuweisung des Zitats 'at-turs' hat sich aber später fortgesetzt und findet sich z.B. auch in der oben angeführten Glosse der Pariser Ṭūsī-Handschrift. Weiter unten, bei Stern Nr. 80, wird 'at-turs' nur aus der syrischen Version zitiert, so daß hier dieser Fehler nicht auftritt. Hier fügt Ibn aṣ-Ṣalāḥ außerdem noch hinzu, daß 'der Schild' ('at-turs') auf Syrisch 'ṣakrā' heiße, was in der Tat stimmt (cf. C. BROCKELMANN, Lexicon Syriacum, ²Halle 1928, S. 475a).

[28] Ein weiteres textliches Zitat aus der syrischen Version bei Ibn aṣ-Ṣalāḥ (dort Stern Nr. 20) wird hier nicht ausführlich behandelt. Es macht zwar eine uns nicht bekannte Variante im griechischen Text sichtbar, trägt aber nichts Wesentliches für die uns hier hauptsächlich beschäftigende Scheidung der Versionen bei. Dabei geht es um den 14. Stern des Fuhrmanns (eventuell = Fl. 5 Aurigae), der nach Ptolemäus (Opera, wie oben in Anm. 13, S. 66,19) über (= ὑπέρ) dem linken Fuß steht (ohne Varianten im Apparat). Diese Präposition müssen auch der syrische und die 'maʾmūnischen' Übersetzer gelesen haben, für die Ibn aṣ-Ṣalāḥ a.a.O. zitiert: 'aṣ-ṣaġīr allaḏī fī aʿlā ar-riǧl al-yusrā' 'der kleine (Stern), der oberhalb des linken Fußes ist'. Auch die Ḥaǧǧāǧ-Überlieferung folgt dieser Lesart: 'al-kawkab aṣ-ṣaġīr allaḏī ʿalā (über) ar-riǧl al-yusrā'. Isḥāq und Ṯābit dagegen scheinen stattdessen in ihrer griechischen Vorlage ὑπό 'unter' gelesen zu haben, denn sie schreiben: 'aṣ-ṣaġīr allaḏī taḥta r-riǧl al-yusrā' 'der kleine, der unter dem linken Fuß ist' (hiernach weiter Gerhard von Cremona, jedoch mit Einsetzung der Präposition selbst nach al-Ḥaǧǧāǧ: 'Minor que est | super | pedem sinistrum').

[29] KUNITZSCH [1], S. 8, 23 f., 33, 59 f.

[30] Bei Ibn aṣ-Ṣalāḥ die Sterne Nr. 12, 21, 42, 54, 61, 62, 70, 81 und 84 (vgl. KUNITZSCH [2], S. 82, Anm. 29).

[31] Z.B. Ibn aṣ-Ṣalāḥ bei Stern Nr. 24 (der 1. Stern des Pfeils [= γ Sagittae]); cf. im übrigen KUNITZSCH [1], S. 51, und KUNITZSCH [2], S. 101, Anm. 16. Dasselbe gilt nachweislich ebenfalls für die Koordinatenwerte in seinem Sternkatalog, vgl. P. KUNITZSCH in: Centaurus 18 (1974), 270-274, und 'Anhang II' bei KUNITZSCH [2].

[32] Vgl. z.B. 'Bär, Ἄρκτος': syr. 'debbā', arab. 'dubb'; 'Drache, Δράκων': syr. 'tan(n)īnā', arab. 'tinnīn'; 'Krone, Στέφανος': syr. 'klīlā', arab. 'iklīl'; 'Fluß, Ποταμός': syr. 'nahrā', arab. 'nahr'; 'Hund, Κύων': syr. 'kalbā', arab. 'kalb'; 'Hase, Λαγωός': syr. 'arnᵉbā', arab. 'arnab'; 'Heck, πρύμνα': syr. 'kōtlā', arab. 'kawtal'; 'Steuerruder, πηδάλιον': syr. 'saukānā', arab. 'sukkān', usw. Hier könnte man schwerlich von einer arabischen 'Übernahme' des syrischen Terminus sprechen.

[33] Vgl. z.B. arab. 'ᶜurqūb' 'Achillessehne' (für griech. σφυρόν [Fuß-] Knöchel), oben in Fall 1, und syr. 'ᶜarqūbā' (genu [?], calx, tendo Achillis), BROCKELMANN, Lex. Syr. S. 551 a. Daneben steht das passendere arab. 'kaᶜb' (Fuß-)Knöchel, das dann später tatsächlich Isḥāq wählt. Ein anderes Beispiel ist älteres arab. 'katif' für griech. ὦμος 'Schulter' (und syr. 'kaṯpā' humerus; BROCKELMANN, Lex. Syr. S. 353 a) neben dem später aufkommenden besser passenden 'mankib' (cf. dazu ausführlich KUNITZSCH [1], S. 71 ff.). Freilich fehlt hierfür ein entsprechendes direktes Belegzitat, und bei Severus Sēbōkt können wir sogar beobachten, daß er für 'épaule' die syrischen Wörter 'rapšā' (wie oben Anm. 5, S. 375,19 [bei Ursa Minor] und 376,9 [bei Pegasus]) und 'kaṯpā' (ib. S. 376,18 bei Orion) durcheinander verwendet. Die arabische Paraphrase 'al-ǧabbār' für 'Orion' könnte ebenfalls auf ein syrisches Vorbild zurückgehen, cf. BROCKELMANN, Lex. Syr. S. 103 a (s.v. gabbārā) und KUNITZSCH [1], S. 194 f. mit Anm. 145; die einheimische altarabische Bezeichnung war dagegen 'al-ǧawzā''.

[34] Eventuell auch Kūšyār ibn Labbān und einige kleinere sehr alte Sternverzeichnisse; cf. KUNITZSCH [1], S. 57 f., und KUNITZSCH [2], Anhang II.

[35] Cf. KUNITZSCH [1], S. 64 und die daselbst in Anm. 164 genannten weiteren Stellen.

[36] Siehe F.J. CARMODY, 'Al-Biṭrûjî, De motibus celorum', Berkely and Los Angeles 1952, S. 31 f., und KUNITZSCH [1], S. 36 f., Anm. 87.

[37] Cf. dazu KUNITZSCH [2], S. 32 und 80.

[38] Cf. KUNITZSCH [1], S. 71 sowie 47, 51, 56. Eine Sonderstellung nimmt al-Bīrūnī ein, cf. KUNITZSCH [1], S. 52 f., und KUNITZSCH [2], Anhang IV.

[39] Cf. KUNITZSCH [1], S. 334 mit Anm. 189; dazu ebenfalls ders., 'Arabische Rückgriffe auf griechische Quellen in der Spätzeit ...', in: Akten des VII. Kongresses für Arabistik und Islamwissenschaft, Göttingen 1976 (Abh. Ak. d. Wiss. in Göttingen, 3. F., Nr. 98), S. 236 ff., speziell 237-240.

IV

DIE ASTRONOMISCHE TERMINOLOGIE IM ALMAGEST

Die wissentschaftliche Terminologie der Astronomie im Arabischen – d.h. bei den Arabern und sodann im weiteren Sinn in der arabisch-islamischen Kultur – entstand nicht auf einmal, aus dem Nichts heraus. Wie viele andere Wissenschaften in der islamischen Welt bildete sich auch die Astronomie nach fremden Vorbildern: die "wissenschaftliche" Astronomie entstand in Anlehnung an die antike griechisch-hellenistische Wissenschaft, die den Arabern und Muslimen durch Übersetzungen einschlägiger Werke ins Arabische bekannt wurde. Aber die arabische Sprache war nicht gänzlich unvorbereitet für die Aufnahme des neuen, fremden Wissensgutes. Bereits in altarabischer, vor- und frühislamischer Zeit hatte es bei den Arabern eine volkstümliche, auf Beobachtung und Erfahrung beruhende Himmelskenntnis mit zugehöriger Nomenklatur und Terminologie gegeben.

Zeugnisse der volkstümlichen altarabischen Himmelskenntnis finden sich häufig im Koran. Später, vor allem im neunten und zehnten Jahrhundert, haben arabische Philologen und Lexikographen das in der Überlieferung verstreute Material zur altarabischen Himmelskenntnis in speziellen Monographien, den sogenannten *anwā'*-Büchern, gesammelt[1].

Zahlreich sind im Koran vertreten die Begriffe Himmel, *al-samā'*, Sonne, *al-šams*, und Mond, *al-qamar*[2]. Häufig kommen auch generell die Sterne vor, *naǧm* – pl. *nuǧūm* und *kaukab* – pl. *kawākib*; einmal (53, 49) wird speziell der Sirius genannt, *al-ši'rā*. Ungeklärt ist, ob mit *al-ḫunnas al-ǧawārī al-kunnas* (81, 15-16; "die rückläufigen [Planeten ?], die [am Himmelsgewölbe] dahinziehen und [immer wieder] ihr Versteck aufsuchen [und unsichtbar werden] [?]"[3]) wirklich die Planeten gemeint sind oder andere Himmelsobjekte oder überhaupt etwas anderes. Auch *burǧ* – pl. *burūǧ* (15, 16; 25, 61; 85, 1; hier wohl noch allgemein "Zeichen, Sternbilder", später in der "wissenschaftlichen" Astronomie

1. Cf. EI², Artikel *Anwā'* (C. Pellat) und *al-Manāzil* (P. Kunitzsch; nachgedr. in P. Kunitzsch, *The Arabs and the Stars*, Northampton 1989, Teil XX). Die *anwā'*-Autoren mischten gelegentlich auch Elemente der zu ihrer Zeit bereits eingeführten fremdbestimmten "wissenschaftlichen" Astronomie mit unter ihre Informationen.

2. S. die Korankonkordanz von M.F. ᶜAbdalbāqī, *al-Muᶜǧam al-mufahras li-alfāẓ al-qur'ān al-karīm*, Kairo ¹1364 [1945]; verbesserte Auflage 1378 [1958-59].

3. Koranstellen werden zitiert nach der deutschen Übersetzung von R. Paret, *Der Koran*, Stuttgart etc. 1962.

terminologisch "Tierkreiszeichen"[4]) und *manāzil* kommen vor (10, 5; 36, 39; dies ziemlich eindeutig als die bekannten "Mondstationen"[5] zu verstehen). Besonders interessant in unserem Zusammenhang erscheinen die beiden auf Arabisch gleichlautenden Stellen 21, 33 und 36, 40: (*wa-*) *kullun fī falakin yasbaḥūna*, Paret: "Alle (Gestirne) schweben an einem Himmelsgewölbe". Da das Verb im Plural steht (und nicht im Dual - somit also schwerlich nur auf die zuvor genannten beiden Gestirne Sonne und Mond bezogen[6]), erscheint die Auffassung des Subjekts als "alle (Gestirne)" gerechtfertigt; das weitere bei Paret ist aber missverständlich. Eher wäre zu übersetzen: "Ein jedes (Gestirn) schwebt in einer Sphäre". Hiermit käme die Aussage des Korans der bekannten Lehre der antiken Astronomie und Kosmologie nahe: jeder Planet schwebt, bewegt sich in/auf seiner Sphäre. Freilich zögert man, die Kenntnis der fremden, nichtarabischen Kosmologie bei Muḥammad vorauszusetzen, und man wird statt streng terminologisch "jeder Planet in seiner Sphäre" eher volkstümlich-diffus interpretieren müssen "jedes (Gestirn) in einer (eigenen) Sphäre".

Die fünf der alten Welt bekannten Planeten waren offensichtlich auch den alten Arabern bekannt, da sie dafür echte alte, in historischer Zeit in ihrer Bedeutung nicht mehr genau fassbare Namen besassen: *ʿuṭārid* (Merkur), *al-zuhara* (Venus), *al-mirrīḫ* (Mars), *al-muštarī* (Jupiter) und *zuḥal* (Saturn)[7]. Dagegen scheint es keinen besonderen Terminus für sie als "Planeten" gegeben zu haben. Die beiden Wörter *naǧm* und *kaukab* wurden unterschiedslos auf alle Arten von Sternen angewendet. Später, in der "wissenschaflichen" Astronomie und in der Astrologie (die im Altertum und Mittelalter - gewissermassen als die "praktische Anwendung" - untrennbar mit der Astronomie verbunden war), wird im Sinne von "Planet" bevorzugt *kaukab* verwendet.

Für Fixsterne überliefern die *anwāʾ*-Bücher mehrere hundert altarabische Namen[8]. Bei den bekanntesten, hellsten Sternen haben später auch die "wissenschaftlichen" Astronomen diese altarabischen Namen weiterbenutzt, neben neuen, aus dem griechischen Fundus abgeleiteten Formen.

4. CF. EI², Artikel *Minṭakat al-Burūdj* (P. KUNITZSCH).

5. CF. EI², *al-Manāzil* (wie oben, Anm. 1).

6. So fasst W. Hartner die Aussage auf: "each of which [scil. the Sun and the Moon] moves in its own sphere", in EI², Artikel *Falak*.

7. Zur Etymologie der Namen siehe W. EILERS, *Sinn und Herkunft der Planetennamen*, München 1976 (Bayerische Akademie der Wissenschaften, Phil.-Hist. Kl., Sitzungsberichte, Jahrg. 1975, Heft 5).

8. Hierzu P. KUNITZSCH, *Untersuchungen zur Sternnomenklatur der Araber*, Wiesbaden 1961; ders., *Über eine anwāʾ-Tradition mit bisher unbekannten Sternnamen*, München 1983 (Bayerische Akademie der Wissenschaften, Phil.-Hist. Kl., Sitzungsberichte, Jahrg. 1983, Heft 5); EI², Artikel *al-Nudjūm* (P. KUNITZSCH).

Auch eine Reihe von Grundbegriffen der Himmelskunde, wie die Namen der vier Haupthimmelsrichtungen, Bezeichnungen für Auf- und Untergang usw. gehörten seit alters der Allgemeinsprache an und wurden später in der "wissenschaftlichen" Astronomie beibehalten.

Wenn wir uns nun der wissenschaftlichen astronomischen Terminologie im Almagest zuwenden, so ist es auch hier so, dass mit einem diffusen Bestand an "Vorauskenntnissen"[9] zu rechnen ist, die sich im Arabischen vor den direkten Übersetzungen eingebürgert haben müssen. Während die verschiedenen Übersetzungen des Almagest etwa in das Jahrhundert vom Ende des achten bis zum Ende des neunten Jahrhunderts fallen, gibt es sowohl erhaltene ältere Übersetzungen (meist astrologischer Texte, vor allem über das Persische gekommen) als auch andere historische Hinweise auf fremdbestimmte astronomisch-astrologische Praxis (z.B. die astronomischen Schriften - auf Syrisch - von Severus Sebokht, Bischof von Qinnasrīn, um 660[10]; die Kuppel im Bad des Wüstenschlosses von Quṣair ᶜAmra mit einer Abbildung des antiken Sternhimmels, um 711-15[11]; die Zitierung der Beschreibung eines griechischen Himmelsglobus für eine Epoche ca. 738 bei Ibn al-Ṣalāḥ, gest. 1154[12]; die Berechnung des günstigen Zeitpunkts für die Grundsteinlegung der Kalifenresidenz Bagdad durch ein Kollegium von Astrologen 762[13]). Man muss also voraussetzen, dass bereits vor den eigentlichen, uns noch erhaltenen Textübersetzungen im arabisch-islamischen Raum Formen von Beschäftigung mit antiker Astrologie (und damit verbunden: Astronomie), mindestens teilweise wohl in mündlichem Kontakt mit entsprechenden Kennern der spätantiken Überlieferung, die ihrerseits natürlich die einschlägige Literatur besassen und benutzten, abliefen, die frühe Formen einer astronomisch-astrologischen Terminologie im Arabischen begründeten und es später den Übersetzern erleichterten, das dem Arabischen gänzlich fremde, neue Material mit einer gewissen Leichtigkeit und Eleganz zu übertragen.

9. Vgl. P. KUNITZSCH *Über das Frühstadium der arabischen Aneignung antiken Gutes*, in *Saeculum* 26 (1975), 268-282.

10. *Le traité sur les constellations*, ed. u. übers. F. NAU, in *Revue de l'Orient Chrétien* 27 (1929-30), 327-410, und 28 (1931-32), 85-100; *Le traité de Severus Sebokt sur l'astrolabe plan*, ed. u. übers. F. NAU, in *Journal Asiatique* 1899, S. 58-101 und 238-303; dazu noch F. NAU, *Notes d'astronomie syrienne*, in *Journal Asiatique* 1910, S. 209-228.

11. F. SAXL, *The zodiac of Quṣayr ᶜAmra*, in K.A.C. CRESWELL, *Early Muslim Architecture*, I, Oxford 1932, 289-295; A. BEER, *The Astronomical Significance of the Zodiac of Quṣayr ᶜAmra*, ib. 296-303; A. BEER, *Astronomical Dating of Works of Art*, in *Vistas in Astronomy* 9 (1967), 177-187 (mit einer Ergänzung von W. HARTNER, ib. 225); M. ALMAGRO et al., *Qusayr ᶜAmra, Residencia y baños omeyas en el desierte de Jordania*, Madrid 1975.

12. IBN AL-ṢALĀḤ, *Zur Kritik der Koordinatenüberlieferung im Sternkatalog des Almagest*, ed. u. übers. P. KUNITZSCH, Göttingen 1975 (Abh. d. Akademie der Wissenschaften in Göttingen, Phil.-Hist. Kl., 3. Folge, Nr. 94), 18, 72f., 29 (arab.).

13. Cf. die Stellen bei Kunitzsch 1989 (wie Anm. 1), Teil XXV, S. 212 Anm. 15.

Wie aus der biobibliographischen Literatur (vor allem Ibn al-Nadīm, *Fihrist*) und anderen Hinweisen zu entnehmen ist, wurde der Almagest dreimal ins Arabische übertragen (davor lag noch eine syrische Übersetzung). Es gab eine "alte" Übersetzung, gegen Ende des achten/Anfang des neunten Jahrhunderts, dann die Übersetzung von al-Ḥaǧǧāǧ ibn Yūsuf ibn Maṭar in Zusammenarbeit mit Sarǧūn ibn Hilīyā (Sergius, Sohn des Elias) al-Rūmī, datiert 212/827-28 (künftig: H), und schliesslich die Übersetzung von Isḥāq ibn Ḥunain, um 890; letztere wurde bald darauf von Ṯābit ibn Qurra (gest. 901) "überarbeitet" (*iṣlāḥ*; diese überarbeitete Fassung künftig: I/T)[14].

Von diesen Versionen haben sich bis heute Handschriften von H (vor allem: Ms. Leiden, or. 680, vor 615/1218-19, vollständig) und I/T (vor allem: Ms. Tunis 07116, dat. 478/1085, ebenfalls vollständig) erhalten[15]. Die "alte" Version wurde – wie aus Vergleichen bei den Koordinatenwerten im Sternkatalog hervorgeht[16] – von al-Battānī (gest. 929) bei der Abfassung seines *al-Zīǧ al-ṣābi'*[17] zugrunde gelegt. Der 1154 gestorbene Arzt und Wissenschaftskritiker Ibn al-Ṣalāḥ kannte noch fünf damals existierende Textfassungen des Almagest und zitiert daraus laufend Koordinatenwerte in seiner Kritik der Überlieferung des ptolemäischen Sternkatalogs im Almagest[18]: die syrische Version, die "alte" arabische Fassung, H, Isḥāq (sogar, wie eigens mehrmals vermerkt, im Autograph!) und I/T.

Es wird nun nicht möglich sein, wie es eigentlich wünschenswert wäre, hier eine vollständige Übersicht und Interpretation der astronomischen Terminologie in den beiden erhaltenen arabischen Versionen H und I/T des Almagest zu geben, denn diese Texte sind nach wie vor (im ganzen) nicht ediert. In kritischer Edition (nach beiden Versionen) liegt lediglich der Sternkatalog aus Buch VII-VIII des Almagest vor[19], und hierzu gibt es Analysen zur Terminologie sowie

14. Nähere Einzelheiten hierzu bei P. KUNITZSCH, *Der Almagest. Die Syntaxis Mathematica des Claudius Ptolemäus in arabisch-lateinischer Überlieferung*, Wiesbaden 1974, 15ff. – Das Ausmass der Eingriffe Ṯābits in Isḥāqs Text lässt sich in den erhaltenen Handschriften von I/T nicht erkennen; cf. auch KUNITZSCH, loc. cit. 68.

15. Cf. KUNITZSCH 1974 (wie Anm. 14), 34ff.; ders. 1986 (wie Anm. 19), 3f., 5ff., 10ff.

16. Cf. KUNITZSCH 1989 (wie Anm. 1), Teil V; ders., Ed. Ibn al-Ṣalāḥ (wie Anm. 12), Anhang II, S. 97-105.

17. Herausgeg. von C.A. NALLINO, *Al-Battānī sive Albatenii Opus astronomicum*, I-III, Mailand 1899-1907.

18. Vgl. Anm. 12.

19. CLAUDIUS PTOLEMÄUS, *Der Sternkatalog des Almagest. Die arabisch-mittelalterliche Tradition*, I: *Die arabischen Übersetzungen*, Hgb. u. übers P. KUNITZSCH, Wiesbaden 1986; II: *Die lateinische Übersetzung Gerhards von Cremona*, Hgb. von P. K., ib. 1990; III: *Gesamtkonkordanz der Sternkoordinaten*, Bearb. von P. K., ib. 1991.

vollständige Verzeichnisse der Nomenklatur der 48 ptolemäischen Sternbilder[20] und der 1025 Einzelsterne[21].

Vorzüglich erschlossen ist bereits der *Zīǧ* von al-Battānī, dessen Edition der Herausgeber C.A. Nallino ein ausführliches Glossar der Terminologie (in arabisch und lateinisch) beigegeben hat[22]. Eine weitere Vorarbeit hat M. Klamroth geleistet, der ein Glossar der Termini in den Kapitelüberschriften sämtlicher Bücher des Almagest in der Version H (nach der Leidener Handschrift) erstellt hat[23]. Als weiteres Hilfsmittel aus jüngster Zeit ist ein Aufsatz von R. Lorch zu nennen, der freilich nur die mittelalterliche lateinische Terminologie behandelt, darunter aber viele aus dem Arabischen übersetzte Texte[24]. An dieser Stelle soll zumindest darauf hingewiesen werden, dass die astronomische Fachterminologie im Almagest (wie natürlich darüber hinaus in vielen anderen Texten) durch die Übersetzungen ins Lateinische in Spanien im 12. Jahrhundert auch für die Entwicklung der europäischen Terminologie von grosser Bedeutung war. In die moderne Astronomie ist davon freilich nicht viel übergegangen, da die ptolemäische Astronomie – insbesondere mit ihren Theorien von Sonne, Mond und Planeten – durch die kopernikanische Revolution ausser Kraft gesetzt wurde, so dass die ältere arabisch-lateinische astronomische Terminologie heute weitgehend obsolet geworden ist.

Entsprechend der geschilderten Situation hinsichtlich der Aufarbeitung der Texte soll im folgenden zuerst gesondert über den Sternkatalog referiert werden, danach über die anderen Teile des Almagest.

Der Sternkatalog mit seiner stereotypen Auflistung der 1025 Sterne in 48 Sternbildern ist für astronomisch-terminologische Beobachtungen wenig ergiebig[25]. Der Begriff D Sternbild (E constellation), ἀστερισμός,

20. KUNITZSCH 1974 (wie Anm. 14), 169-212.

21. Ibid., 217-370. *Terminologisches* ib. 71ff., 171f., 214ff.

22. Vgl. Anm. 17.

23. M. KLAMROTH, *Ja'qûbî's Auszüge aus griechischen Schriftstellern*, in *Zeitschrift der Deutschen Morgenländischen Gesellschaft* 42 (1888), Ptolemäus: 17-34, spez. das Glossar: 28-34.

24. R. LORCH, *Astronomical Terminology*, in *Méthodes et instruments du travail intellectuel au moyen âge, Etudes sur le vocabulaire*, éd. Olga WEIJERS, Turnhout 1990 (CIVICIMA, *Etudes sur le vocabulaire intellectuel du moyen âge*, III), 182-196.

25. In den kommenden Zitaten finden neben den bereits eingeführten H und I/T noch folgende Abkürzungen Verwendung: G = griechischer Text der Μαθηματικὴ σύνταξις, ed. von J.L. HEIBERG: *Ptolemaeus, Opera*, vol. I, pars I-II, Leipzig: Teubner 1898-1903; D = deutsche Übersetzung aus dem Griechischen von K. MANITIUS: *Ptolemäus, Handbuch der Astronomie*, I-II, Leipzig: Teubner, ¹1912-13; ²1963; E = englische Übersetzung aus dem Griechischen von G.J. TOOMER: *Ptolemy's Almagest, Translated and annotated*, London bzw. New York etc. 1984; L = lateinische Übersetzung aus dem Arabischen von Gerhard von Cremona, Toledo, um 1150-80 (für den Sternkatalog nach der Ed. KUNITZSCH, vol. II - vgl. Anm. 19, für andere Textstellen nach Handschriften).

wurde im Arabischen analog als Verbalsubstantiv[26] *kaukaba* (wie ein Infinitiv eines vierradikalen arabischen Verbs *kwkb*) übertragen (so auch in L nachempfunden als *stellatio*). Diese Form der Übertragung könnte man als "etymologisierende Übersetzung" kennzeichnen, da sie Bildungskomponenten oder - prinzipien griechischer Wörter aufspürt und daraus entsprechende Übersetzungsformeln konstruiert. In jüngeren Handschriften und in abgeleiteten Texten findet man vereinfacht *ṣūra* "(Stern-)Bild" sowie die Formel *kawākib* + Genit., "die Sterne von N.N.".

Häufig zählt Ptolemäus unter einem Sternbild Sterne mit auf, die zwar in enger räumlicher Nähe zu dem betreffenden Bild stehen, die aber nicht konstituierend an seiner Ausgestaltung teilhaben; solche Sterne bezeichnet Ptolemäus mit dem Ausdruck ἀμόρφωτος (D nicht in das Bild miteinbezogen, von anderen oft "externe Sterne" genannt; E outside the constellation); hierfür schwankt die arabische Wiedergabe, am häufigsten findet sich *laisa(t) fī ṣūra* = L *non est/sunt in forma.*

Neun Sterne bezeichnet Ptolemäus, anstatt ihnen eine der von ihm aufgestellten sechs Grössenklassen (G μέγεθος, A *ʿiẓam*) zuzuordnen, als ἀμαυρός "dunkel" = H *muẓlim* (= L *tenebrosus*), I/T *ḫafīy* (= L *occultus*).

Astronomisch interessant ist im Sternkatalog vor allem noch die Farbangabe ὑπόκιρρος (D rötlich, E reddish), die Ptolemäus sechs Sternen[27] beigegeben hat. H überträgt (infolge eines sprachlichen Missverständnisses: ὑπόκιρρος falsch mit κηρός "Wachs" in Verbindung gebracht) den Terminus an allen Stellen als *šamʿī* "wächsern" (so auch L: *cerea,* aber bei α Boo mit abweichender Punktierung als *audiens*); I/T hat eine Formel des Typus *yaḍribu ilā l-...*, "er schlägt in Richtung, neigt hin zu...", wobei das eigentliche Substantiv der Farbangabe in den erhaltenen Handschriften an allen Stellen zu einem unkenntlichen

26. Gleichsam "Gestirnung, Gestaltung als Sternbild", zu dem Verb ἀστερίζειν.

27. Baily Nr.110, α Boo; cf. Kunitzsch 1974 (wie Anm. 14), S. 230, Nr. 71;
393, α Tau; 267, 279;
425, β Gem; 271, 300;
553, α Sco; 291, 386;
735, α Ori; 310, 491;
818, α CMa; 320, 531.

Die Bezeichnung von α CMa, Sirius, als "rötlich" bzw. "orangefarbig" ist seit langem ein bekanntes astronomisches Skandalon, da dieser Stern weder heute noch vor zweitausend Jahren ähnlich wie die anderen fünf Sterne diese Farbe gezeigt haben kann. Eine befriedigende Erklärung für das Vorkommen der Farbangabe ὑπόκιρρος bei α CMa im Almagest konnte - trotz zahlreicher Hypothesen - bis heute nicht vorgebracht werden.

Wort mit dem *rasm al-ḥwsy* (mit leichten Varianten) entstellt ist (= L *tendit ad aerem / aerem clarum ualde / rapinam / ceram* [→ *terram*]).

Philologisch und lexikalisch bietet der Sternkatalog mit den Positionsbeschreibungen der Sterne innerhalb der Bilder viel interessantes, teilweise abgelegenes Material, dessen Verfolgung durch die Versionen für Philologen sehr reizvoll sein kann. Für die reine Astronomie und ihre Terminologie ist daraus jedoch kein Gewinn zu ziehen[28].

Was nun letztlich die astronomische Terminologie des Almagest im eigentlichen Sinne angeht, so lassen sich daraus gegenwärtig nur willkürliche Proben vorführen; eine umfassende systematische Darstellung ist noch nicht möglich, da die einschlägigen Texte (H, I/T, L) noch nicht ediert sind[29]. Wir folgen dabei dem Verfahren von J. Klamroth (vgl. Anm. 23) und führen im folgenden einige Stichwörter nach griechischen Lemmata (aus G) vor, zu denen wir die arabischen Entsprechungen in H und I/T (nach den Handschriften von Leiden und Tunis) sowie in Auswahl auch deren lateinische Wiedergabe in L[30] setzen. Wie bei Klamroth sind diese Termini aus den Texten der Kapitelüberschriften entnommen, die den einzelnen Büchern vorangestellt sind[31]. Auch dieses Verfahren ist unzulänglich, da durchaus damit zu rechnen ist, dass im Text selbst gelegentlich andere Termini verwendet werden. Die arabische Fachliteratur ist für das Schwanken und die Unbeständigkeit im Gebrauch der Fachsprache leidig bekannt. Doch bleibt angesichts des Mangels kritischer Editionen im Augenblick kein anderer Weg, als dies verkürzte Verfahren anzuwenden, um wenigstens einige Beispiele vorführen zu können.

28. Wer Näheres über die Nomenklatur der Sternbilder und der Einzelsterne sucht, sei auf die in Anm. 20-21 genannten Verzeichnisse und Analysen sowie auf die in Anm. 19 genannten Texteditionen verwiesen.

29. Der alte Druck Venedig 1515 des Almagest zeigt Gerhards Text in einer "Mischversion" (aus den beiden Rezensionen 𝔄 und 𝔅; cf. die in Anm. 19 erwähnte Edition des Sternkatalogs) und kann daher nicht zur Ermittlung des genauen Wortlauts einzelner Stellen herangezogen werden.

30. Um Zeugnisse beider Rezensionen von Gerhards Übersetzung einzubeziehen, stützen wir uns einmal auf die Exzerpte bei Kunitzsch 1974 (wie Anm. 14), 131ff., und darüber hinaus auf Mikrofilme der Handschriften Paris, B.N. lat. 16200 (dat. 1213, eine Abschrift aus der ältesten erhaltenen Handschrift B.N. lat. 14738, Ende 12. Jh., von der kein Film vorlag – als Vertreter der 𝔄-Klasse, welche eine noch unvollkommene Frühfassung von Gerhards Version darstellt; Sigle in der in Anm. 19 genannten Edition: *pa*) sowie Vat. lat. 2057 (13. Jh. – als Vertreter der 𝔅-Klasse, welche die ausgereifte überarbeitete endgültige Fassung darstellt; Sigle in der Edition: *vl*).

31. Mit ziemlicher Gewissheit stammen diese Überschriften nicht von Ptolemäus selbst. Sie sind – bereits im griechischen Bereich – hinzugewachsen und gehörten fortan zum Textcorpus. Sowohl alle uns vorliegenden griechischen Handschriften (die beiden frühesten darunter aus dem 9. Jh.) enthalten sie wie sie auch in den Vorlagen enthalten waren, die die arabischen Übersetzer benutzten. Für den von uns behandelten Zeitraum und in unserem thematischen Zusammenhang erscheint es daher zulässig, sie für die interkulturellen Text- und Terminologievergleiche heranzuziehen.

ἀνωμαλία (III, 4 etc.; D Anomalie, E anomaly): H, I/T *iḫtilāf*: L *diuersitas*.

ἀπόγειον (IX, 7 etc.; D Apogeum, E apogee): H, I/T *al-buʿd al-abʿad:* L *longitudo longior*. Das Gegenstück hierzu ist περίγειον (z.B. am Anfang von X, 1; D Perigeum, E perigee): H, I/T *al-buʿd al-aqrab:* L *longitudo propinquior*[32].

ἀστρολάβον ὄργανον (V, 1; D das Astrolab, E an 'astrolabe' instrument): H *āla min ḥalaq tu'ḫaḏu bihā al-kawākib wa-tuʿrafu mawāḍīʿuhā bi-l-ṭūl wa-l-ʿarḍ* = L *instrumentum armillarum quo accipiuntur stelle et sciuntur loca earum in longitudine et latitudine*, I/T *āla yuqāsu bihā al-kaukab* [var. *al-kaukab al-ṯābit, al-kawākib al-ṯābita*].

περιοδικαὶ ἀποκαταστάσεις (IX, 3; D periodische Wiederkehren, E periodic returns): H *al-ʿaudāt al-dauwārīya* = L *reditiones uolubiles* [Ms. Paris *reditiones reuolutionum*], I/T *ʿaudāt adwār* + Genit.

διάκρισις (V, 19 etc.; D Berechnung, E determination): H *tamyīz* = L *cognitio*, I/T *taqwīm*. – διευκρίνησις (XI, 11; D Berechnung, E determination): H *tamyīz*, I/T *taʿdīl* = L *equatio*.

διάστασις (XIII, 9; D Elongation, E elongation): H *buʿd*: L *longitudo*.

διόρθωσις (IV, 7 etc., X, 4 etc.; D Korrektion, E correction): H *taqwīm* = L [in IV] *equatio*, I/T *taṣḥīḥ* = L [in X] *uerificatio* [bei X, 9 beide Hss. in der vorangestellten Kapitelliste: *certificatio*, aber in der Kapitelüberschrift im Inneren des Textes wieder: *uerificatio*].

ἐκκεντρότης (X, 3 etc.; D Exzentrizität, E eccentricity): H *falak* +

32. Es gab auch eine arabische Transkription: ἀπόγειον = *'fyǧywn* [*afīǧiyūn*] und περίγειον = *'fryǧywn* [*afrīǧiyūn*] (cf. z.B. MUḤAMMAD IBN AḤMAD AL-ḪWĀRIZMĪ, *Mafātīḥ al-ʿulūm*, ed. G. VAN VLOTEN, Leiden 1895, 221, 5), die ihrerseits ins Lateinische weitertranskribiert wurde: *efegion* und *efregion*, mit vielen Varianten, wobei auch häufig beide Wörter ineins zusammenfielen als *effegion* u. ä.Cf. auch NALLINO (wie Anm. 17) I 282 (dazu noch Anm. 8 von S. XXXVI, Forts. auf S. XXXVII), der dort die Kolumnentitel einer Tabelle bei al-Battānī bespricht, die aus Almagest XIII, 5 übernommen ist. Ergänzend zu Nallino sei hier mitgeteilt: G hat in diesen Kolumnentiteln βόρειον / νότιον πέρας: H *al-buʿd al-abʿad al-šamālī / al-buʿd al-aqrab al-ǧanūbī*; besser in I/T *al-buʿd al-abʿad wa-huwa fī l-šamāl / al-buʿd al-aqrab wa-huwa fī l-ǧanūb* = L, Ms. Vat. *efegion – septentrio / efregion – meridies* [in Ms. Paris schlecht: *effegion* bzw. *effegun – septentrio / effegun – meridies*]; um zu entscheiden, wie die Begriffe *efegion / efregion* in Gerhards Version hineingekommen sind, müsste die Überlieferung in allen arabischen und lateinischen Handschriften geprüft werden (Nallino hatte nur den unzuverlässigen Druck von 1515 herangezogen).

Genit. *al-ḫāriǧ al-markaz*, I/T *ḫurūǧ al-falak al-ḫāriǧ al-markaz... ʿan al-markaz* = L *egressio orbis egredientis centri... a centro; al-ḫurūǧ ʿan al-markaz* = L *egressio a centro.*

ἔκλειψις, σεληνιακή / ἡλιακή (VI, 9. 10; D Mondfinsternis - Sonnenfinsternis, E lunar eclipse - solar eclipse): H *al-kusūf al-qamarī / al-šamsī*, I/T *kusūf al-qamar / al-šams.*

ἐπίκυκλος (V, 5 etc.; D Epizykel, E epicycle): H, I/T *falak* (*al-*) *tadwīr* = L *orbis reuolutionis.*

ἐποχή (III, 7 etc.; D Epoche[33], E epoch): H *mauḍiʿ* (aber in IV, 9: *ibtidāʾ*), I/T *ḥāṣil* (aber im Text von X, 1 und Überschrift XI, 4, im Plural: *taḥṣīlāt*).

ὁ ἰσημερινός (II, 2; D Äquator, E equator): H [*falak*] *muʿaddil al-nahār* = L *orbis equationis diei*, I/T *dāʾirat muʿaddil al-nahār.*

κίνησις [μέση] (III, 2 etc.; D [gleichförmige] Bewegung, E [mean] motion): H, I/T *ḥarakat* + Genit. [*al-wusṭā*].

κατὰ κορυφήν [γίνεται] (II, 4; D [in den] Zenit [kommt], E [reaches the] zenith): H [*aiy ahl bilād yakūnu maǧrā* + Genit. *ʿalā*] *samt ruʾūsihim* = L [*quas terras inhabitent illi super*] *summitatem capitum* [*quorum ... transit*], I/T [*man allaḏīn taṣīr ... ʿalā*] *samt ruʾūsihim.*

ὁ μεσημβρινὸς κύκλος (II, 10; D Meridian, E meridian): *falak*[34] *niṣf al-nahār* = L *orbis meridiei*, I/T *dāʾirat niṣf al-nahār.*

ὁ λοξὸς κύκλος (II, 2; D Ekliptik, E ecliptic): H *al-falak al-māʾil*, I/T *al-dāʾira al-māʾila.* Für "Ekliptik" G auch (II, 7 etc.): ὁ διὰ μέσων τῶν ζῳδίων κύκλος: H *falak al-burūǧ*, I/T *al-dāʾira allatī tamurru bi-ausāṭ al-burūǧ.*

μετάπτωσις (IX, 7; D Weiterbewegung, E displacement): H *intiqāl*, I/T *naqla.*

ὁρίζων (II, 2; D Horizont, E horizon): H *dāʾirat al-ufq*, I/T *al-ufq.*

33. Zu "Epoche" gibt D I 182 Anm. a folgende Erklärung: Unter Epoche ist der in Ekliptikgraden ausgedrückte Ort (τόπος) zu verstehen, welchen die Sonne zu einem bestimmten Zeitpunkt, der als Ausgangspunkt ihrer gleichförmigen Bewegung gilt, innehat (ἐπέχει) oder innegehabt hat.

34. KLAMROTH (wie Anm. 23) 31 schreibt abweichend: *ḫaṭṭ*.

παράλλαξις (V, 11 etc.; D Parallaxe, E parallax): H, I/T *iḫtilāf al-manẓar* = L *diuersitas aspectus*.

παραχώρησις (XIII, 6; D Abweichung, E deviation): H *tabāʿud*, I/T *ḥaraka*.

οἱ ε̄ πλανώμενοι (IX, 1; D die fünf Wandelsterne, E the five planets): H, I/T *al-kawākib al-ḫamsa al-mutaḥaiyira* = L *stelle quinque erratice*[35].

προήγησις (XII, 1 etc.; D Rückläufigkeit, E retrogradation): H *taqaddum al-kawākib allaḏī huwa ruǧūʿuhā*, I/T *taqaddum* = L *precessio* [nur bei XII, 6 Ms. Vat.: *antecessio*][36].

στηριγμός (XII, 7 etc.; D Stillstand, E station): H (Plur.) *maqāmāt al-kawākib wa-ruǧūʿuhā*, I/T *al-wuqūfāt* = L *stationes*.

αἱ συζυγίαι (VI, 2. 4; D Syzygien, E syzygies): H, I/T *al-ittiṣālāt* = L *applicationes* [H in V, 10. 14: *al-iǧtimāʿ wa-l-istiqbāl*].

αἱ σύνοδοι (VI, 1 etc.; D Konjunktionen, E conjunctions): H, I/T *al-iǧtimāʿāt* = L *coniunctiones*.

Es ist zu beobachten, dass H und I/T zum Teil die gleichen Termini benutzen - hier legte vielleicht das griechische Original nur die eine Wiedergabe nahe, oder beide Übersetzer befolgten bereits vorher geprägte Muster; zum Teil sind aber auch H und I/T verschieden - vermutlich strebte der (spätere) Isḥāq mit seinen neu geprägten Formulierungen grössere Genauigkeit in der Sache oder im sprachlichen Ausdruck gegenüber seinem Vorgänger al-Ḥaǧǧāǧ an. Ganz allgemein verhält sich al-Ḥaǧǧāǧ in seiner Übersetzung des Almagest relativ frei[37], während Isḥāq alle Einzelheiten im Wortlaut des komplizierten griechischen Textes pedantisch (natürlich nicht mit Beibehaltung der griechischen Wortstellung) nachgestaltet, so dass seine Version zwar sachlich zuverlässiger, dafür aber sprachlich schwerer verständlich ist[38]. In der Überlieferung hat freilich Isḥāq (mit 9 erhaltenen Handschriften[39]) die

35. S. hierzu KUNITZSCH 1974 (wie Anm. 14) 141f. (Anm. b zu IX, 1 - L).

36. Hier wird "Rückläufigkeit" etymologisierend (zu προήγησις) mit *taqaddum* (= L *precessio*) wiedergegeben; der eingebürgerte klarere Ausdruck dafür ist sonst *ruǧūʿ*, cf. NALLINO (wie Anm. 17) II 333 und *Mafātīḥ al-ʿulūm* (wie Anm. 32) 221, 9ff.

37. Cf. KUNITZSCH 1974 (wie Anm. 14) 66, 72f.

38. Ibid., 69f., 72.

39. Cf. die Edition (wie in Anm. 19) II 4; die dort unter Nr. 10 erwähnte Handschrift aus Jaipur ist zu streichen. Wie eine Untersuchung durch Prof. G. Saliba und Dr. R. Lorch *in loco* (anlässlich eines Kongresses im Dezember 1991) ergab, enthält sie vielmehr

Version von al-Ḥaǧǧāǧ (mit nur zwei erhaltenen Handschriften[40]) ausgestochen. Immerhin hatten sich al-Farġānī, al-Yaᶜqūbī und al-Kindī in ihren Bearbeitungen bzw. Auszügen des Almagest auf H gestützt[41], während I/T die Ausgangsbasis für die weitverbreitete Bearbeitung (*taḥrīr*) von Naṣīr al-Dīn al-Ṭūsī (vollendet 1247) stellte[42].

Gerhard von Cremona ist unter allen mittelalterlichen arabisch-lateinischen Übersetzern derjenige, der am engsten und wörtlichsten an der arabischen Vorlage klebt. Man kann seine Übersetzung von Fall zu Fall sogar zur Emendation des arabischen Textes heranziehen. Für mittelalterliche westliche Leser ohne Arabischkenntnisse dürfte seine Version nur sehr schwer, in Teilen vielleicht gar nicht verständlich gewesen sein. Auch moderne Bearbeiter könnten (im Falle einer Edition) seinen Text nur unter ständiger Beiziehung des Arabischen etablieren. Als praktischer Hinweis sei hinzugefügt, dass Gerhard den Text von Buch I-IX nach H und den Rest von Buch X-XIII nach I/T übertragen hat; der Sternkatalog (in VII-VIII) erscheint auf der Basis von I/T, aber mit vielen eingearbeiteten Einzelheiten aus H (die sich in gleicher Weise in der arabischen Handschrift in Tunis, als mit Siglen gekennzeichnete Randnotizen, finden)[43]. Die Zitate aus L sind daher in den Beispielen oben je nachdem H oder I/T zugeordnet (sofern diese beiden verschiedene Termini oder Formulierungen haben).

Die oben vorgestellten Beispiele bieten nur eine erste kurze oberflächliche Übersicht. Ein vollständiges, rundum erschlossenes Glossar der astronomischen Terminologie des arabischen Almagest von der Güte desjenigen, das Nallino seiner Edition des *Zīǧ* von al-Battānī beigegeben hat, bleibt bis auf weiteres ein Desiderat. Es kann erst erstellt werden, wenn einmal der arabische Almagest nach den beiden Versionen H und I/T vollständig ediert ist.

dieselbe anonyme Bearbeitung des Almagest, die auch in Ms. Teheran, Naṣīrī 789 vorliegt (zu dieser vgl. die zitierte Edition, III 200: Nachtrag).

40. Ibid., II 3; die eine dieser beiden (London, Brit. Libr. Add. 7474) enthält nur die Bücher I-VI.

41. Cf. Kunitzsch 1974 (wie Anm. 14) 48f., 16, 38 Anm. 89.

42. Ibid., 47f.

43. Ibid., 104ff., und P. Kunitzsch, *Gerhard von Cremona als Übersetzer des Almagest*, in M. Forstner (Hgb.), *Festgabe für Hans-Rudolf Singer, Zum 65. Geburtstag am 6. April 1990...*, Frankfurt am Main etc. 1991, I 347-358.

V

A HITHERTO UNKNOWN ARABIC MANUSCRIPT OF THE *ALMAGEST*

In my edition of Ptolemy's star catalogue[1] I have registered 14 manuscripts containing the Arabic translations of the *Almagest*, complete or in major sections: a) version of al-Ḥajjāj, two manuscripts (nos. 1-2 in the list: Leiden, cod. or. 680, complete; London, BL Add. 7474, only Books I-VI)[2]; and b), version of Isḥāq ibn Ḥunayn emended by Thābit ibn Qurra, ten manuscripts (nos. 3-12: Tunis, BN 07116, complete; Tehran, Sipahsalar 594, defective; London, BL Add. 7475, only Books VII-XIII; Paris, BNF ar. 2482, only Books I-VI; Paris, BNF ar. 2483, only Books I,10 - VII,4; Escorial 914, only Books V-IX; Escorial 915, only Books VII-XIII; Jaipur, Maharaja Mansingh II Library – see below; Paris, BNF hebr. 1100, complete, in Hebrew script; Cambridge, UL Mm.6.27, in Hebrew script, defective)[3]. To these were later added Tehran, Naṣīrī 789, and Jaipur, Maharaja Sawai Man Singh II Museum 38 (348)[4].

After the publication of the star catalogue, I obtained new pieces of information which require to remove three manuscripts from the above list. The total of major Arabic manuscripts of the *Almagest* thus amounts to eleven manuscripts.

Of MS Tehran, Naṣīrī 789 I obtained a (badly made) microfilm from which I recognized that this appears to be no direct translation of the *Almagest*, but rather some sort of a recension, uncertain whether by Quṭb al-Dīn al-Shīrāzī himself, who seems to have written the manuscript, or by someone else[5].

[1] Claudius Ptolemäus, *Der Sternkatalog des Almagest. Die arabisch-mittelalterliche Tradition*, ed. P. Kunitzsch, I-III, Wiesbaden 1986-1991.

[2] *Sternkatalog* I, p. 3.

[3] *Sternkatalog* I, p. 4.

[4] *Sternkatalog* II, p. 171 (additions to I, p. 4). Other manuscripts containing only scattered, small fragments of the work were not registered there.

[5] *Sternkatalog* III, p. 200.

MS Jaipur 20 (= no. 10 in my list, above) seems to be another exemplar of the text contained in MS Naṣīrī 789 [6].

MS Jaipur 38 (348) turned out to be a copy of Naṣīr al-Dīn al-Ṭūṣī's *Taḥrīr* of the *Almagest*[7] (just as also MS Jaipur 19[8]).

Now, a new, hitherto unknown, Arabic manuscript of the *Almagest* turned up in 1998. It was offered for auction by Christie's, London, on April 28, 1998[9]. The illustration on p. 31 of Christie's Catalogue shows fol. 160r of the text, which I could easily identify, by comparison with microfilms of the known Arabic versions of the *Almagest*, as the end of XII,1 and beginning of XII,2 in the Isḥāq-Thābit version. This is, then, the twelfth Arabic manuscript of the *Almagest* in the numbering of the revised form of my earlier list.

The manuscript was acquired by Mr. Lawrence J. Schoenberg, Longboat Key, Florida, in whose collection it now has the number LJS 268. Mr. Schoenberg kindly allowed me to inspect the manuscript in Philadelphia, where it was transferred to be shown in an exhibition of Schoenberg manuscripts, March to June, 2001[10]. In the following I give a description of the manuscript which is, in several details, updated and more correct than the provisional descriptions in the Christie's Catalogue and in an internet catalogue of the Schoenberg manuscripts.

The manuscript contains 185 folios. It was written, in Andalusian Maghrebi script (black, with elements in red such as chapter headings and diagrams etc.), by Aḥmad ibn Aḥmad ibn Salāma al-Ṣ.n[...]ānī al-Hājirī in Saragossa, completed on Monday, the 10th of Jumādā al-ākhir 783 = 13 Alūl 5141 of the Creation (*khalīqa*, i.e., in the Jewish calendar) = 2 September 1381 AD, when the Sun was at an altitude of eighty (or rather two - written in a word, not clearly readable) degrees above the

[6] I owe this information to Prof. G. Saliba who kindly copied for me some chapter beginnings in the manuscript in Jaipur in December 1991.

[7] This information is also owed to the kind endeavours of G. Saliba.

[8] On this latter one, see D.A. King, A Handlist of the Arabic and Persian Astronomical Manuscripts in the Maharaja Mansingh II Library in Jaipur, in: *Journal for the History of Arabic Science* 4 (1980), 82 (under no. 1).

[9] Cf. the catalogue: Christie's, London, *Islamic Art and Indian Miniatures, Tuesday, 28 April 1998*, lot no. 54 (pp. 30ff., with illustrations).

[10] I express my sincere gratitude to Mr. Schoenberg for this generous permission as well as to Dr. M.T. Ryan, Director of the Annenberg Rare Book and Manuscript Library, University of Pennsylvania, Philadelphia, PA, and his staff, who extended to me all kinds of help during my visit there. Mr. Schoenberg stresses that he welcomes all researchers interested in the manuscripts of his collection; he may be contacted by e-mail under: <schoenbriz@home.com>.

Western horizon of Saragossa. The scribe wrote it for his master and teacher, Abū Isḥāq Yaʿqūb ibn Isḥāq ibn Yaʿqūb known as Ibn al-Qursunuh (Yacob ben Isaac al-Corsono) al-Isrāʾīlī, astronomer of King Pedro (IV) of Aragon and Count (*wālī*) of Barcelona[11].

In the course of the text there appear three more datings which are all in harmony with the colophon: f. 132v, at the end of Book IX (*Jumādā al-awwal* [sic] *sanat* 783); f. 145r, at the end of Book X (9 *Jumādā al-awwal ... sanat* 783); and f. 157v, at the end of Book XI (29 *Jumādā al-awwal alladhī min sanat 783 li-taʾrīkh al-hijra*).

On f. lr the title is given as: *Kitāb al-majisṭī li-Baṭlaymūs* [sic][12] *al-Qalūdhī al-mansūb ilā 'l-taʿālīm fīhi thalāthat ʿashar* [sic] *maqāla*. Underneath, three successive later owners are mentioned: Aḥmad ibn Mubārak (*laṭafa 'llāhu bihi* [not: Luṭf Allāh!] *āmīn*), Muḥammad al-Raʾīs, and Muḥammad ʿAbd al-Ḥayy ibn ʿAbd al-Kabīr al-Kattānī al-Ḥasanī[13].

On the verso of the front fly-leaf there is a long anonymous notice on the importance of the work and the value of the present manuscript. It is in careful modern Maghrebi script and quotes at the beginning (without naming the source) Ḥājjī Khalīfa's explanation of the Greek

[11] All this is given in the colophon, f. 185r. On the addressee, the Jewish court astronomer of King Pedro IV of Aragon, Yacob ben Isaac Corsono, who played a major role in the establishing of the king's astronomical tables, cf. J.M. Millás Vallicrosa (ed.), *Las tablas astronomicas del Rey Don Pedro el Ceremonioso*, Madrid - Barcelona 1962, esp. pp. 11ff., 63ff. (Millás spells the name as Corsino, but more correctly it seems to be Corsuno or Corsono); J. Chabás, Astronomía andalusí en Cataluña: Las Tablas de Barcelona, in: *From Baghdad to Barcelona. Studies ... in Honour of Prof. Juan Vernet*, Barcelona 1996, II, 477ff., esp. p. 521; J. Samsó, An outline of the history of Maghrebi zijes from the end of the thirteenth century, in: *Journal for the History of Astronomy* 29 (1998), 93ff., esp. p. 94 (where apparently the same Jewish astronomer is quoted, from an anonymous text from Tilimsān, ca. 1515 AD, as Abū Kursūm al-Yahūdī). Cf. also *The Jewish Encyclopaedia*, 3 (New York – London 1903), 593, s.v. Carsono; *Encyclopaedia Judaica*, 3 (Berlin 1929), 606, s.v. Astronomie, 14. Jh.: Jakob Karsi (Jakob Al-Korsono). Another member of the Corsono family seems to be the Moroccan theological scholar Judah ben Joseph, 14th c., mentioned in *Encyclopaedia Judaica*, 2 (Jerusalem 1971), col. 551, s.v. Alcorsono. I express my thanks to Mme. Judith Leifer and her staff in the library of the Center for Judaic Studies, Philadelphia, PA, for helping me to trace these latter sources.

[12] This spelling and vocalization is given here and in the title and colophon of Book I. In all the other Book titles and colophons it is *Baṭlamyūs*, in harmony with the usual medieval Arabic spelling of the name.

[13] On the last one, cf. *EI*², IV, 774f. He died, in exile in France, on 28 September 1962 at the age of 77.

title[14] and mentions the ownership by Muḥammad ʿAbd al-Ḥayy. It further quotes the date of the manuscript's writing and states that therefore the manuscript is at present 547 years old. From this it is clear that the notice was written in the year 1928.

Under this lengthy text there follow some lines, apparently in the same hand, but written with another pen. In translation, they say: "I have now presented it as a souvenir to the well-known politician (or: diplomat), the trustworthy and important man, Excellence Monsieur R-y-n-y-w-ʾ, ambassador (*safīr*) of the grand nation of France, because the best that a man can present to bis beloved friend is books of science; the noblest of these is what is like this present precious volume. It is hoped that his Excellency will most gladly accept it".

The riddle behind this passage is (partly) solved by a pencilled notice, in French, on top of the verso of the back fly-leaf, which runs: "Almagesté de Ptolemée sur l'Astronomie, traduit en Arabe – 1381. Contient les 13 chapitres alors que l'exemplaire (de 1321) de le [sic] Bibliotheque nationale ne contient que les 7 premiers chapitres. Offert par le Cherif à M. Regnaul avec dedicace" (throughout in this spelling)[15]. "Le Cherif" can be no other than the Sultan of Morocco who was, in the year 1928 (and after), the young Muḥammed ibn Yūsuf (Mohammed V). The manuscript thus seems to have come, after Muḥammad ʿAbd al-Ḥayy, into the possession of the Sultan of Morocco who gave it as a noble present to a certain M(onsieur) Regnaul(t)[16].

The manuscript consists of 185 folios numbered, at the bottom of each recto page, in pencil, in modern European numerals from 1 to 185. The two fly-leaves, at the beginning and the end, are not numbered. In the middle there is a section where the individual who put down the folio numbers realized that several folios were wrongly bound. He then tried to restore the correct order by crossing out the folio numbers and adding, to their right, new numbers[17]. In its present state the

[14] See Ḥājjī Khalīfa, *Kashf al-ẓunūn*, ed. Yaltkaya - Bilge, II (Istanbul 1943), col. 1594; cf. also P. Kunitzsch, *Der Almagest. Die Syntaxis Mathematica des Claudius Ptolemäus in arabisch-lateinischer Überlieferung*, Wiesbaden 1974, 121. The Greek words *mājistūs*, *mājistī* are miswritten in the notice es *fāḥistūs*, *fāḥistī*.

[15] The reference seems to be to MS Paris, BNF ar. 2483, which indeed contains only I,10 - VII,4 of the *Almagest*. But in its present form it is not dated; it is only generally ascribed to about the 15th century.

[16] It has not been possible to trace the identity of Monsieur Regnaul(t). If in the future I succeed in obtaining more information, I shall publish it in a short notice in a following volume of this journal.

[17] In the following quotations, in such cases both numbers will be given separated

leather-bound manuscript has the thirteen Books of the Almagest in the following order:

I:	lv-14r (the beginning of the text of ch. 1 is missing)
II:	14v-30v
III:	31r-37v (breaks off in ch. 3, the rest of III is missing)
IV:	50r (end of ch. 9, former parts missing) - 52r
V:	52v-55v (breaks off in ch. 5), 56r-v (ch. 8), 57r (ch.10) - 66v
VI:	67r-84r
VII:	84v-101v
VIII:	132/102r (star catalogue breaks off on 139/109r, after the 18th star of Centaurus), 139/109v blank, cont. - 145/115v
IX:	145/115v-149/119v; 102/124r-110/132v
X:	111/133r-124/145^{bis}r
XI:	124/145v-131/152v (beg. ch. 5); 150/120r (cont. ch. 5) - 158/157v
XII:	159/158r-171r
XIII:	171v-184v.

Folios 38r-49v (III,4 - IV, beg. of ch. 9) are missing. Also some text from V is absent, but without interruption in the folio counting. Further, several tables were not, or not completely, written down (without interruption in the folio counting). At the beginning, on lv, there is the title of the work and the list of chapter headings for Book I. But the list breaks off in the heading of ch. 13, after the word *al-nahār*. The rest of the page is blank, the text of the beginning of ch. 1 is missing. On 2r, ch. 1 is continued without regard to the missing portion on 1v.

The manuscript has the Arabic text of the *Almagest* in the version of Isḥāq ibn Ḥunayn emended by Thābit ibn Qurra. As observed in other manuscripts of the Isḥāq-Thābit version, however, some portions of text appear in the Ḥajjāj version: the star catalogue (VII,5 - VIII,1); several chapter headings in Books IX and X (some headings also show a mixture between the two versions of al-Ḥajjāj and Isḥāq-Thābit); the first few lines of text in ch. 1 of Book IX[18].

The Arabic text is fully dotted. Vowels are added in the more important portions of the text such as Book titles, colophons, and the foreign names in the star catalogue. But the scribe's vocalization often seems to echo colloquial usage rather than the requirements of classical Arabic grammar. So, e.g., the name of Ptolemy, Baṭlamyūs, is always written

by a stroke / . To the left of the stroke appears the crossed-out number, to its right the new number.

[18] This corresponds to the observations in other manuscripts registered in Kunitzsch, *Der Almagest* (note 14), 131-149. For IX,1 especially, see there, p. 142.

with *ḍamma* above the *sīn*, also in the genitive case, where a *fatḥa* would be required. The article is often given as *il-* instead of *al-*; similarly there is *illatī* for *allatī*. In the colophon, at the end, we find *muwāfiqah* (partic., with *sukūn* above *h*, no *tā' marbūṭa*), and *bi-muwāfiqa* (infinitive, in statu constructo, with dotted *tā' marbūṭa*, but *sukūn* above it).

Finally, I mention some variant readings in the star catalogue different from the readings in the other known manuscripts[19].

4 *al-labina*: *al-l.d.ma* – 9 *al-'unuq*: *al-'ayyūq* – IV *qīfāwus*, *al-multahib*: *fanqāwus*, *al-multahab* (with *sukūn* above *b*) – V *bu'ūṭis*, *aṣ-ṣayyāḥ*: *f.rūṭ.s*, *al-ṣabāḥ* – VI *al-iklīl*: *al-aklīl* – VII *al-sulaḥfāh*: *al-salḥafāt* – 96, 99 *al-khazaf*: *al-jawf* – XI *al-nhlb*, *barsūs*: *al-mtlhb* (corrected from *al-mlthb*, as in the name of IV), *bwsws* (without dot under *b*) – XII *anīkhus*: *al-ḥisi* (sic vocalized) – 161 *al-sakhlatayn*: *al-s.l.ḥayn* – XV *al-nawl*: *al-nawq* – XVII *al-dulfīn*: *al-dilfīn* (sic, with *sukūn* above *nūn*) – 332 *al-mlyk*: *al-mlbk* (perhaps rather *al-lbk*) – 351 *al-ḍafīra*, *al-dhu'āba*: *al-ṣghra*, *al-dāyir* – 372 *al-khimār*: *al-ḥy'r* – 363 *awwal muqaddam al-'ṭ'f*: *mutaqaddim awwal al-'.ṭāb* – XXXI *al-jady*: *al-jiddī* (sic vocalized) – 518 *bhbn*: *nahr* – 558 *al-sukkān*: *al-yasār* (*y* without dots) – 573 *qānubus*: *fān.s* (*n* without dot) – XLIV *qanṭūris*, *al-ẓulmān*: *qanṭūris* (sic vocalized, with *sukūn* above *s*), *al-ẓalmān* (sic vocalized, with *sukūn* above *n*).

The zodiacal sign Aries is always called, in all places in the column for the signs in the longitudes, *al-kabsh* (but the constellation itself is called *al-ḥamal*).

As specific characteristics of the Schoenberg manuscript it can be mentioned that it has, like the Isḥāq-Thābit manuscripts b, d, a[20], the star catalogue in the Ḥajjāj version. Common with MS d (Escorial 914) alone, it has the first lines of IX,1 in the Ḥajjāj version[21]. Further, like most of the Isḥāq-Thābit manuscripts[22], it has several chapter headings in Books VII, IX and X in the Ḥajjāj version; others show a mixture between the Ḥajjāj and Isḥāq-Thābit versions. Minor variants of one or a few words only do not show a specific relationship to the

[19] The references here are to Kunitzsch, *Der Almagest* (note 14), Roman numerals for the constellation names (pp. 172-203, there) and Arabic numerals for individual star descriptions or names (pp. 217-348, there). The full spectrum of Arabic variants is displayed in *Sternkatalog* (note 1), vol. I.

[20] The sigla are the same as those used in Kunitzsch, *Der Almagest* (note 14), and *Sternkatalog* (note 1).

[21] Cf. *Der Almagest* (note 14), 142.

[22] Cf. *Der Almagest* (note 14), 132-149.

other Isḥāq-Thābit manuscripts. The 'Books' are always called *maqāla* (with the exception of the colophon of Book X, where it is *qawl*, corresponding to the Ḥajjāj version), and the chapters are *naw*ʿ[23].

In sum, the Schoenberg manuscript is a most valuable and welcome addition to the stock of surviving manuscripts of the Arabic *Almagest*. It shows the form in which the *Almagest* was actually known and used by astronomers near the end of the 14th century in al-Andalus.

Appendix

I use the occasion of this article to give a short description of a fragment of the Arabic *Almagest*, comprising 16 folios, in the famous Qarawīyīn Library at Fes. Prof. Bennesser Alaoui Abdallah of the Faculté des Lettres in the Université Sidi Mohamed Abdellah, Fes, kindly sent me the information and copies of the relevant sources. This piece of text was first mentioned in an article by al-ʿĀbid al-Fāsī, *Khizānat al-Qarawīyīn wa-nawādiruhā*, in: *Majallat al-Makhṭūṭāt al-ʿArabīya* 5,1 (Dhū l-Qaʿda 1378/May 1959), p. 13, under no. 49. A page from this fragment is shown, as photograph, in the Moroccan journal *Daʿwat al-Ḥaqq* 9 (Shawwāl 1397/October 1977), in an article by Dr. ʿAbd al-Hādi al-Tāzī, *Ṣadā al-Fārābī fī 'l-Maghrib*, pp. 92ff. The fragment is listed in *Fihris Makhṭūṭāt al-Qarawīyīn*, ed. Muḥammad al-ʿĀbid al-Fāsī, II (Al-Dār al-Bayḍāʾ, 1400/1980), p. 238, as no. 654[24]. The photograph in *Daʿwat al-Ḥaqq* shows the first page of the fragment, which contains the beginning of ch. 2 of Book XII of the *Almagest*, written in Andalusian Maghrebi script, in the Isḥāq-Thābit version.

[23] Cf. *Der Almagest* (note 14), 130f.

[24] In the caption of the illustration in *Daʿwat al-Ḥaqq* it is given as no. 644/40, which may be a misprint for 654. The entry in the catalogue of 1400/1980 was also supplied to me by Dr. M. Comes of Barcelona. To both, Prof. Bennesser Alaoui and Dr. Comes, go my thanks for their generous help.

VI

THE SECOND ARABIC MANUSCRIPT OF PTOLEMY'S *PLANISPHAERIUM*

The Greek text of Ptolemy's *Planisphaerium* seems to be lost[1]. Until recent times, the work was known in the West only in the Latin version translated by Hermann of Carinthia from the Arabic (Tolosa, 1143)[2]. Of the Arabic version three manuscripts are known to exist[3]: Istanbul, Ayasofya 2671, 76^v-97^r (dated 621 H = A.D. 1224); Tehran, private library of Khān Malik Sāsānī (dated 607 H = A.D. 1210); and Kabul, Maktabat Ri'āsat al-Maṭbū'āt (sic *RIMA* II, 19). Lately, C. ANAGNOSTAKIS has published the Istanbul MS in fac-simile accompanied by an English translation and comments[4].

The translator of the Arabic version from the Greek is unknown. For a long time it was assumed that the translator was the Spanish-Arabic astronomer Maslama al-Majrīṭī (d. 1007). The first to make this assumption was Hermann himself, at the end of his preface (cf. HEIBERG, CLXXXVI, 28-29). This wrong assumption may have been provoked by the fact that the Arabic text translated by Hermann was accompanied by a number of notes and an "additional chapter" all of which are explicitly attributed to Maslama ("*Dixit Maslem...*", etc.)[5].

[1] PAULY-WISSOWA-KROLL, *Real-Encyclopädie der classischen Altertumswissenschaft*, art. Klaudios Ptolemaios, 23,2 [46. Halbband], Stuttgart 1955, col. 1788ff.; esp. 1829-31 (*Planisphaerium*, B. L. VAN DER WAERDEN); G. J. TOOMER, art. Ptolemy, in: *DSB* XI (New York 1975), 186ff.; esp. 197f., 205.

[2] Ed. J. L. HEIBERG, Ptolemaeus, *Opera*, vol. II (Leipzig 1907), 227-259 (cf. also the Prolegomena, pp. XIIf.; CLXXX-CLXXXIX); German translation by J. DRECKER, in: *Isis* 9 (1927), 255-278. HEIBERG used six manuscripts; until now we found seven more copies.

[3] Cf. F. SEZGIN, *GAS* V (Leiden 1974), 170.

[4] C. ANAGNOSTAKIS, *The Arabic Version of Ptolemy's Planisphaerium*. Ph.D. thesis, Yale University 1984 (published in book form by University Microfilms International, Ann Arbor, Michigan, in 1986). For a recent discussion of the *Planisphaerium* see J. L. BERGGREN, Ptolemy's Maps of Earth and the Heavens: A New Interpretation, in: *Archive for History of Exact Sciences* 43,2 (Oct. 1991), 133-144.

[5] A corpus of notes on the *Planisphaerium* by Maslama was found by G. VAJDA

It is, however, quite improbable that a Greek mathematical work could have been translated from Greek into Arabic in Spain in the second half of the tenth century. Fortunately, there is plain evidence that the Arabic text of the *Planisphaerium* (which is not listed in the *Fihrist* of Ibn al-Nadīm) existed in the Arabic East already in the first half of the tenth century, i.e. about half a century before Maslama's activities. The grandson of Thābit ibn Qurra, Ibrāhīm ibn Sinān ibn Thābit (909-946), refers to Ptolemy's *Kitāb tasṭīḥ al-kura* three times in his *Risāla fī 'l-asṭurlāb*[6]; on one of these occasions (p. 312) he quotes – "from memory", as he says – the *Incipit* of the *Planisphaerium* in a wording which is basically identical to that in the Istanbul MS. Thus, it can now be accepted for sure that the *Planisphaerium*, like all the other material translated from Greek etc. into Arabic, was translated in the East, around or before 900 so that it was ready to be used by Ibrāhīm ibn Sinān some time before 946[7]. Maslama's contribution seems to be restricted to the "notes" and the "additional chapter" added to the basic text which had reached him somehow, like others (e.g., al-Khwārizmī's Tables), from the East. Since the "notes" and the "additional chapter" are explicitly introduced under his name (in the Arabic as well as in Hermann's version) it appears justified to assume that the text itself did not suffer any interventions by Maslama.

The Arabic text in MS Istanbul is not very well preserved, The hand is an elegant *naskhī*, but there are many defects, often in the letters intended for geometrical demonstration, and elsewhere[8]. All

in MS Paris, B.N. ar. 4821 (dated 11 Sha'bān 544 = 14 December 1149); cf. *Rivista degli studi orientali* 25 (1950), 7-9. Of these, the "additional chapter" added to the *Planisphaerium* and the succeeding "Chapters indispensable for whosoever wants to construct an astrolabe" were edited and translated into Spanish by J. VERNET – M. A. CATALÁ, Las obras matematicas de Maslama de Madrid, in: *Al-Andalus* 30 (1965), 16-45. The "notes" on the *Planisphaerium* (eleven in number), which precede in the manuscript, were edited by P. KUNITZSCH and R. LORCH, *Maslama's Notes on Ptolemy's* Planisphaerium *and Related Texts*, Munich 1994 (Bayerische Akademie d. Wissenschaften, Phil.-hist. Kl., Sitzungsberichte, Jahrg. 1994, Heft 2).

6 See *Rasā'il ibn Sinān*, ed. A. S. SAIDAN, Kuwait 1403/1983, *risāla* VII (pp. 305-318), pp. 309, 312, 315.

7 Cf. also ANAGNOSTAKIS (note 4), 28, 55; P. KUNITZSCH, The Role of Al-Andalus in the Transmission of Ptolemy's *Planisphaerium* and *Almagest* (to be published in the next volume of this journal).

8 ANAGNOSTAKIS (pp. 98-101) has proposed numerous emendations, mostly for geometrical letters and for some numbers.

the diagrams are omitted; their places are left blank. The correspondence to Hermann's Latin is often loose; this may be partly due to Hermann's style of translating which is extremely free, in sharp contrast to the extreme literalness of Gerard of Cremona for example. For the second half of Chapter 18 Hermann's Arabic text was apparently different from that in MS Istanbul[9].

Of the other two Arabic manuscripts the one in Kabul is inaccessable at present – if it has survived the deplorable events of the last decade in that city at all.

On the other hand, a good xerox-copy of the *Planisphaerium* in the Tehran manuscript has recently become available to us[10].

According to the description in the *Nashrīye* [cf. Sezgin, as in note 3], vol. 5 (1968), p. 114f., the manuscript in the library of Khān Malik Sāsānī is a *majmūʿa* and contains three texts:

1: *Kitāb fī ʿamal musaṭṭaḥ yaqūmu maqāma 'l-asṭurlāb [fī] jamīʿ al-ʿurūḍ wa 'l-buldān*, "Book on the construction of a projected [plate] that can take the place of an astrolabe [in] all latitudes and places";

2: the *Planisphaerium*;

3: the treatise on the construction of the astrolabe by Kūshyār [ibn Labbān] [al-]Jīlī.

It is written in *naskhī* script and dated 607 H = A.D. 1210 (that is, about fourteen years before the Istanbul MS)[11]. There are owners' marks from 910 H = A.D. 1504-05 and 1158 H = A.D. 1745-46.

The present copy comprises section no. 2, the *Planisphaerium*. It consists of 24 folios, i.e. 48 pages which, for practical use, we numbered from page 0 (the first recto page, blank, with an owner's notice, a geometrical note and another short note, on the text, in Persian) to page 47 (the last verso page with the end of the *Planisphaerium*; the colophon bears no date). Each page carries seventeen lines of text.

The *naskhī* script in the Tehran manuscript (henceforth: T) is a bit

[9] Hermann's reading is, however, confirmed by Maslama's citation of the text in his note to this chapter. This is shown in detail in the edition of Maslama's "notes" (cf. above, note 5).

[10] The copy was procured by Dr. A. Djafari Naini for Prof. M. Folkerts, Institut für Geschichte der Naturwissenschaften, Universität München. I express my gratitude to both of them for making the text available for this study.

[11] On the copy of page 0 the utter left margin of the preceding page is visible showing the last ends of three lines of oblique writing; in it, in numerals, the number 617 appears. If this refers to the date of copying, the MS would date from 617 H = A.D. 1220-21, not 607. But this is not sure.

stiff, less elegant than that of the Istanbul manuscript (henceforth: I). On pp. 17, 37 and 47 there are collation notes: *qūbilat bi-aṣlihā/aṣlihi.* There are some marginal notes in various hands. An Oriental reader has added – mostly in the former part of the text – marginal corrections to the geometrical letters, or even corrected the letters in the text itself and in some diagrams, in a rather recent hand as it seems.

MS T contains basically the same text as I, but with better readings in many places. Also for Chapter 18 (cf. above) it is identical with I and belongs to the same branch of transmission; another branch is represented by Maslama's note (to this chapter) and Hermann's version.

In T, all the fourteen diagrams are represented (though sometimes a bit carelessly) in the same places where I had left blanks. The diagrams and their letters are basically identical in T and Hermann/HEIBERG, except for some parts of the construction which are occasionally reversed in HEIBERG (left to right, etc.). A synopsis of the diagrams in all the sources is as follows:

Diagram no.	MS I, blank on fol.	MS T, diagram on p.	Corresponding diagram Hermann/ HEIBERG p.	Chapter acc. to HEIBERG	Diagram reconstructed by ANAGNOSTAKIS, Fig.
1	77^{v}	3	227	1	1
2	79^{r}	7	230	2	2
3	80^{r}	8	232	3	3
4	80^{v}	10	233	4	4
5	85^{v}	21	237	8	5
6	88^{v}	28	242	10	6
7	89^{v}	30	245	11	–
8	90^{v}	33	246	12	7
			repet. 248	13	7
9	92^{v}	38	250	14	8
10	93^{v}	39	251	15	9
11	94^{r}	40	253	16	10
12	95^{r}	43	254	17	11
13	95^{v}	44	256	18	12
14	96^{r}	45	257	19	13

This first description of the text of the *Planisphaerium* in T is now concluded by a list of better readings in T (against I). In fifteen cases the new readings confirm the emendations of ANAGNOSTAKIS (which will be noted below). The modern reader's corrections (cf. above) are not considered. As a reference, folio and line numbers of I are given which can be easily followed in ANAGNOSTAKIS' fac-simile edition. Dots etc. are supplied according to modern usage.

Place in MS I	MS I		MS T
76^v, 2	بطليموس	–	بطلميوس
ult.	فكان	–	مكان
77^r, 13	بعداً	–	بعد (*in ras.*)
ult.	قوس آر	–	قوس آن [= *emend.* ANAGN.]
77^v, 2	طر	–	طم [= *emend.* ANAGN.] /
	ثم سم	–	تمر
78^r, 7	القسمة	–	قسمة
78^v, 6	ربحك	–	زبحد
79^r, 3	*post* فزاوية زطل T *add.* إذاً		
9	وهي النقطة	–	وهي النقط
79^v, 1	يتّصل	–	يصل
3	حيطك	–	*modern corr.* (*marg.*) ح ب ط د; *ante illud* T *add* دائرة
7	فصلنا	–	وصلنا
12	*ante* ح ط ا جـ T *add.* خطا		
80^r, 1	وهذه دائرة	–	وهي دائرة
6	سقاطعتان	–	يتقاطعان
12	منهما	–	منها
14	نقطة وَ...	–	نقطة زَ و... [= z] [= *emend*. ANAGN.]
ult.f.	وببعد هر هك	–	هر *deest* [= *emend.* ANAGN.]
81^r, 14	وكذلك	–	ولذلك
15	فالأجزاء	–	*dto., corr. in* بالأجزاء
ult.	وسبعون	–	وتسعون [= *emend.* ANAGN.]
81^v, 5	التي بها تكون بها	–	التي تكون بها
82^v, 2	دقيقة, *supra* درجة	–	درجة

Place in MS I	MS I		MS T
83^{v}, 3	فكان	–	مكان
14	نَفْسَهُ	–	يقسمه
ult.	ع ر	–	ع ن [= *emend.* ANAGN.]
84^{r}, 11	مساوياً	–	مساوٍ
84^{v}, 8	حدت	–	حدث [= *emend.* ANAGN.]
85^{r}, ult.	وثلثمائة	–	ثلثمائة
85^{v}, 3		*idem*	
86^{v}, 9	*post* وخمسة عشر T *add* درجة		
87^{r}, 12	وأربعون	–	وأربع وأربعون
ult.	القطبا	–	القطب
87^{v}, 8	ظاهراً	–	ظاهر
88^{r}, 9	مساوية	–	متساوية /
	أيضاً الذى	–	أيضاً التى
14	نفسها	–	بعينها
88^{v}, 8	على الاستواء	–	على نهار الاستواء
12	أو أقصر	–	أو أقصر ما
14	ر كل	–	ر كجل
89^{r}, 1	دائرة الأفق	–	مركز دائرة أفق
9	وثلثمائة	–	ثلثمائة
89^{v}, 10	من البيّن	–	ومن البيّن
14	كم التى	–	كم
90^{r}, 13	عليها	–	عليه
90^{v}, 1	التى بها يكون بها	–	التى بها تكون
91^{r}, 13	هكم	–	كم [= *emend.* ANAGN.]
91^{v}, 2	من خمس وأربعين درجة	–	من سبع وخمسين درجة [= *emend.* ANAGN.]
3	وإحدى وأربعين	–	واحداً وأربعين
92^{r}, 1	وهى خمسة وثلثين	–	و هى خمسة وثلثون زماناً
4	ذكرناه	–	بيّنّاه
5	ذكرناه	–	ذكرنا /
	يتهيأ لنا أن	–	يتهيأ لنا فيه أن

Place in MS I	MS I		MS T
92^{r}, 8	دوائر	–	بدلَ دوائر
13	وهو خط خط	–	وهو خط حط [= *emend.* Anagn.]
92^{v},11	مساوياً	–	مساوٍ
15	فقوس سكن	–	فقوس سلن [= *emend.* Anagn.]
93^{r}, 7	بالنقطتين	–	بالقطبين
93^{v}, 7	خط خط	–	خط رحط [= *emend.* Anagn.]
94^{r}, 3	محيط	–	خط محيط
94^{v}, 8	خط كع	–	خط لع
15	خط دل	–	خط دم
16	خط بم	–	خط نم
95^{r}, 16	خط أخر	–	خط احن [= *emend.* Anagn.] /
	خطى دحر	–	خطى دحن [= *emend.* Anagn.]
95^{v}, 5	دائرة قرس	–	دائرة قرش [= *emend.* Anagn.]
9	بنقط س ف	–	بنقط ع س ف
13	للغرض	–	للعرض
96^{r}, 1	هى زوا قائمة	–	هى على زوايا قائمة
97^{r}, 1	معها	–	معما
3	فلا يكون	–	ولا يكون
5	بطليموس	–	بطلميوس

VII

THE ROLE OF AL-ANDALUS IN THE TRANSMISSION OF PTOLEMY'S *PLANISPHAERIUM* AND *ALMAGEST**

The major role in the transmission of Arabic-Islamic science to the West was played by Muslim and Christian Spain. In comparison, much less Arabic works were translated in Salerno and in Sicily and only two in the lands of the Crusaders in the Near East.

Among the works translated into Latin in Spain in the twelfth century were two fundamental texts in astronomy authored by the Alexandrian astronomer and scientist Claudius Ptolemaeus (2nd cent. A.D.), *viz.* his handbook of astronomy, the *Almagest*, which was the most important source for astronomers for centuries both in the Orient and in Europe, and his *Planisphaerium*, which was understood as presenting the theoretical foundations for the construction of the astrolabe, the most popular astronomical instrument in the Middle Ages in East and West.

Let us first look at the *Almagest* which has become the object of new interest and has provoked a number of studies, investigations and editions in the last few decades.

The Greek text of the *Almagest* has survived and was edited in two volumes by J. L. Heiberg in the Teubneriana, Leipzig 1899 and 1903. In the late eighth to the late ninth centuries several Arabic versions were produced in Baghdad. I shall not go into details here[1] and content myself mentioning that two of these became of greater historical influence, the translation of al-Ḥajjāj b. Yūsuf b. Maṭar (212 H = A.D. 827-28) and the translation of Isḥāq b. Ḥunayn (around A.D. 880) with emendations by Thābit b. Qurra (d. A.D. 901). It can be taken for sure that it was from one, or both, of these Arabic versions

* Revised and updated version of a paper given at the Fifth International Symposium on the History of Arabic Science, Granada, March 30 - April 4, 1992. A résumé of this paper, in Arabic, was published in: *Proceedings of the Fifth International Symposium for the History of Arabic Science held at the University of Granada, Granada - Spain 30 March - 4 April 1992*, ed. by M. Mawaldi, vol. I, Aleppo 1415/1995, p. 91.

[1] See the details in Kunitzsch 1974: 15-82.

that the Spanish-Arabic astronomer Maslama al-Majrīṭī studied the *Almagest* in the second half of the tenth century[2]. In the early twelfth century the Spanish-Arabic astronomer Jābir b. Aflaḥ cited the two versions by name and used them in his *Iṣlāḥ al-Majasṭī*[3]. A few decades later Gerard of Cremona followed the same two Arabic versions in his Latin translation of the *Almagest*. It can be added that all the surviving Arabic manuscripts of the *Almagest* represent these two versions (two for al-Ḥajjāj, and nine for Isḥāq-Thābit)[4].

In the course of recent work on the transmission of the *Almagest* and the edition of the star catalogue from Books VII-VIII, I made some observations with special reference to the role of Spain which I shall shortly describe here.

One of the surviving Arabic manuscripts is in Hebrew characters (Paris, B.N. hebr. 1100). It contains the *Almagest* in the Isḥāq-Thābit version, but the star catalogue is in the Ḥajjāj version. The first part of it was copied in Calatayud in A.D. 1380, the second part (including the star catalogue) also in Calatayud in A.D. 1475, evidently by a grandson of the first writer[5].

Another manuscript, also containing the Isḥāq-Thābit version (the star catalogue again being in the Ḥajjāj version), in Arabic script, in the Maghrebi style, is MS Escorial 914. It is incomplete and contains Books V-IX only. No date of copying appears. On the initial folio there are some owners' notes. The second of these was Rabbi Ṭodrōs

[2] Cf. Ṣā'id al-Andalusī 169,2-3. The assumption brought forward by Vernet 1965: 17 that Maslama might also have used the Greek text for understanding difficult passages in the Arabic appears to us quite improbable (see also below, on the *Planisphaerium*).

[3] Cf. Kunitzsch 1974: 36 note 87; Lorch 1975: 97.

[4] Cf. Ptolemy 1986: 3f. MS no. 10 listed here (Jaipur, Maharaja Man Singh II Library, no. 20) has to be removed from the list of manuscripts of the Arabic text of the *Almagest*. Recent investigation of the MS *in loco* by Dr. R. Lorch, Munich, and Prof. G. Saliba, New York, showed that the Jaipur MS contains the same text as MS Teheran, Naṣīrī 789 (described in Ptolemy 1991: 200), *viz.* a paraphrase or epitome of the *Almagest* by an unknown author made on the basis of the Isḥāq-Thābit version. The total number of known manuscripts of the Arabic text of the *Almagest* is thus reduced to eleven. A further manuscript mentioned in Ptolemy 1990: 171 (Jaipur, Maharaja Man Singh II Museum, no. 348) was found, on the same occasion, to contain al-Ṭūsī's *Taḥrīr* of the *Almagest*, copied in Sūrat, Gujerat, in 1023 H = A.D. 1614. (I am most grateful to Dr. Lorch and Prof. Saliba for supplying me with this valuable information collected during a congress in Jaipur in December 1991.)

[5] For the details see Ptolemy 1986: 6ff.

b. Moshe al-Qusṭanṭīnī, the same who in 1380 copied the first part of MS hebr. 1100. The manuscript then seems to have remained in the possession of the Qusṭanṭīnī family until, near the end of the 15th century, it came into the hands of a member of the Calavera family in Zaragoza. Rabbi Ṭodrōs has also noted, in Hebrew, on the same page that the manuscript contained the five Books from V to IX (i.e. the same fragmentary part that still exists today).

Here we have the rare example in the history of text transmission that a manuscript used by later copyists has survived. We thus now have both, the source manuscript (Escorial 914) and the later copy in Hebrew characters made by two members of the Qusṭanṭīnī family in Calatayud (Paris, hebr. 1100). The Escorial manuscript was already fragmentary when Rabbi Ṭodrōs obtained it, in the same way as we have it today. Apart from the direct notes in the Escorial manuscript, also many copying errors in the Hebrew manuscript testify that the Hebrew copyists used that Arabic source whose Maghrebi script they were not able correctly to read in all places. On the other hand, the Hebrew copyists must have used at least one more Arabic manuscript because occasionally they have better, or more complete, readings than the Escorial manuscript.

The second example is formed by a striking relationship between Gerard's Latin translation and the Arabic manuscript kept in the National Library of Tunis (no. 07116). In Gerard's translation we have found that for Books I-IX he followed the Arabic version of al-Ḥajjāj, while for the rest of the work, Books X-XIII, he followed the Isḥāq-Thābit version[6]. The star catalogue (in Books VII-VIII, i.e. in the Ḥajjāj portion of the translation) was rendered from the Isḥāq-Thābit version. But Gerhard's source appears to have been a special form of Isḥāq-Thābit, because in the verbal descriptions of the stars' locations and in the coordinates in many cases Ḥajjāj material was inserted instead of the genuine Isḥāq-Thābit text[7].

Now, by chance, the Tunis manuscript, which is a pure Isḥāq-Thābit manuscript, also for the star catalogue, contains in the margins of the star catalogue a great number of parallel words or formulas or coordinate values quoted from the Ḥajjāj version and distinctly marked as such by certain abbreviations added to each of them. The manuscript is dated 478 H = A.D. 1085 and is written in the Maghrebi style. Most

[6] See Kunitzsch 1974: 97-102.

[7] In the edition (Ptolemy 1990) the Ḥajjāj elements are printed in italics.

certainly it was copied in Muslim Spain. The marginal notes from al-Ḥajjāj in the Tunis manuscript completely coincide with Gerard's insertions of Ḥajjāj material in the course of his Isḥāq-Thābit version of the star catalogue.

It is therefore evident that Gerard, who worked on the translation of the *Almagest* in Toledo in the years roughly between 1150 and 1180[8], must have used an Arabic manuscript of the Isḥāq-Thābit version very closely related to the Tunis manuscript. We do not assume that it was the Tunis manuscript itself that was used by Gerard; in this case some traces of permanent use, and perhaps some handwritten notes by him, should exist in the manuscript. This is not the case. Rather we assume that in the intervening seventy or more years between the dates of the copying of the Tunis manuscript and of Gerard's translation more Arabic copies were made of the Isḥāq-Thābit version in the style of MS Tunis, i.e. including those quotations from al-Ḥajjāj, and that in his time Gerard used a manuscript belonging to that special branch of transmission. This example again illustrates the living interest and the ample activities developing in the Spanish area in the eleventh and twelfth centuries concerning the sciences, both in the Muslim and the Christian regions.

Next we come to the *Planisphaerium.* Its Greek text seems to be lost. Since long time the work was best known in the Latin translation made from the Arabic by Hermann of Carinthia (Tolosa, 1143)[9]. Of the underlying Arabic version not much was known nor has the text been edited. A common opinion[10] ascribed the Arabic translation to the Spanish-Arabic astronomer Maslama al-Majrīṭī (d. 398 H = A.D. 1007-08[11]).

This seemed highly improbable, because all the translations of Greek scientific texts were made in the East, mostly at Baghdad; no specimens are known that were produced in Spain. It even can be doubted that there existed in Muslim Spain any person knowing Greek at all or sufficiently to carry out scientific translations. Quite conclusive in

[8] Cf. Ptolemy 1990: 2f.

[9] Cf. Carmody 1956: 18 (no. 9). Hermann's Latin version was edited by J. L. Heiberg in Ptolemaeus, *Opera*, vol. II, Leipzig 1907, pp. 225-259; a German translation of this, by J. Drecker, is in: *Isis* 9 (1927), 255-278.

[10] Cf., e.g., Sarton 1931: 174 (item 5); Carmody 1956: 18 (no. 9); Vernet-Catalá 1965: 17f.; Neugebauer 1975: 871; Burnett 1978: 108.

[11] Cf. Ṣā'id al-Andalusī 169,11-12.

this concern is the report transmitted by the Andalusian historian Ibn Juljul. According to him, in the year 337 H = A.D. 948-49 the Spanish Omaiyad caliph ʿAbd al-Raḥmān al-Nāṣir received from the Byzantine emperor a number of valuable gifts among which an illustrated copy of Dioscurides' book on the *materia medica*. Since al-Nāṣir found nobody in his territory who was in command of Greek so as to read and translate Dioscurides' work, he asked the emperor to send him a man of that quality. The emperor indeed sent him the monk Nicolaus (Nīqūlā) who arrived in Cordoba in 340 H = A.D. 951-52 and who, in collaboration with several local scholars, established the identity of a number of Greek plant names (it is not mentioned that he translated the entire book)[12]. Nicolaus is reported to have died near the beginning of the reign of Caliph al-Ḥakam II (reg. 961-976).

This may suffice to demonstrate the improbability of Maslama being the translator of the *Planisphaerium* from Greek into Arabic. The erroneous modern opinion to that effect may have been provoked by a remark in this sense in the preface of Hermann's Latin translation and by a number of marginal notes and additions to the Latin text which are explicitly attributed to Maslama.

Now, in one Arabic manuscript (Paris, B.N. ar. 4821, foll. 69v-81r) a collection of Maslama's notes and additions to the *Planisphaerium* was really found. So Maslama's authorship for the notes and the additions can be regarded as safely established[13].

But what about the Arabic text of the *Planisphaerium* itself? When and where may it have originated? Presently three manuscripts of it are known[14] two of which are accessible (Istanbul, Aya Sofya 2671, foll. 76v-97r, and Tehran, Khān Malik Sāsānī)[15].

Surprisingly, Ibn al-Nadīm's *Fihrist*, our best and most complete bibliographical source on the Arabic translations from the Greek, does not list the *Planisphaerium* among Ptolemy's works. It is only

[12] Cf. Ullmann 1970: 260; a German translation of Ibn Juljul's report is in Rosenthal 1965: 265-268.

[13] Maslama's "additional chapter" and the notes referring to the astrolabe were edited by Vernet-Catalá 1965; the notes to the *Planisphaerium* were edited, in Arabic and Latin, by Kunitzsch-Lorch 1994.

[14] Cf. Sezgin 1974: 170.

[15] The text of the Istanbul MS was published in facsimile by Anagnostakis 1984. A collation of the text in the Tehran MS is given in Kunitzsch 1995. The third manuscript, in Kabul, has remained inaccessible.

under the entry of Pappus that the title occurs, where Pappus' commentary on the *Planisphaerium* is mentioned: *Kitāb tafsīr Kitāb Baṭlamyūs fī tasṭīḥ al-kura* (adding that it was translated by Thābit b. Qurra)[16]. Does that mean that the Arabic translation of the *Planisphaerium* did not exist by the time of the completion of the *Fihrist* in 377 H = A.D. 987-88? This appears highly doubtful.

The Arabic title given in the *Fihrist* is not complete. In full wording it runs: *Kitāb fī tasṭīḥ basīṭ al-kura*, "Book on the Projection of the Surface of the Sphere". In this form it is found in the Istanbul and Tehran manuscripts, in the Paris manuscript with Maslama's notes[17] and in a treatise by Ibn al-Ṣalāḥ (d. 1154) on the projection of the sphere[18].

Luckily, there is yet another author who mentions the *Planisphaerium*, Ibrāhīm b. Sinān b. Thābit, a grandson of the famous Thābit b. Qurra. His lifetime is fixed to A.D. 909-946. In his treatise (*risāla*) on the astrolabe he underlines that he wishes to treat matters in a short and concentrated form, otherwise it would come close to copying Ptolemy's lengthy and more detailed *Planisphaerium.* And then he even quotes, from memory as he says (*fī-mā aḥfaẓuhu*), the initial lines of Ptolemy's text.

To our great satisfaction it is evident that the words he quotes are basically, and partly *verbatim*, identical to the text of the *Planisphaerium* as found in the Istanbul and Tehran manuscripts: *Innahu lammā kāna min al-mumkin yā Sūrī wa-mimmā yuntafaʿu bihi fī abwāb kathīra* etc., "Since it is possible, o Syrus, and useful in many aspects..."[19].

So now we have definite proof that the Arabic version of the *Planisphaerium* existed in the Arabic East in the first half of the tenth century, before 946.

[16] Ibn al-Nadīm 269,9.

[17] Vernet-Catalá 1965: 18, 21 read *basṭ* instead of *basīṭ* (in the Paris MS). But on a film of the manuscript it becomes visible that the Paris MS, in two instances (on foll. 69v and 75v), also has *basīṭ* (the undotted *rasm* of the word contains four strokes, three for the *sīn* and one for the *yā*ʾ). Cf. also Kunitzsch-Lorch 1994: 13 note 1.

[18] Ibn al-Ṣalāḥ, Saray 3342, fol. 55v,21 = Carullah 2060, fol. 143v,16. In the title of his own treatise Ibn al-Ṣalāḥ uses the formula *tasṭīḥ al-basīṭ al-kurī*.

[19] Ibrāhīm b. Sinān 312.

It remains unclear why it was not registered by Ibn al-Nadīm. The most obvious reason would be that no copy of it had come into his hands and none of his numerous acquaintances among the learned circles of Baghdad had mentioned it in front of him. Pappus' commentary, on the other hand, in the translation of Thābit b. Qurra, was known to him and was also cited by Ibrahīm b. Sinān and still later by Ibn al-Ṣalāḥ. Of this no text has come to light until now, as far as we know[20].

In conclusion, with regard to the *Planisphaerium* we can now safely say: an Arabic translation of it was produced at some point by an unknown translator in the Arabic East. This version was known to and used by Ibrāhīm b. Sinān who died in 946. A few decades later it was also available in Muslim Spain where in the second half of the tenth century it was studied by Maslama al-Majrīṭī[21]. Maslama then wrote notes and additions to it, but he was not at all involved in the translation of the text itself[22]. These notes seem to have been transmitted in some manuscripts together with the text, so that they could be included by Hermann of Carinthia in his Latin translation of 1143.

The endeavours of Maslama and his disciples on the astrolabe and the *Planisphaerium* seem to have radiated into Northeastern Spain, where in the last decades of the tenth century for the first time bits of Arabic astronomical knowledge, notably the astrolabe, were received by interested Western circles, first attempts were made to compose texts and perhaps the first Western astrolabe was prepared after Arabic models (the so-called "Carolingian Astrolabe", formerly in the collection of the late Marcel Destombes, Paris, and now kept in the Museum of the Institut du Monde Arabe in Paris).

Among the material that reached Northern Spain there seems to have been the Arabic version of the *Planisphaerium*. Two fragments from it have recently been found, in Latin translation, among scattered portions of texts belonging to the oldest Arabo-Latin text corpus on

[20] Cf. Sezgin 1974: 175 (item 2). In Sezgin 1978: 95 (item III) the same text is erroneously attribluted to Theon. In Ibn al-Ṣalāḥ (*loc. cit.* in note 18) the author is called *B.b.s al-Iskandarānī*, whereas the *Fihrist* (*loc. cit.* in note 16, line 8) calls him *B.b.s al-Rūmī*.

[21] This parallels his study of the *Almagest* mentioned by Ṣāʿid al-Andalusī (cf. above, with note 2).

[22] In Ṣāʿid al-Andalusī's entry on Maslama (pp. 168f.) only his study of the *Almagest* is mentioned, but not his endeavours on the *Planisphaerium* and the astrolabe.

the astrolabe of that time[23]. This may have been the point of departure for the formation of the (fabulous) idea that Ptolemy was the inventor of the planispheric astrolabe, an idea that later spread in many medieval Western writings on the astrolabe[24].

The history of the text of the *Planisphaerium* once more confirms that the role of Al-Andalus in the transmission of science lay in the reception and assimilation of Arabic material and its handing over to the Latin West. The translation of Greek material, on the other hand, was mainly, if not totally, an achievement of the Arabic East.

Literature

Anagnostakis 1984: C. Anagnostakis, *The Arabic Version of Ptolemy's Planisphaerium*, Ph.D. thesis, Yale University 1984 (published in book form by University Microfilms International, Ann Arbor, Michigan, in 1986).

Burnett 1978: C. S. F. Burnett, Arabic into Latin in Twelfth Century Spain: the Works of Hermann of Carinthia, in: *Mittellateinisches Jahrbuch* 13 (1978), 100-134.

Carmody 1956: F. J. Carmody, *Arabic Astronomical and Astrological Sciences in Latin Translation. A Critical Bibliography*. Berkeley and Los Angeles 1956.

Ibn al-Nadīm: Ibn al-Nadīm, *Kitāb al-Fihrist*, ed. G. Flügel, vol. I: Text, Leipzig 1871.

Ibn al-Ṣalāḥ: Ibn al-Ṣalāḥ, *Kitāb fī kayfīyat tasṭīḥ al-basīṭ al-kurī wa-huwa al-maʿrūf bi-l-asṭurlāb*. MSS Istanbul, Saray 3342, foll. 55vff.; Carullah 2060, foll. 143vff.

Ibrāhīm b. Sinān: Ibrāhīm b. Sinān b. Thābit, *Rasāʾil Ibn Sinān*, ed. A. S. Saidan, Kuwait 1403/1983 (here especially: [*al-Risāla al-sābiʿa*] *fī l-asṭurlāb*, pp. 305-318). The edition is based on MS Bankipur 2468.

Kunitzsch 1974: P. Kunitzsch, *Der Almagest. Die Syntaxis Mathematica des Claudius Ptolemäus in arabisch-lateinischer Überlieferung*, Wiesbaden 1974.

Kunitzsch 1982: P. Kunitzsch, *Glossar der arabischen Fachausdrücke in der mittelalterlichen europäischen Astrolabliteratur*, Göttingen 1982 (Nachrichten der Akademie der Wissenschaften in Göttingen, Phil.-Hist. Kl., Jahrg. 1982, Nr. 11).

[23] Cf. Kunitzsch 1993.

[24] See the texts cited in Kunitzsch 1982: 518 (no. 3); cf. also Kunitzsch 1993: 98 and note 11.

Kunitzsch 1993: P. Kunitzsch, Fragments of Ptolemy's *Planisphaerium* in an Early Latin Translation, in: *Centaurus* 36 (1993), 97-101.

Kunitzsch 1994: P. Kunitzsch, The second Arabic manuscript of Ptolemy's *Planisphaerium*, in: *Zeitschrift für Geschichte der arabisch-islamischen Wissenschaften* 9 (1994), 83-89.

Kunitzsch-Lorch 1994: P. Kunitzsch and R. Lorch, *Maslama's Notes on Ptolemy's* Planisphaerium *and Related Texts*, München 1994 (Bayerische Akademie der Wissenschaften, Phil.-Hist. Kl., Sitzungsberichte, Jahrg. 1994, Heft 2).

Lorch 1975: R. P. Lorch, The Astronomy of Jābir ibn Aflaḥ, in: *Centaurus* 19 (1975), 85-107.

Neugebauer 1975: O. Neugebauer, *History of Ancient Mathematical Astronomy*, I-III, Berlin etc. 1975.

Ptolemy 1986: Claudius Ptolemäus, *Der Sternkatalog des Almagest. Die arabisch-mittelalterliche Tradition*, vol. I: *Die arabischen Übersetzungen*, ed. P. Kunitzsch, Wiesbaden 1986.

Ptolemy 1990: idem, vol. II: *Die lateinische Übersetzung Gerhards von Cremona*, ed. P. Kunitzsch, Wiesbaden 1990.

Ptolemy 1991: idem, vol. III: *Gesamtkonkordanz der Sternkoordinaten*, ed. P. Kunitzsch, Wiesbaden 1991.

Rosenthal 1965: F. Rosenthal, *Das Fortleben der Antike im Islam*. Zurich and Stuttgart 1965.

Ṣāʿid al-Andalusī: Ṣāʿid al-Andalusī, *Ṭabaqāt al-Umam*, ed. Ḥ. Bū-ʿAlwān, Beirut 1985.

Sarton 1931: G. Sarton, *Introduction to the History of Science*, vol. II, part 1, Washington 1931 (repr. 1950).

Sezgin 1974: F. Sezgin, *Geschichte des arabischen Schrifttums*, vol. V: *Mathematik bis ca. 430 H.*, Leiden 1974.

Sezgin 1978: idem, vol. VI: *Astronomie bis ca. 430 H.*, Leiden 1978.

Ullmann 1970: M. Ullmann, *Die Medizin im Islam*, Leiden/Köln 1970 (Handbuch der Orientalistik, I. Abt., Ergänzungsbd. VI, 1. Abschnitt).

Vernet-Catalá 1965: J. Vernet and M. A. Catalá, Las obras matematicas de Maslama de Madrid, in: *Al-Andalus* 30 (1965), 15-45 (with an addition by M. A. Catalá, pp. 46-47).

VIII

Fragments of Ptolemy's *Planisphaerium* in an Early Latin Translation

The Greek text of the *Planisphaerium*, on the projection of the circles on the sphere onto a plane, by Ptolemy (2nd century AD) seems to be lost[1]. But an Arabic translation is extant, presumably made in Baghdad around or before AD 900[2]; it survived in three manuscripts, two of which[3] are accessible and have recently been used for research[4]. The Arabic text afterwards reached Muslim Spain, where it was studied and commented upon by Maslama al-Majrīṭī (d. 1007–08)[5]. In 1143, in Tolosa in Northern Spain, the Arabic text of the *Planisphaerium* was translated into Latin by Hermann of Carinthia[6]. Probably induced by the comments added to the text explicitly under Maslama's name, Hermann erroneously thought that the text of the *Planisphaerium* itself had been translated into Arabic by Maslama; this wrong assumption was retained by most modern scholars until recently[7]. But it is now clear – especially because of Ibrāhīm ibn Sinān's quotations (cf. note 2, above) – that the Arabic text of the *Planisphaerium* existed in the Arabic East half a century or more before Maslama.

Now, Professor A. Borst, Konstanz, has recently directed our attention to two pieces of text found in MS Paris, B.N. lat. 7412 (11th cent.), 11r–v, inserted among scattered portions of texts centered around the *Sententie astrolabii* as edited by Millás[8] and Bubnov[9]. These two pieces, however, do not form part of any of those texts.

Surprisingly, it turned out that the two pieces are a primitive Latin rendering of the beginnings of Chapters 2 and 3 of Ptolemy's *Planisphaerium.* This is of considerable historical interest, because it proves that the Latin scholars in late tenth-century north-eastern Spain who made the first Western European attempt to study and understand the astrolabe based on Arabic material[10] must also have had at hand the Arabic version of the *Planisphaerium*, or parts of it. At the same time it provides additional evidence for the fact that the first Western attempts at using the astrolabe were stimulated by the achievements of contemporary Spanish-Arabic astronomers, among whom Maslama was the most prominent. Maslama had also known, and commented on, the *Planisphaerium* (see above). Though found uniquely in the Paris manuscript, this first Latin rendering of parts of the *Planisphaerium* – roughly 150 years before Hermann of Carinthia's translation in 1143 – obviously belongs to the voluminous astrolabe *corpus* produced by those Latin scholars in late tenth century. Perhaps more of it was translated and is still hidden in remote places in other manuscripts.

The evidence here presented for the knowledge of the *Planisphaerium*, or parts of it, by some Latin scholars of late tenth century, most happily produces the 'missing link' in the development of the wide-spread medieval theory that Ptolemy was the 'inventor' of the planispheric astrolabe. This is historically wrong, because what Ptolemy describes is not yet the fully developed plane astrolabe as existing in the Middle Ages, in the Orient and, subsequently, in Europe. Nevertheless, the theory of projection demonstrated in the *Planisphaerium*, which is at the base of the construction of the plane astrolabe, induced Western authors to regard Ptolemy as the inventor of the instrument[11].

Now follows the edition of the two portions of text found in MS Paris, B.N. lat. 7412[12].

<11r> Dixit ptolomeus: exient[a] toti circuli de alafac[b] qui terminant[c] super simile[d] circuli[1] de alfelec[e] id est de rota inclinata uel

circumuoluta, non est[f] inneme[g] partiuntur[2] circulum[3] equalitatis[4] peisatu <?> diei[5,h]. || Et dicam quia si alium circulum faciemus declinatum a circulo[6] aequalitatis diei loco[7] cir- <11v> culi alofc[i], donec iste circulus diuidat circulum diei aequalitatis singulariter per medium, quia duo truncabunt istum circulum cum circulo qui andat in medio casarum, qui se contemplantur super alcoter[k] cum fortitudine[l]. Dico quia linea que inter[8] illos est transit per almerkerz[m] circuli aequalitatis diei[n].

[1] circuli: circulu MS [2] partiuntur: partitur MS [3] circulum: circulu MS [4] equalitatis: equali mente MS [5] diei: die MS [6] a circulo: fiat circulus MS [7] loco: fiet MS [8] inter: infra MS

[a] *exient*: perhaps corrupted from *erunt*, for Ar. *takūnu*.
[b] Ar. MS Istanbul has sing. *al-ufq*, perhaps the translator read the plural, *al-āfāq* ("horizons"); *de* paraphrases the Arabic genitive.
[c] Ar. *tursamu*, "are drawn".
[d] corresponds to Ar. *ʿalā mithli mā*, "in the same way as".
[e] Ar. *al-falak*, "sphere"; for *de*, cf. *sub* b, above.
[f] imitates Ar. *laysa* which here only means "not".
[g] Ar. *innamā*, "only".
[h] This section is the beginning of Chapter 2 of the *Planisphaerium*, corresponding to Arabic, MS Istanbul, Aya Sofya 2671, 78r,ult.–78v,2, and transl. Hermann ed. Heiberg, 220, 4–6.
[i] Ar. *al-ufq*, "horizon".
[k] Ar. *al-quṭr*, "diameter".
[l] Ar. *bi 'l-quwwa*, "in potentia, potentialiter".
[m] Ar. *al-markaz*, "center".
[n] Beginning of Chapter 3, corresponding to MS Istanbul, 79r,12–79v,2, and Hermann ed. Heiberg, 231, 15–20.

NOTES

1. Cf. B. L. van der Waerden, article "Ptolemaios, Klaudios" (works: *Planisphaerium*), in Pauly-Wissowa-Kroll, *Real-Encyclopädie der classischen Altertumswissenschaft*, 46. Halbbd. (Stuttgart, 1955), col. 1829–31; G. J. Toomer, article "Ptolemy", in *Dictionary of Scientific Biography* (ed. C. C. Gillispie), XI (New York, 1975), pp. 186ff., esp. 197f., 205; O. Neugebauer, *History of Ancient Mathematical Astronomy*, I–III (Berlin etc., 1975), pp. 857–879.
2. This version was used and cited by Ibrāhīm ibn Sinān ibn Thābit (909–946) in Baghdad, that is, ca. fifty years before Maslama; cf. P. Kunitzsch, "The Role of Al-Andalus in the Transmission of Ptolemy's *Planisphaerium* and *Almagest*", to appear in the

Proceedings of the Fifth International Symposium on the History of Arabic Science, Granada, 30 March – 4 April 1992.

3. Istanbul, Aya Sofya 2671 (dated 621 H = AD 1224), and Tehran, private library of Khān Malik Sāsānī (dated 607 or 617 H = AD 1210 or 1220–21). The third manuscript, in Kabul, has remained inaccessible. The Istanbul manuscript was studied and edited in facsimile by C. Anagnostakis, *The Arabic Version of Ptolemy's Planisphaerium*, Ph.D. thesis, Yale University 1984 (published in book form by University Microfilms International, Ann Arbor, Michigan, in 1986).
4. P. Kunitzsch, "The second Arabic manuscript of Ptolemy's *Planisphaerium*", to appear in *Zeitschrift für Geschichte der Arabisch-Islamischen Wissenschaften*; P. Kunitzsch and R. P. Lorch, Maslama's notes on Ptolemy's *Planisphaerium* (edition of texts, in preparation).
5. An edition of the Arabic text and Latin versions of Maslama's notes by P. Kunitzsch and R. P. Lorch is under preparation (cf. note 4).
6. Ed. J. L. Heiberg, in Ptolemaeus, *Opera*, vol. II, Leipzig 1907. For the identification of Tolosa, cf. R. Lemay, "De la scholastique à l'histoire par le truchement de la philologie: itinéraire d'un médiéviste entre Europe et l'Islam", in *La diffusione delle scienze islamiche nel medio evo europeo*, Convegno internazionale, Roma, 2–4 ottobre 1984. Rome, 1987, pp. 399ff., esp. 451. On Hermann's translation, cf. also C. S. F. Burnett, "Arabic into Latin in Twelfth Century Spain: the Works of Hermann of Carinthia", *Mittellateinisches Jahrbuch* 13 (1978), pp. 100ff., esp. 108–112.
7. Cf. Kunitzsch (note 2), note 10; *idem* (note 5), note 8.
8. J. M. Millás Vallicrosa, *Assaig d'història de les idees físiques i matemàtiques a la Catalunya medieval*, Barcelona 1931.
9. N. Bubnov (ed.), *Gerberti postea Silvestri II papae Opera Mathematica*, Berlin 1899, esp. pp. 109ff.
10. For the section on the use of the instrument in the *Sententie astrolabii* (the longest and most important text among those edited by Millás, as in note 8), an Arabic source has been identified in al-Khwārizmī's treatise on the subject; cf. P. Kunitzsch, Al-Khwārizmī as a Source for the *Sententie astrolabii*, in *From Deferent to Equant: A Volume of Studies in the History of Science in The Ancient and Medieval Near East in Honor of E. S. Kennedy* (= *Annals of the New York Academy of Sciences*, vol. 500), New York 1987, pp. 227–236 (repr. in P. Kunitzsch, *The Arabs and the Stars*, Northampton 1989, item IX).
11. See, for example, the texts I, IV, V, XII, HC (= Hermann the Lame), RM (= Raymond de Marseille), AW and NN, discussed and used in P. Kunitzsch, *Glossar der arabischen Fachausdrücke in der mittelalterlichen europäischen Astrolabliteratur* (Nachrichten d. Akademie d. Wissenschaften in Göttingen, Phil.-hist. Kl., 1982, 11), Göttingen 1983. The same allusion in RB (= Rodolfus Brugensis) is of a different character, because – as Hermann of Carinthia's disciple – Rodolfus relied on Hermann's translation of the *Planisphaerium*; the allusion in MC is directly translated from an Arabic treatise on the construction of the astrolabe, presumably by al-Zarqāllu (Azarchel, d. 1100); cf. D. A. King, *Islamic Astronomical Instruments*, London 1987, item III, text no. 9 (p. 52, in English, and p. 69, in Arabic).
12. I express my sincere thanks to Professor Borst for his consent to the present edition. In the survey of contents of MS 7412 by M. Destombes (in *Archives Internationales*

d'Histoire des Sciences 15, 1962, pp. 43–45) our text was not registered. In the outer right margin of fol. 12r, our manuscript contains two notes added to the text called III by Millás (note 8) and by Kunitzsch (note 11) [inc. *Philosophi qui sua sapientia ...*], on the construction of the front of the astrolabe. Both notes are incomplete, the lines being cut off on the right at the edge of the folio. The upper note is explicitly ascribed to *Ptolomeus*. But the two notes are not really related to Ptolemy since they describe details of the practice of constructing the circles on the astrolabe, using such terms as *circinus* and *regula*, which are alien to the *Planisphaerium*.

IX

Das Arabische als Vermittler und Anreger europäischer Wissenschaftssprache*

Summary: The role played by the 'Arabs', *i. e.* the peoples of Arabic-Islamic civilization, in the transmission and development of the sciences – from Antiquity and into medieval Europe – is well known. In the present contribution it is discussed in which way the 'Arabs' formed their own scientific terminology and in which way they contributed to the formation and development of the Western, European scientific terminology. In the translations of scientific works from Arabic into Latin in the middle ages, mainly four ways of rendering the Arabic terminology are observed: simple transliteration; modified Latinized transliteration; literal translation; and free rendering by newly formed or inherited Latin or Greek terms. In the course of time, transliterated Arabic terms were more and more suppressed – though many of them live on among us until today – and supplied by corresponding Western terminology.

Schlüsselwörter: Arabismen, Arabisch (spanisches), Arabisch-islamische Kultur, Fachtermini (arabisch-lateinische), Mittelarabisch, Spanien, Transkription (arabisch-lateinisch), Übersetzungen (griechisch-arabische, arabisch-lateinische), Wissenschaftssprache (arabisch), Wissenschaftssprache (lateinisch); MA (X-XIII Jh.).

Zur Annäherung an unser Thema ist es erforderlich, zunächst den Begriff des Arabischen zu definieren und im historischen Kontext transparent zu machen. Natürlich war das Arabische ursprünglich die Sprache der Araber, also der Bewohner der Arabischen Halbinsel und einiger in die nördlich angrenzenden Gebiete gewanderter Stämme, zumeist Beduinen, und zwar viele Jahrhunderte oder länger vor dem Islam. Entsprechend dem zivilisatorischen Standard der Zeit und Gegend kannte die Sprache noch keine wissenschaftliche Terminologie im engeren Sinne; viele Sachgebiete, die später als ‚Wissenschaften' ausgeformt wurden – Heilkunde, Naturbeobachtung, Zählen und Rechnen usw. –, waren im Ansatz vorhanden und wurden in Form volkstümlicher Traditionen gepflegt. Nach dem Aufkommen und der Ausbreitung des Islam wurden diese beduinischen Araber Herren alter Kulturreiche rund um das Mittelmeer, ihre Sprache wurde zur κοινή im islamischen Reich. Schrittweise begann damit zugleich eine Rezeption der Wissenschaften in den eroberten Regionen, vor allem griechischer und persischer Materialien. Das Frühstadium dieser Übernahme ist uns verborgen, die erhaltenen Übersetzungstexte zeigen bereits ein ausgereiftes Stadium der Assimilation.

Das Arabische machte dabei Wandlungen in zweierlei Hinsicht durch: es wurde nunmehr auch von den Angehörigen der vielen anderen Nationen angewendet, die unterworfen wurden und die im islamischen Reich aufgingen, und es wurde in seinem Wortschatz und seinen Ausdrucksformen erweitert und fortentwickelt hin auf die zahlrei-

* Vortrag, gehalten auf dem 30. Symposium der Gesellschaft für Wissenschaftsgeschichte, 20.–23. Mai 1993 im Deutschen Studienzentrum Venedig.

chen Anforderungen in dem sich neu bildenden islamischen Staatswesen, sei es im Hinblick auf den Kult, auf Verwaltung, auf Literatur und eben auch auf die Wissenschaften. Die Orientalisten nennen diese Form des Arabischen ‚Mittelarabisch'.

Wenn wir also im Rahmen der Wissenschaftsgeschichte und der Überlieferungsfragen von „den Arabern" und „dem Arabischen" sprechen, so ist damit die in arabischer Sprache geschriebene Wissenschaftsliteratur des islamischen Kulturraums gemeint, an der viele Nationen – neben den Arabern auch Perser, Turkvölker, Berber, syrisch-aramäische Bevölkerungen, Inder usw. – ihren Anteil haben. Man spricht daher ganz treffend von der arabisch-islamischen Kultur. Ich werde im folgenden der Einfachheit halber von „den Arabern" und „dem Arabischen" sprechen, dabei sind diese beiden Begriffe aber stets in Anführungszeichen gesetzt zu denken, im Sinne des eben Dargelegten.

Die kulturgeschichtliche Leistung der Araber entfaltete sich in zwei Richtungen: sie übernahmen und entwickelten das griechisch-antike Wissenschaftserbe, und sie übermittelten diese Wissenschaften weiter an das Abendland. Für die Entwicklung der Wissenschaftssprache bedeutet das: die Araber schufen zunächst eine arabische Wissenschaftssprache; anschließend wirkten sie mit an der Ausbildung der lateinischen, europäischen Wissenschaftssprache.

Hier zeichnen sich zwei Überlieferungswege ab, einer im Osten, in den griechischsprachigen Raum von Byzanz hinein – der, auch wegen der Kontinuität des Umgangs mit dem altgriechischen Material, einen etwas anderen Verlauf nahm als im Westen und der seinerseits nur wenig auf den Westen weiterwirkte –, der andere im Westen, vor allem von Spanien, weniger von Sizilien und Süditalien aus, in das lateinische westliche Europa hinein.

Wir durchschauen noch nicht im einzelnen die Umstände, die zu dem überwältigenden Interesse des Westens an der *doctrina arabum* geführt haben. Die Einstellung des Westens zu der neu in die Geschichte eintretenden Welt des Islam war von Anfang und auf Dauer grundsätzlich feindselig. Man lernte die Sarazenen als Eroberer, als Glaubensfeinde, als vielfach überlegene Macht kennen, die die bis dahin unangefochtene Alleingültigkeit christlich-abendländischen Selbstverständnisses erschütterte und relativierte. Vom Tage von Tours und Poitiers, 732, an bis zur Rückgewinnung von Granada, 1492, ja bis zum Sieg über die Türken vor Wien, 1683, zog sich der beständige Kampf gegen die Muslime hin, denen man europäischen Boden, zeitweilig sogar Teile des Heiligen Landes selbst, wieder zu entreißen suchte. Daneben bildeten sich freilich in den gemeinsamen Berührungsgebieten auch Formen eines gemischten, mehr oder weniger toleranten Zusammenlebens aus.

Es mag sein, daß so die Christen in Spanien und Süditalien auf viele Interna des arabischen Lebens, darunter die Pflege der Wissenschaften, aufmerksam wurden. Im westlichen Abendland war um diese Zeit, im 9.–10. Jahrhundert, die Verbindung zu den griechischen Quellen nahezu völlig abgerissen; die Wissenschaften kannte man nur noch in den schwachen, verwässerten Ausläufern, die durch spätlateinische Kompendien und Enzyklopädien weitergetragen wurden, stets eingezwängt in die kirchlich-theologischen Rahmenbedingungen. Demgegenüber hatten sich im muslimischen Spanien nach der Eroberung und der Aufbauzeit die Verhältnisse bis zum 10. Jahrhundert soweit konsolidiert, daß dort – unter Zufluß des entsprechenden Materials aus dem Osten – die Wissenschaften ebenso lebendige Pflege erfuhren wie im Bagdader Kalifenreich. Hier konnte man nun wieder auf die Namen Galenos, Aristoteles, Ptolemaios, Eukleides usw. stoßen.

Wir müssen grundsätzlich festhalten, daß alle lateinischen Übernahmen arabischen Wissenschaftsmaterials in Spanien auf der christlichen Seite der Grenze, in den rückeroberten Gebieten stattfanden. Grenzgänger aller Art müssen die Kenntnis über das Material wie auch dieses selbst immer wieder in den christlichen Norden transportiert haben; Männer aus diesem Umkreis waren dann auch häufig an den Übersetzungen selbst beteiligt, wie wir noch sehen werden.

Eine erste Welle der Beschäftigung mit arabischer Wissenschaft – in diesem Fall Astronomie und Astrologie – kam im Raum um Barcelona gegen Ende des 10. Jahrhunderts in Gang. Vor allem das Astrolab, jenes praktische und vielseitig verwendbare astronomische Handinstrument, das die Araber aus spätgriechischer Tradition übernommen hatten, scheint Interesse erweckt zu haben. Es entstanden hier lateinische Schriften über Form und Gebrauch des Instruments, die teils aus dem Arabischen übersetzt wurden, teils in Anlehnung an arabische Quellen und nach vorliegenden Instrumenten selbständig verfaßt wurden. Aus dieser ältesten Entlehnungsschicht haben einige arabische Sternnamen – Rigel, Altair, Aldebaran – bis in die heutige Zeit überdauert[1].

Die Hauptperiode der Übersetzungen fällt ins 12. Jahrhundert. Unter den bekannten Übersetzern finden sich sowohl Namen aus Spanien selbst, wie Johannes Hispalensis, als auch aus vielen anderen Teilen Europas, wie Adelhard von Bath, Gerhard von Cremona, Plato von Tivoli, Hermann von Kärnten und andere. Spanien wurde so zum Haupteinfallstor der arabischen Wissenschaften ins westliche Europa. Wir wollen hier nicht in eine vollständige Aufzählung von Namen und Orten eintreten.

Um uns näher an die Frage der Wissenschaftssprache heranzuarbeiten, müssen wir noch etwas weiter auf die Begleitumstände dieser lateinischen Übersetzungen eingehen. Es sei am Rande erwähnt, daß, wie etwa im 13. Jahrhundert vor allem unter Alfons X. von Kastilien, auch Übersetzungen aus dem Arabischen ins Altspanische stattfanden; diese blieben jedoch außerhalb des spanischen Sprachgebiets ohne Einfluß. Allerdings wurden mehrere so ins Altspanische übertragene Werke unmittelbar anschließend im selben Milieu weiter ins Lateinische übersetzt, und diese nahmen hernach vollen Anteil an der Verbreitung und Rezeption wie alle anderen lateinischen Übersetzungen.

Wie gut konnten diese Männer, die wir „die Übersetzer“ nennen, Arabisch, wie war ihr Bildungshintergrund, wie weit reichten ihre Fachkenntnisse auf den vielen Sachgebieten, aus denen sie übersetzten? In Bagdad waren es einst nahezu ausschließlich keine echten Araber gewesen, die aus dem Griechischen und Persischen – teils mit syrischer Zwischenstufe – ins Arabische übersetzt hatten, sondern Angehörige christlicher und anderer nichtmuslimischer Gruppen, die die Sprachen von sich aus beherrschten und in Einzelfällen diese Sprachkenntnis in Konstantinopel selbst noch vertieften. Die Männer aus ganz Europa, die, vom Ruf der *doctrina arabum* angezogen, nach Spanien strebten, besaßen alle zunächst überhaupt keine Kenntnis der arabischen Sprache. In den christlich-spanischen Orten, in die sie kamen, war zu dem Zeitpunkt das Arabische als gesprochene Sprache ebenfalls bereits mehr oder weniger längst verdrängt. Sie bedurften also eigener Bemühungen, um Sprachkundige und Lehrer zu finden – wie sodann übrigens auch, um Exemplare der zu übersetzenden Texte aus dem arabischen Raum jenseits der Grenze zu beschaffen.

Das arabische Sprachgebiet ist – damals wie noch heute – vom Phänomen der Diglossie gekennzeichnet: Die gesprochene Umgangssprache, der sogenannte Dialekt, ist erheblich verschieden von der grammatisch korrekten Schriftsprache, die nur ein kleiner Prozentsatz der ja größtenteils analphabetischen Bevölkerung einigermaßen sicher beherrscht. Die Kontakte jener Übersetzer zu den Sprachkennern werden also zumindest in der Anfangsphase, oft wohl auch dauerhaft, über den spanisch-arabischen Dialekt des Arabischen verlaufen sein. Dafür zeugen auch die Schreibungen vieler transkribierter arabischer Fremdwörter in den Übersetzungen, die ganz klar phonetische Erscheinungen des spanisch-arabischen Dialekts erkennen lassen, wie die hier besonders stark ausgeprägte *imāla,* d. h. die nach *ǟ* hin „geneigte" Aussprache des arabischen *ā,* das in den Transkriptionen meist durch lateinisches **e** wiedergegeben wird (in spanischen Ortsnamen kommt es zuweilen sogar als **i** vor). Semantische Eigenheiten des spanischen Arabisch fanden ebenfalls ihren Niederschlag in den Übersetzungen und damit in der europäischen Wissenschaftssprache. Daß etwa das ptolemäische Sternbild Θηρίον, das Raubtier, in der Astronomie bis heute Lupus, der Wolf, heißt, beruht darauf, daß das von den Bagdader Übersetzern als Äquivalent von Θηρίον eingesetzte Wort *sabu*ᶜ im spanischen Arabisch spezifisch den Sinn von ‚Wolf' angenommen hatte[2].

Die arabischen Sprachkenntnisse der Übersetzer werden also von unterschiedlicher Güte gewesen sein. Wer wie Gerhard von Cremona vermutlich mehrere Jahrzehnte lang in Spanien wirkte[3], wird sich im Laufe so langer Zeit merklich vervollkommnet haben. Bei kürzerer Tätigkeit – und in der Anfangsphase auch bei Gerhard – werden die Sprachkenntnisse dagegen eher bescheiden gewesen sein. Man darf hier durchaus Vergleiche zur Gegenwart ziehen mit Beobachtungen, wie schnell oder langsam Europäer das Arabische erlernen können; im Mittelalter kommt dabei noch das Fehlen all der unzähligen Hilfsmittel hinzu, die dafür heutzutage zur Verfügung stehen.

Es ist daher nicht zu verwundern, wenn bei manchen Übersetzungen Helfer namentlich erwähnt werden, die dem ‚Übersetzer' beim Übersetzungsvorgang zur Seite standen. Für Gerhard von Cremona kennen wir die Darstellung von Daniel von Morley in seiner *Philosophia,* daß jenem bei der Übersetzung des *Almagest* von Ptolemaios ein Mozaraber namens Galippus (Ġālib) assistiert habe[4]. In einem anderen Fall wird die Prozedur noch näher beschrieben: der Helfer habe aus dem Arabischen mündlich ins spanische Umgangsidiom übertragen, der ‚Übersetzer' dann daraus weiter ins Lateinische[5]. Es ist strittig, wie weit sich solche Angaben auf alle Übersetzer und auf deren sämtliche Werke ausdehnen lassen. Immerhin genügen diese Hinweise, um uns vor Augen zu führen, daß die Übersetzungen oft – und vielleicht gar überwiegend – nicht unter ausgesprochen günstigen Umständen zustande kamen.

Was die jeweilige Fachkenntnis der Übersetzer angeht, so kann diese kaum ausgereicht haben, um Texte aus den verschiedensten Gebieten mit gleicher Kompetenz zu übertragen. Gerhard von Cremona hat im Sternkatalog des *Almagest* zunächst den grotesken Fehler begangen, die Breitenwerte bei den nördlichen Sternen mit 300° (und Zusammensetzungen) statt mit 60° wiederzugeben[6]; in der revidierten endgültigen Fassung[7] ist dieser absurde Fehler berichtigt. (Hier könnte das Ideal eines *fidus interpres* einmal die Sachkenntnis zunächst überwogen haben, bis später dann doch der Sachverstand siegte.)

Der westlich-lateinische Bildungshintergrund der Übersetzer scheint besser fundiert gewesen zu sein. In den Vorworten, die sie oft ihren Übersetzungen voranstellen (bei Hermann von Kärnten gelegentlich sogar innerhalb der Übersetzung selbst), stellen sie ihre Bildung zur Schau. Sie kennen und zitieren nicht nur klassische und spätantike lateinische Autoren, sondern sie kennen auch die zeitgenössische, bereits übersetzte Fachliteratur und beziehen sich darauf.

Hier liegt ein für unsere Thematik ganz wesentliches Problem, nämlich die bisher von der Forschung überhaupt noch nicht aufgeworfene, geschweige denn untersuchte Frage, wieweit diese Übersetzer auf eine bereits vorhandene ältere lateinische Fachterminologie zurückgreifen konnten und wieweit sie in jedem Einzelfall davon Gebrauch gemacht haben. Die Aufarbeitung der gesamten lateinischen Übersetzungsliteratur steckt ja noch in den Anfängen. Nur sehr wenige Texte liegen in modernen kritischen Editionen vor. Weitergehende Untersuchungen in größerem Rahmen fehlen noch völlig und lassen sich bei dem unvollkommenen Stand der Erschließung des Materials auch noch gar nicht vornehmen. Immerhin gibt es erste Ansätze, sich dem Problem zu nähern[8].

Bei der Wiedergabe wissenschaftlicher termini aus dem Arabischen lassen sich in den lateinischen Übersetzungen etwa folgende Hauptformen beobachten: bloße Transkription des arabischen Wortes; assimilierte Transkription, wobei das arabische Wort latinisiert, also einfach mit lateinischen Flexionsendungen versehen wird; wortgetreue Übersetzung; und nichtwörtliche Wiedergabe durch einen, vermutlich älteren oder sonstwie eingebürgerten, lateinischen – und zuweilen auch griechischen – terminus.

Die Transkriptionen arabischer Wörter sind besonders zahlreich in den ältesten Übersetzungen. Die nordspanischen Astrolabtexte vom Ende des 10. Jahrhunderts wimmeln von arabischen Wörtern; stellenweise werden komplette Zeilen transkribiert[9]. Immerhin wird bereits hier den arabischen Wörtern meist eine lateinische Übersetzung oder zumindest eine – oft umständliche – sachliche Definition beigefügt. Im 12. Jahrhundert finden wir auch bei Adelhard von Bath zahlreiche derartige Arabismen. In den 40er Jahren des Jahrhunderts hat Johannes Hispalensis beispielsweise in seiner Übersetzung der Astrolabschrift von Ibn aṣ-Ṣaffār ebenfalls zahlreiche Fremdwörter beibehalten. Gerhard von Cremona verwendet im *Almagest* fast keine Transkriptionen arabischer termini, nur im Sternkatalog macht er gelegentlich von dieser Möglichkeit Gebrauch, bei einigen *appellativa* und öfter bei unübersetzbaren Eigennamen. Im ganzen läßt sich eine mit fortschreitender Zeit abnehmende Tendenz dieses Vorgehens beobachten. Der Grund für die Benutzung solcher Transkriptionen ist nicht ohne weiteres zu erkennen. Oft handelt es sich bei den Wörtern um einfache, leicht zu übersetzende Begriffe; und sie werden ja meist zusätzlich auch mit einem lateinischen Interpretament versehen. Es scheint, als habe man den arabischen Wörtern eine ganz besondere systemimmanente Aussagekraft zuerkannt, die dem behandelten Gegenstand erst die rechte Weihe höherer Wissenschaft verleihe.

In der sich in der Folge weiterentwickelnden lateinischen Fachliteratur werden die meisten arabischen Wörter nach und nach abgelegt und nur noch die lateinischen Äquivalente weiterbenutzt. Einige aber überleben, wie beim Astrolab zum Beispiel die *almucantarat* (Parallelkreise zum Horizont)[10] oder die *alidade* (der drehbare Beobachtungszeiger auf der Rückseite)[11]. Drei arabische Begriffe aus den Astrolabtexten des 12. Jahrhunderts haben die Zeiten bis heute überdauert, sind in alle europäischen Nationalsprachen eingedrungen und werden von den Astronomen und Nautikern der ganzen Welt gebraucht: *Zenit, Nadir* und *Azimut*[12]. Unabhängig vom praktischen Gebrauch haben es die meisten lateinischen Astrolabautoren bis hin zu Johannes Stoeffler, 1512, und noch späteren für nötig befunden, bei der obligaten einleitenden Beschreibung des Instruments immer wieder all jene arabischen Namen der Bestandteile, die einst am Ende des 10. Jahrhunderts und dann wieder von den Übersetzern des 12. Jahrhunderts in die Literatur eingeführt worden waren, zumindest einmal aufzulisten[13]. Für schwierige Begriffe, wie etwa viele Namen aus der *materia medica*, für die keine lateinischen

Äquivalente hatten gefunden werden können, lebten die arabischen Formen ebenfalls weiter. Übrigens befinden sich unter den so übernommenen Fremdwörtern manche, die gar nicht arabisch sind, sondern persischer, indischer, ja selbst griechischer Herkunft; die Araber hatten sie nicht übersetzen können, sie behielten sie ihrerseits in Transkription bei. Hierzu gehören unter anderen eine Reihe astrologischer termini, wie *alcocoden* und *hyleg*[14], das mißratene *effegion* (für Apogäum und Perigäum) und viele Wörter aus der *materia medica*.

Für die assimilierten Transkriptionen nenne ich als Beispiele den Begriff *aux* (Genitiv *augis*; ‚Apogäum') aus arabisch *auǧ*, das seinerseits auf sanskrit *ucca* zurückgeht[15], oder Adelhard von Baths *dacaicae* (Genitiv *dacaicarum*) aus dem arabischen Plural *daqāʾiq* (Singular *daqīqa*), ‚Minuten' (bei Koordinatenwerten)[16], oder den astrologischen terminus *hylegium*, aus arabisch *hīlāǧ* oder arabisiert *hailāǧ*, das wiederum dem mittelpersischen *hīlāk* nachgebildet ist, welches seinerseits das griechische ἀφέτης ‚Loslasser', ‚Entsender' wiedergab (das weitergetragene persische Material war seinerseits vielfach ebenfalls griechisch beeinflußt)[17].

Die Gruppe der wörtlich aus dem Arabischen übersetzten termini ist fast ebenso problematisch wie die dem Europäer unkenntlichen Transkriptionen aus dem Arabischen; denn die hier geprägten Ausdrücke lassen die dahinter stehende Sache oft nur schwer erkennen. Ein bekanntes, bis heute geläufiges Beispiel aus dieser Gruppe ist etwa das mathematische Wort *sinus*, das wörtlich ein arabisches *ǧaib* wiedergibt; ursprünglich handelte es sich dabei jedoch nicht um das arabische Appellativ dieser Bedeutung, sondern um eine mit denselben Buchstaben (unvokalisiert) geschriebene Transkription eines Sanskritwortes *jīvā* ‚Sehne'[18]. Die Begriffe Apogäum und Perigäum umschreiben die Araber umständlich als *al-buʿd al-abʿad* und *al-buʿd al-aqrab*, „die weiteste/die nächste Distanz", was Gerhard von Cremona ganz wörtlich genauso umständlich übertrug als *longitudo longior* und *longitudo propinquior*. Daneben hatte es aber auch arabische Transkriptionen der beiden griechischen Wörter gegeben, die als *efegion* und *efregion* ins Lateinische übernommen wurden, wo sie dann unter der Feder von unkundigen Schreibern vielfach in ein Wort als *effegion* zusammenfielen. Gerhard von Cremona benutzt diese Transkription auch an Stellen, wo im Arabischen gar nicht die entsprechenden arabischen Formen stehen; die unkenntlichen Transkriptionen hatten sich also im lateinischen Bereich bereits verfestigt[19].

In der vierten Gruppe schließlich finden wir freie terminologische Übersetzungen, wobei ohne Rücksicht auf den arabischen Wortlaut inhaltlich korrespondierende lateinische Äquivalente (die zuweilen auch aus griechischen Fachausdrücken bestehen können) eingesetzt werden. So finden wir bei einigen Übersetzern für den umständlichen, aus drei Wörtern bestehenden arabischen terminus das kurze, direkte lateinische *equator*, oder für die ebenfalls dreigliedrige arabische Bezeichnung kurz *ecliptica*. Ein Viereck kann wörtlich heißen *habens quatuor latera* oder auch kurz *quadrilaterum* und noch „echter" *quadratum*. In den verschiedenen Versionen der *Elemente* von Eukleides kommen neben den besonders bei Adelhard von Bath noch vorherrschenden Transkriptionen später auch termini aus der eigenen europäischen Tradition zur Anwendung, wie *hypotenusa*, *diametros*, *rectangulus* und *orthogonius*, *piramis* usw.[20].

Im ganzen haben die lateinischen Übersetzungen aus dem Arabischen den mittelalterlichen westlichen Gelehrten eine Masse von Texten übermittelt, die oft schwer verständlich, wenn nicht gar unverständlich waren. Es gibt Übersetzer wie etwa Hermann von Kärnten, die relativ frei übersetzten, sich um eine Art von elegantem Latein bemühten und dabei vielfach eigene Interpretationen und Verständnishilfen einflochten – die nicht immer richtig gewesen sein müssen –, im Gegensatz zu anderen wie vor allem Gerhard von Cremona und in abgeschwächter Form auch Johannes Hispalensis, die den arabischen Wortlaut so eng nachahmten, daß es für dürchschnittliche Leser größte Schwierigkeiten bereitet haben muß, den sachlichen Inhalt überhaupt zu verstehen. Mit der Terminologie unserer Gruppen drei und vier – wörtliche und freiere lateinische Übertragung – konnten sich die Abendländer immerhin auf der Basis des Lateinischen mit einem gewissen Verständnis auseinandersetzen; dagegen müssen die termini der ersten beiden Gruppen – transkribierte arabische Wörter – für sie bloße Chiffren geblieben sein, die sie blindlings als schweres Gepäck weiterschleppten, ohne deren Struktur und genaue semantische Formation zu durchschauen. Auch der modernen Forschung bieten diese Arabismen dasselbe Hindernis; zu ihrer sachgerechten Erklärung reichen bloße lexikalische Recherchen nicht aus, man muß in jedem Einzelfall auf die Texte selbst zurückgreifen – unter Einschluß des Arabischen –, um die kontextgebundene sachgerechte Erklärung zu finden.

Ungeachtet all dessen bleibt die Bedeutung der arabischen Wissenschaft für die Entwicklung der mittelalterlichen europäischen Wissenschaft und ihrer Terminologie als ein wesentliches historisches *movens* bestehen – arabisch nebst all dem, was dahinter an griechischem, persischem und indischem Einfluß steht.

Kennzeichnend für die mittelalterliche Auseinandersetzung mit der Fachterminologie ist, daß sehr früh schon die Abfassung von Glossaren einsetzt, worin die arabischen, aber auch lateinische termini gesammelt werden, denen man als Interpretament üblicherweise die in den Texten selbst vorkommenden Definitionen hinzufügt. Aus den Astrolabtexten vom Ende des 10. Jahrhunderts hat Fulbert von Chartres (gest. 1028) ein Glossar von 28 arabischen termini kompiliert[21]. Andere Glossare liegen vor zur *Sphaera* von Johannes Sacrobosco[22], zur *Astronomie* von Geber (Ǧābir ibn Aflaḥ)[23] und zur *Theorica planetarum*-Literatur[24]. Aus dem medizinischen und pharmakologischen Bereich kennen wir die sogenannten *sinonima*, so zum *Canon* des Avicenna, zum Werk des Rasis und anderer[25]. Zwei astronomisch-astrologische Glossare habe ich 1977 ediert[26].

Die Aufarbeitung des überlieferten arabo-lateinischen Materials setzte in der Renaissance ein. Astronomisch-astrologische, medizinische und andere Texte der Übersetzungsliteratur wurden analysiert, die Nomenklatur und Terminologie sprachlich auf die Originale zurückgeführt, natürlich mit all den Fehlern und Unzulänglichkeiten, die die noch unvollkommene Kenntnis der Sprachen und der Quellentexte mit sich brachte. Daneben aber setzte jene Bewegung ein, die man als Anti-Arabismus bezeichnet, die die schwer verständlichen mittelalterlichen Übersetzungstexte als „barbarisch" verurteilte und die die gesamte arabische Wissenschaft an sich als orientalisches Geschwätz denunzierte und verwarf. Jetzt konnte man ja auch wieder auf die griechischen Originale selbst zurückgreifen und sein lateinisches Sprachgefühl wieder an den klassischen Autoren schulen.

Aus heutiger Sicht sind wir in der Lage, dem arabischen Beitrag zur Entwicklung der Wissenschaften und ihrer Terminologie größere Gerechtigkeit widerfahren zu lassen.

Abschließend sei noch hinzugefügt, daß die Araber selbst sich in den letzten Jahrzehnten, angeregt durch die Resultate internationaler Forschung, zunehmend ihrer großen Kulturleistung im Mittelalter bewußt geworden sind. Der Begriff des *turāṯ*, des Kulturerbes, stellt heut einen bedeutenden Faktor dar im Prozeß der Gewinnung eines neuen Selbstwertgefühls unter den Arabern und darüber hinaus in der gesamten islamischen Welt.

1 S. hierzu P. Kunitzsch: Arabische Sternnamen in Europa. Wiesbaden 1959.
2 Vgl. P. Kunitzsch (Hrsg.): Claudius Ptolemäus, Der Sternkatalog des Almagest. Die arabisch-mittelalterliche Tradition, II: Die lateinische Übersetzung Gerhards von Cremona. Wiesbaden 1990, S. 161, Anm. 1.
3 Vgl. P. Kunitzsch (wie Anm. 2), S. 2f.; derselbe: Gerhard von Cremona als Übersetzer des Almagest. In: M. Forstner (Hrsg.): Festgabe für Hans-Rudolf Singer. Frankfurt am Main etc. 1991, S. 347–358, speziell S. 347f.
4 Daniel von Morley, ‚Philosophia'. Hrsg. von G. Maurach. *Mittellateinisches Jahrbuch* 14 (1979), 204–255; siehe § 192.
5 O. Bardenhewer: Die pseudo-aristotelische Schrift Über das reine Gute. Freiburg 1882, S. 123f. (Zitat aus dem Vorwort der Übersetzung von Avicennas *Liber de anima* von Dominicus Gundisalvus und Johannes Avendehut).
6 Vgl. P. Kunitzsch (wie Anm. 2), S. 5f.
7 P. Kunitzsch (wie Anm. 2), S. 6f.
8 Vgl. etwa O. Weijers (Hrsg.): Méthodes et instruments du travail intellectuel au moyen âge. Etudes sur le vocabulaire. Turnhout (Belgien) 1990. Ein Band über die Bildung des arabischen wissenschaftlichen Vokabulars in derselben Reihe (CIVICIMA), hrsg. von D. Jacquart, ist im Druck. An kleineren Arbeiten aus jüngster Zeit nenne ich J. D. Latham: Arabic into Medieval Latin. *Journal of Semitic Studies* 17 (1972), 30–67; 21 (1976), 120–137; 34 (1989), 459–469; C. Vázquez de Benito/M. Teresa Herrera (a): Problemas de la transmisión de arabismos. *Al-Qanṭara* 4 (1983), 151–181; dieselben (b): Los arabismos de Ruices de Fontecha en Dubler. *Al-Qanṭara* 6 (1985), 103–117; J. J. Barcia Goyanes: Los términos árabes en la osteología de Vesalio. *Al-Qanṭara* 5 (1984), 293–327.
9 Zum Beispiel eine Tafel der sieben Klimata aus dem Umkreis der ältesten nordspanischen Arbeiten, herausgegeben von J. M. Millás Vallicrosa: Assaig d'història de les idees físiques i matemàtiques a la Catalunya medieval, I. Barcelona 1931, S. 290–292; auch übernommen in einige Handschriften von *De utilitatibus astrolabii* von [Pseudo- (?)] Gerbert, siehe: Gerberti postea Silvestri II papae Opera Mathematica, edidit N. Bubnov. Berlin 1899, S. 141; vgl. hierzu noch E. Honigmann: Die sieben Klimata und die ΠΟΛΕΙΣ ΕΠΙΣΗΜΟΙ. Heidelberg 1929, S. 190f.
10 Vgl. P. Kunitzsch: Glossar der arabischen Fachausdrücke in der mittelalterlichen europäischen Astrolabliteratur. (Nachrichten der Göttinger Akademie der Wissenschaften, Phil.-hist. Klasse, Jg. 1982, Nr. 11) Göttingen 1983; hier Nr. 31.
11 P. Kunitzsch (wie Anm. 10), Nr. 19.
12 P. Kunitzsch (wie Anm. 10), Nr. 43a; 36; 43b/44.
13 Zu den übernommenen arabischen Astrolabtermini siehe P. Kunitzsch (wie Anm. 10).
14 Vgl. P. Kunitzsch: Mittelalterliche astronomisch-astrologische Glossare mit arabischen Fachausdrücken. (Sitzungsberichte der Bayerischen Akademie der Wissenschaften, Phil.-hist. Kl., Jg. 1977, Heft 5) München 1977; zu *alcocoden* speziell S. 35f., Nr. 4; zu *hyleg* speziell S. 49f., Nr. 32.
15 P. Kunitzsch (wie Anm. 14), S. 40, Nr. 9.
16 Die astronomischen Tafeln des Muḥammed ibn Mūsā al-Khwārizmī..., hrsg. von H. Suter. (Kgl. Danske Vidensk. Selsk. Skrifter, 7. R., Hist. og Filos. Afd. III. 1) Kopenhagen 1914, Index S. 242.
17 Vgl. P. Kunitzsch (wie Anm. 14), S. 49f., Nr. 32.
18 Vgl. C. A. Nallino (Hrsg.): Al-Battānī sive Albatenii Opus astronomicum. Bd. 1, Mailand 1903, S. 155.
19 Vgl. P. Kunitzsch: Die astronomische Terminologie im Almagest (in dem im Druck befindlichen von D. Jacquart herausgegebenen Band, siehe oben Anm. 8), sub ἀπόγειον, mit Anm. 32.
20 Vgl. die Tabellen bei H. L. L. Busard (Hrsg.): The First Latin Translation of Euclid's Elements Commonly Ascribed to Adelard of Bath. Toronto 1983, S. 397–399.
21 Ediert von M. McVaugh/F. Behrends: Fulbert of Chartres' Notes on Arabic Astronomy. *Manuscripta* 15 (1971), 172–177; dazu auch P. Kunitzsch (wie Anm. 10), S. 481f.
22 Ediert bei L. Thorndike: The Sphere of Sacrobosco and its Commentators. Chicago/London 1949, S. 472–475.
23 Ms. Paris, B. N. lat. 7292, fol. 49r–51r; vgl. R. Lorch: Astronomical Terminology (in dem von O. Weijers herausgegebenen Band, siehe oben Anm. 8), S. 182, Anm. 2 (dort Druckfehler 5292 statt 7292).
24 Ediert bei O. Pedersen: Classica et Mediaevalia, Dissertationes IX. Kopenhagen 1973, S. 584–594.
25 Vgl. M. Ullmann: Die Medizin im Islam. (Handbuch der Orientalistik, 1. Abt., Ergänzungsband VI, 1) Leiden/Köln 1970, S. 237–241 (dort dazu auch weitere Forschungsliteratur).
26 P. Kunitzsch (wie Anm. 14).

X

Erfahrungen und Beobachtungen bei der Arbeit mit Texten der arabisch-lateinischen Übersetzungsliteratur (Mathematik/Astronomie)

Die Arbeit auf dem Gebiet der Geschichte der Mathematik - und der Naturwissenschaften allgemein - in Antike und Mittelalter ist zu einem wesentlichen Teil philologische Arbeit. Das Quellenmaterial, mit dem der Forscher es zu tun hat, sind überwiegend Texte. Ihre zuverlässige Erschließung bildet die Grundlage für alle weiteren Studien über die Inhalte. Für die Medizingeschichte äußerte sich in diesem Sinne ähnlich W. Artelt bereits 1949: „Ihr Gegenstand ist medizinisch, aber ihre Methoden sind die des Historikers und Philologen, sind geisteswissenschaftlich"[1]. Diese untrennbare intime Verbindung zwischen Natur- und Geisteswissenschaft wird von den Vertretern der aktuellen Naturwissenschaft bei gelegentlichen Exkursen in historische Gefilde nicht immer erkannt oder hinreichend beachtet. Ja sogar bei Historikern der Naturwissenschaften selbst kann gelegentlich das Verständnis dafür fehlen. Bei einem Gastvortrag im Hamburger Institut für die Geschichte der Naturwissenschaften über die arabisch-lateinische Überlieferung des *Almagest* (1967) lautete anschließend die Frage des ersten Fragestellers: „Was ist denn daran Naturwissenschaft?"

Für die Antike ist die Erschließung des Materials erfreulich fortgeschritten, wenn auch noch längst nicht abgeschlossen. Die wichtigsten griechischen Texte sind ediert, es gibt zuverlässige Übersetzungen in moderne Sprachen und vielfältige Untersuchungen und Gesamtdarstellungen. Dennoch bergen Handschriften allerorten noch viel nicht identifiziertes sowie spätgriechisch-byzantinisches Material. Im *Corpus des Astronomes Byzantins* ist hier eine weitläufige Editionsarbeit begonnen. Klassisch-lateinische und spätlateinische Texte aus Astronomie und Astrologie wurden in letzter Zeit in größerem Stil von Frankreich aus ediert. A. Le Bœuffle hat die lateinische Terminologie und Nomenklatur von Astronomie und

1 W. Artelt: *Einführung in die Medizinhistorik. Ihr Wesen, ihre Arbeitsweise und ihre Hilfsmittel,* Stuttgart 1949, S. V (hier zitiert nach B. D. Haage: „Vom Nutzen interdisziplinärer Forschung bei der Interpretation mittelalterlicher Literatur ...", in: *Mediaevistik* 1, 1988, 39-59, spez. S. 39).

Astrologie zusammenhängend untersucht[2]. In Deutschland widmet sich besonders W. Hübner (Münster) der Erforschung antiker Astronomie und Astrologie. Wir sehen: trotz der vermeintlich abgeschlossenen Erfassung des Materials hat sich die Arbeit gerade in den letzten Jahrzehnten wieder stark belebt, auch die Naturwissenschaften der Antike sind noch immer im Stadium der Erschließung von Grundlagen und der stetig verfeinerten Analyse des Überlieferten. Von einem abschließenden Gesamtbild kann auch hier noch lange nicht die Rede sein. Wie weit fortgeschritten auch immer die klassische Philologie - im Vergleich etwa zur Orientalistik oder zur Mediävistik - sein mag: auch sie leidet darunter, daß der Philologe üblicherweise die Texte der „Fachliteratur", in unserem Falle also der Naturwissenschaften, links liegen läßt, einmal weil er ihnen anscheinend im Vergleich zu den Werken der Dichter, Philosophen und Historiker keinen so hohen geistesgeschichtlichen Rang beimißt, zum anderen vermutlich auch, weil auf seiner Seite die nötigen Spezialkenntnisse zur adäquaten Behandlung des Stoffes fehlen.

Was das Mittelalter betrifft, so ist die Situation, aber auch die Ausgangslage, wesentlich ungünstiger. Hier hat man es mit zwei sprachlich und kulturmorphologisch stark verschiedenen Räumen zu tun: einmal mit der arabisch-islamischen Kultur in ihrer gesamten Ausdehnung einschließlich des spanisch-arabischen Bereiches, und zum anderen mit Europa, sowohl in seinem westlich-lateinischen wie in seinem östlich-byzantinischen Teil. Für die betroffenen Philologien, die arabische und die mittelalterlich-lateinische - wenn wir die mittelgriechisch-byzantinische einmal beiseite lassen -, gilt das gleiche wie für die zuvor erwähnte klassische Philologie: ihre Vertreter wenden sich im allgemeinen stets den geistesgeschichtlich bedeutsamen und charakteristischen Texten aus Religion, Poesie, Literatur, Historiographie und Philosophie zu; die „Fachliteratur" bleibt Außenseitern überlassen.

Abgesehen davon haben die orientalische und die mittellateinische Philologie eine wesentlich schlechtere Ausgangslage. Als eigenständige Wissenschaften existieren sie erst seit verhältnismäßig kurzer Zeit, die Arabistik im eigentlichen Sinne erst seit ca. 200 Jahren und die Mediävistik noch viel kürzer, etwa seit der Wende vom 19. zum 20. Jahrhundert[3]. Beide müssen mit viel kürzerer fachlicher Vor-

2 A. Le Bœuffle: *Les noms latins d'astres et de constellations,* Paris 1977; ders.: *Astronomie, Astrologie. Lexique latin,* Paris 1987.

3 Vgl. den Vortrag über das Mittellateinische Wörterbuch in der Vortragsreihe „Lexikographie an der Bayerischen Akademie der Wissenschaften" (München, 14. Mai 1990), S. 2-4 des Manuskripts. Ich danke Frau

auserfahrung und mit viel weniger Personal auskommen als die klassische Philologie, dafür sind sie auf der anderen Seite mit einem um ein Vielfaches reicheren überlieferten Material konfrontiert, das eigentlich einen entsprechend größeren Aufwand verlangte. Entsprechend groß ist in beiden Bereichen der Anteil des noch gar nicht oder bisher unzulänglich Erforschten.

Was endlich speziell die Bearbeitung der mittelalterlichen arabisch-lateinischen Übersetzungsliteratur in den Naturwissenschaften angeht, so komplizieren sich hier die Voraussetzungen um ein weiteres Element. Wer Texte dieser Gattung bearbeiten will, muß neben dem Stoff des fachlichen Gehalts zugleich die beiden Ausgangssprachen, das Arabische und das Lateinische, hinlänglich beherrschen, um dem überlieferten Material in genügender Weise gerecht werden zu können. Für diese Kombination von Kenntnissen und Voraussetzungen gibt es bis heute keine etablierten Institutionen, und so bleibt dieser Forschungsbereich weiterhin auf die Leistungen von interdisziplinären Einzelgängern angewiesen.

Blicken wir auf die Erforschung der Naturwissenschaften im arabisch-islamischen Bereich, so müssen wir feststellen, daß sie weitgehend noch im Stadium der bio-bibliographischen Vorarbeiten steht. Nach den nach heutigen Maßstäben nicht mehr ausreichenden Arbeiten von G. Wenrich, F. Wüstenfeld, M. Steinschneider, H. Suter, C. Brockelmann usw. kann der Erforscher der arabisch-islamischen Naturwissenschaften heute auf eine Reihe von guten Hilfsmitteln zurückgreifen, die in den Jahrzehnten nach dem Zweiten Weltkrieg geschaffen wurden: M. Ullmanns Bände über die *Medizin* sowie die *Natur- und Geheimwissenschaften im Islam*[4]; F. Sezgins *Geschichte des arabischen Schrifttums,* die aber nur die Zeit bis ca. 430 H = 1040 erfaßt[5]; die neue Ausgabe der *Encyclopaedia of Islam* mit ihren ausführlichen Sach- und Personalartikeln[6]; und das *Dictionary of Scientific Biography*[7]. Hierzu kommen noch die bekannten arabischen Bio- und Bibliographien wie Ibn an-Nadīm (*al-Fihrist*), Ibn al-Qifṭī, Ibn Abī Uṣaybiʿa, Ṣāʿid al-Andalusī, Ibn Ǧulǧul usw. Mit all diesen Hilfsmitteln bewegt man sich jedoch nur in einem Kreis sekundärer Informationen, die der Verifikation anhand des erhaltenen primären Textmaterials bedürfen.

Dr. Th. Payr für die Überlassung eines Exemplars des Vortragsmanuskripts.

4 Im Rahmen des *Handbuchs der Orientalistik,* 1. Abt., Ergänzungsbd. VI, 1 und 2 (Leiden/Köln, 1970 und 1972).

5 Bd. I ff., Leiden 1967 ff.

6 Bd. I ff., Leiden [1954–]1960 ff.

7 16 Bde., New York 1970–1980.

Will man mit einem der einschlägigen arabischen Texte arbeiten, so begegnet man einer Reihe von Problemen, die es zunächst zu überwinden gilt. Da ist einmal die mehr technische Seite der Identifikation und Beschaffung der betreffenden Handschriften. Die Angaben - auch in den genannten neueren Hilfsmitteln - beruhen meist nicht auf eigenem Augenschein der Verfasser, sondern stammen aus älteren Katalogen und können fehlerhaft sein, sowohl was den Autor wie die Identität des Textes und Nummer und Folioangaben der Handschrift angeht. Die Beschaffung von Filmen oder Kopien von Handschriften in orientalischen Bibliotheken ist in den letzten Jahren immer schwieriger und langwieriger oder oft ganz unmöglich geworden. Viele Arbeitsprojekte leiden unter der Schwierigkeit oder Aussichtslosigkeit, das nötige Textmaterial zu beschaffen. Hat man endlich die gewünschte Handschrift erhalten, so gilt es, sich der Identität des gesuchten Textes zu vergewissern. Im günstigsten Fall kann man dazu zuverlässige Kataloge wie die von W. Ahlwardt[8] oder M. Krause[9] heranziehen, die ihre Texte ausführlich mit Incipit und Explicit verzeichnen. Ohne solche Vergleichsmöglichkeit bleibt eine gewisse Unsicherheit über die Authentizität des Textes bestehen, da entsprechende Notizen in den Handschriften selbst, zumal wenn sie von anderer Hand später hinzugefügt wurden, nur eine relative Zuverlässigkeit besitzen.

Die Probleme der Textidentifikation vergrößern sich erheblich, wenn es sich um arabische Übersetzungen antiker griechischer Texte handelt. Ein merkwürdiges Charakteristikum der arabischen Übersetzungsliteratur ist, daß in vielen Fällen derselbe Text mehrfach übersetzt wurde. Die Gründe hierfür sind nicht unmittelbar zu erkennen. Vielleicht erschienen gewisse frühe Übersetzungen späteren Fachleuten als unzulänglich, so daß sie erneute Übersetzungen veranlaßten bzw. selbst vornahmen. Auch mögen neue, als „besser" angesehene griechische Handschriften aufgetaucht sein, die daraufhin eine erneute Übersetzung nach sich zogen, oder von älteren Übersetzungen waren keine, oder keine guten bzw. vollständigen Handschriften mehr zugänglich, so daß man sich zu einer neuen Übersetzung genötigt sah. Möglich ist endlich auch, daß der persönliche Ehrgeiz von Mäzenen oder einzelnen Übersetzern den Anlaß gab, vor allem gewisse berühmte Standardwerke erneut in einer eigenen Übersetzung vorzulegen.

8 (Kgl. Bibliothek zu Berlin): W. Ahlwardt: *Verzeichnis der arabischen Handschriften,* 10 Bde., Berlin 1887-1899.

9 M. Krause, „Stambuler Handschriften islamischer Mathematiker", in: *Quellen und Studien zur Geschichte der Mathematik Astronomie und Physik,* Abt. B, Bd. 3, Heft 4 (Berlin 1936), S. 437-532.

Als Beispiel kann ich hier den *Almagest* nennen, mit dessen Überlieferungsgeschichte[10] und teilweiser Edition[11] ich mich seit vielen Jahren befaßt habe. Die bio-bibliographische Literatur erwähnt davon fünf verschiedene arabische Textfassungen: eine „erste“ oder „alte“ oder „ma'mūnische“ Übersetzung, eine Übersetzung von al-Ḥaǧǧāǧ ibn Yūsuf (datiert 212 H = 827/28) und eine Übersetzung von Isḥāq ibn Ḥunayn (um 880–90). Zu der ersten und der dritten von diesen soll Ṯābit ihn Qurra (gest. 901) je eine Bearbeitung (*iṣlāḥ*) geliefert haben. Tatsächlich erhalten sind elf bisher bekannt gewordene Handschriften, davon zwei mit der Ḥaǧǧāǧ- und neun mit der Isḥāq-Fassung, die letztere in Ṯābits Bearbeitung. Die Isḥāq/ Ṯābit-Fassung wurde auch von Naṣīr ad-Dīn aṭ-Ṭūsī als Grundlage für seine Rezension (*taḥrīr*) des *Almagest* (1247) verwendet und ist darin klar wiederzuerkennen. Von den drei anderen Fassungen besitzen wir, außer den bibliographischen Notizen, nur einige Zitate aus der „alten“ Version sowie aus Isḥāqs Urfassung (ohne Ṯābits Bearbeitung). Besonders wichtig ist hierzu die kritische Abhandlung von Ibn aṣ-Ṣalāḥ (gest. 1154) über die Koordinatenüberlieferung im Sternkatalog des *Almagest*[12]. Ibn aṣ-Ṣalāḥ dokumentierte zu 88 Sternen die Koordinatenwerte aus fünf ihm vorliegenden Textfassungen des *Almagest,* darunter Isḥāqs Urfassung sowie, separat, die Isḥāq-Version in Ṯābits Bearbeitung. Diese Zitate sind von äußerster Wichtigkeit, da sie das einzige Hilfsmittel darstellen, um die Überlieferung in den Handschriften zu beurteilen und einzuordnen. Was die „Bearbeitungen“ von Ṯābit betrifft, so kennen wir von derjenigen der „alten“ Übersetzung außer der bibliographischen Notiz im *Fihrist* gar nichts. Andererseits liegt uns die Isḥāq-Version nur in der von Ṯābit bearbeiteten Form vor (abgesehen von Ibn aṣ-Ṣalāḥs Zitaten von Koordinatenwerten aus dem Sternkatalog).

Es ist uns nicht möglich zu erkennen, worin Ṯābits Eingriffe in den Text bestehen, wie weit und wie häufig er „Verbesserungen“ angebracht hat. Einmal (am Ende von Buch VI, Kap. 13) findet sich

10 P. Kunitzsch: *Der Almagest. Die Syntaxis Mathematica des Claudius Ptolemäus in arabisch-lateinischer Überlieferung,* Wiesbaden 1974.

11 Claudius Ptolemäus: *Der Sternkatalog des Almagest. Die arabisch-mittelalterliche Tradition,* I: *Die arabischen Übersetzungen,* P. Kunitzsch, Wiesbaden 1968; II: *Die lateinische Übersetzung Gerhards von Cremona,* P. Kunitzsch, 1990; III: *Gesamtkonkordanz der Sternkoordinaten,* von P. Kunitzsch, 1991.

12 Ediert von P. Kunitzsch, Göttingen 1975 (Abh. Akad. d. Wiss. in Göttingen, Phil.-hist. Kl., 3. Folge, Nr. 94).

in einer Handschrift[13] am unteren Seitenrand eine separate Anmerkung, die namentlich von Ṯābit stammt. Diese einzige Anmerkung könnte es schwerlich sein, deretwegen dem Ṯābit eine eigene „Bearbeitung" zugeschrieben wurde. In den Koordinatenzitaten bei Ibn aṣ-Ṣalāḥ stimmen die Werte aus Isḥāqs Urfassung bis auf drei Ausnahmen[14] immer mit denen in Ṯābits „Bearbeitung" überein.

Wenn wir so die Unterschiede zwischen Isḥāqs Urfassung und der von Ṯābit bearbeiteten Fassung heute nicht mehr greifen können, so sind wir andererseits in der glücklichen Lage, immerhin die Isḥāq/Ṯābit-Fassung und die Version von al-Ḥaǧǧāǧ genau zu unterscheiden, von denen je eine vollständige Handschrift erhalten ist. Dabei wird erkennbar, daß es sich wirklich um verschiedene Übersetzungen handelt. Neben Passagen, die völlig verschieden formuliert sind, gibt es aber auch solche, die sehr ähnlich lauten und sich nur in einzelnen termini unterscheiden. Unsere bisherige Kenntnis der gesamten Übersetzungsvorgänge reicht nicht aus, um diese Befunde plausibel zu erklären.

Hinzu tritt ein weiteres Phänomen: die Kontamination und Substitution der Versionen in den Handschriften. In einigen Isḥāq-Handschriften gibt es Stellen, wo im laufenden Text einzelne Sätze oder Absätze im Ḥaǧǧāǧ-Wortlaut erscheinen. Von fünf Isḥāq-Handschriften, die den Sternkatalog enthalten, haben ihn nur zwei im Isḥāq-Wortlaut, dagegen drei im Wortlaut von al-Ḥaǧǧāǧ. Hier kann man noch nicht von Kontamination im eigentlichen Sinn reden, hier handelt es sich eher um Substitution von Textpartien, vermutlich wegen fehlender Stellen in der Vorlage. Ganz stark ausgeprägt ist die Kontamination dagegen bei den Zahlenwerten der Koordinaten im Sternkatalog. Als verbürgt dürfen wir nur die Direktzitate bei Ibn aṣ-Ṣalāḥ ansehen; im Vergleich mit diesen Zitaten, die sich leider nur auf 88 von den 1025 Sternen des Katalogs beziehen, wird das volle Ausmaß der Kontamination in den erhaltenen Handschriften sichtbar. Wo diese Gegenkontrolle fehlt, können wir die Angaben der Handschriften lediglich zur Kenntnis nehmen, sie aber nicht definitiv bestimmten Versionen zuordnen.

Bei den arabischen Übersetzungen griechischer Texte kommt noch ein weiteres Element der Unsicherheit hinzu. Manche Werke wurden zunächst ins Syrische und dann erst von da aus weiter ins Arabische übersetzt, teils von demselben Übersetzer, teils von verschiedenen Übersetzern. Die möglichen Einflüsse solcher syrischen Zwischenstufen auf die arabischen Fassungen ließen sich gelegentlich

13 Ms. Tunis 07116 (dat. 478 H = 1085), fol. 109^{v}; vgl. Kunitzsch (s. Anm. 10), S. 68 mit Anm. 178.

14 Ibn aṣ-Ṣalāḥ (s. Anm. 12), Sterne Nr. 6, 35, 83.

in Einzelfällen bereits nachweisen. Im großen Ganzen aber ist es noch nicht gelungen, allgemein gültige Kriterien für syrischen Einfluß in arabischen Übersetzungstexten herauszuarbeiten. Dazu gibt es bisher noch viel zu wenig einschlägige Texteditionen. Beim *Almagest* scheint das Syrische keine Rolle gespielt zu haben; nach den vorliegenden sekundären Angaben haben sowohl al-Ḥaǧǧāǧ als Isḥāq den *Almagest* direkt aus dem Griechischen ins Arabische übertragen. Andererseits gehört Isḥāq seiner Herkunft nach bekanntermaßen dem christlich-syrischen Milieu an, und so mag sein Übersetzungsarabisch aus diesem Grunde Züge von syrischer Verfremdung aufweisen. Hierzu wären noch langwierige Untersuchungen an seinem gesamten Übersetzungswerk und an seinen eigenen Schriften nötig, um über diese Problematik begründete Aufschlüsse zu gewinnen.

Ich kann hier nicht auf weitere Beispiele eingehen, wie etwa auf die ähnlich gelagerte, aber zugleich viel kompliziertere Situation bei Euklids *Elementen*[15]. Worauf es mir ankommt, ist, nachdrücklich festzustellen, daß die Bearbeitung jedes einzelnen der unzähligen arabischen Übersetzungstexte immer aufs neue alle die Probleme aufwirft, die ich soeben umrissen habe. Zunächst unscheinbar wirkende Varianten in mehreren Handschriften können sich jederzeit als Elemente separater Übersetzungen erweisen. Es ist, selbst bei wenig prominenten, kleineren Texten stets mit der Möglichkeit der Existenz mehrerer, verschiedener Übersetzungen zu rechnen. Dazu kommen jene Bearbeitungen (*iṣlāḥ*) und Rezensionen (*taḥrīr*), an denen die arabische Übersetzungsliteratur so reich ist. Die Bearbeitung arabischer Übersetzungstexte wird also in jedem Fall von vornherein mehr Aufmerksamkeit, Sorgfalt und Arbeitsaufwand erfordern als diejenige von eigenen Schriften arabischer Autoren.

Die Kompliziertheit und Komplexität der arabischen Verhältnisse überträgt sich im ganzen sodann auf das weite Feld der mittelalterlichen lateinischen Übersetzungen aus dem Arabischen. Für eine schlüssige Bearbeitung eines lateinischen Übersetzungstextes muß eben in jedem einzelnen Fall die Abhängigkeit von einer ganz bestimmten arabischen Ausgangsversion fixiert und durch den gesamten Text hindurch kontinuierlich nachgewiesen werden.

15 Vgl. P. Kunitzsch: „Findings in some texts of Euclid's *Elements* (Mediaeval transmission, Arabo-Latin)", in: *Mathemata. Festschrift für Helmuth Gericke,* Stuttgart 1985 (Boethius, Bd. XII), S. 115–128; ders.: „Letters in geometrical diagrams, Greek-Arabic-Latin", in: *Zeitschrift für Geschichte der Arabisch-Islamischen Wissenschaften* 7 (1991/92), 1–20.

Natürlich steht man bei der Arbeit mit lateinischen Übersetzungstexten ungefähr vor den gleichen Problemen wie bei der arabischen Übersetzungsliteratur. Die Identität eines Textes, die Zuschreibung an einen bestimmten Übersetzer - auch hier gibt es wie im Arabischen Mehrfachübersetzungen von Texten! -, die Authentizität des Wortlauts sind zunächst nicht gesichert; auch hier sind, wie im arabischen Bereich, die Angaben von Handschriftenkatalogen oft unvollständig oder falsch. Außerdem fehlt im mittelalterlich-lateinischen Raum - mit wenigen Ausnahmen - jene interne Sekundärliteratur, die im arabisch-islamischen Bereich wenn schon nicht absolute Sicherheit verschafft, so doch immerhin wertvolle Hinweise auf Übersetzer, Texte, Versionen und Bearbeitungen gibt, die für den modernen Bearbeiter eine nicht zu unterschätzende Hilfe darstellen. Darüber hinaus ist das gesamte Feld der mittelalterlich-lateinischen Übersetzungsliteratur auch von Seiten der Mediävistik noch viel weniger erschlossen als die arabische Übersetzungsliteratur in der Orientalistik. Für den Bearbeiter ist dieses Gebiet daher in viel größerem Ausmaß *terra incognita* als für den Bearbeiter arabischer Übersetzungen.

Bevor ich auf Einzelbeispiele komme, will ich bereits hier vorweg hervorheben, daß der Bearbeiter lateinischer Übersetzungstexte Philologe im doppelten Sinne sein muß: er muß nicht nur sein Latein, sondern zugleich auch das Arabische beherrschen. Bei allen Varianten und Unsicherheiten, auf die er in seinen Quellenhandschriften stößt, wird er seine Entscheidung über „richtig“ und „falsch“ stets an der arabischen Übersetzungsvorlage zu orientieren haben. Wird diese methodische Voraussetzung außer acht gelassen, können sich zahllose Fehler und Mißverständnisse aller Art in sein Vorhaben einschleichen.

Auch von den lateinischen Übersetzungen sind die meisten Texte noch nicht ediert (von purgierten und daher unbrauchbaren Frühdrucken der Humanistenzeit einmal abgesehen). Urteile allgemeiner Art sind daher hier zunächst ebenso wenig möglich wie bei den arabischen Übersetzungen. Dennoch zeichnen sich anhand vorliegender Beispiele, die teilweise auch erst aus Handschriften gesammelt sind, für manche Übersetzer charakteristische Eigenschaften ab. So sehe ich eine Skala, an deren einem Ende Gerhard von Cremona als derjenige Übersetzer steht, der den arabischen Wortlaut in allerengster Anlehnung ins Lateinische überträgt, während am anderen Ende Hermann von Kärnten rangiert, dessen Übersetzungen so frei sind, daß es zuweilen schwerfällt, überhaupt noch die Zusammengehörigkeit mit dem zugrunde liegenden arabischen Text festzustellen. In der Mitte, aber mehr zu Gerhard hin, befindet sich Johannes Hispalensis, der dem Arabischen zwar auch sehr eng folgt, aber nicht so extrem wörtlich wie Gerhard. Diese Beobachtungen lassen sich aus einer Reihe von Texten ableiten, die zweimal übersetzt wurden, wie etwa die *Astronomie* von al-Farġānī, die zuerst 1135 von Johannes Hispalensis und später noch einmal von Gerhard von Cremona übersetzt wurde, oder das astrologische *Introductorium maius* von Abū Maʿšar, das zuerst Johannes Hispalensis und danach erneut Hermann von Kärnten übertrug. Mit Hermanns hier dokumentiertem Stil ist gut seine Übersetzung des *Planisphaeriums* von Ptolemäus (1143) zu vergleichen, von dem es zwar keine Parallelübersetzung gibt, die aber im Vergleich zum arabischen Text dieselben Stilelemente aufweist.

Einen gewissen Fortschritt in der Fertigkeit der Übersetzer beim Übersetzen des fremdartigen semitischen Idioms ins Lateinische zwischen der Frühzeit (etwa Ende des 10. Jahrhunderts) und der Blütezeit (12. Jahrhundert) kann man eindeutig beobachten. Die *Sententie astrolabii* vom Ende des 10. Jahrhunderts, aus Nordostspanien, lassen noch deutlich die Mühsal und Unbeholfenheit erkennen, mit der ein Übersetzer - vielleicht Lupitus von Barcelona - Partien aus dem Astrolabtraktat von al-Ḫwārizmī übertrug[16].

Wie es um die arabischen Sprachkenntnisse der Übersetzer des 12. Jahrhunderts stand, ist nur schwer zu ermessen. Der bekannte Befund der „Diglossie" im arabischen Sprachraum legt es nahe, daß auch die Übersetzer in Spanien zunächst mit der mündlichen Umgangssprache, dem spanisch-arabischen Dialekt, Bekanntschaft machten. Wie weit die einzelnen Übersetzer es vermochten, sich daneben auch Kenntnisse der arabischen Schriftsprache, der *fuṣḥā*, anzueignen und wie tief diese gingen, ist schwer abzuschätzen. Bekannt ist das strenge Urteil von Roger Bacon (1214-1292): „Not one of these translators had any true knowledge of the languages or of the sciences"[17]. Für einige Übersetzer bzw. für die Übersetzungen bestimmter Text wird in den Quellen angegeben, daß der (westliche) Übersetzer dabei die Hilfe eines sprachgewandten Einheimischen in Anspruch nahm, so Gerhard von Cremona bei der Übersetzung des *Almagest* einen Mozaraber namens *Galippus* (= Ġālib), nach dem Zeugnis von Daniel von Morley[18], oder Dominicus Gundisalvi bei der Übersetzung von Avicennas *Liber de anima* den Juden Io-

16 Vgl. P. Kunitzsch: „Al-Khwārizmī as a Source for the *Sententie astrolabii*", in: *From Deferent to Equant: A Volume of Studies ... in Honor of E. S. Kennedy,* New York 1987 (Annals of the New York Academy of Sciences, vol. 500), S. 227-236.

17 Zitiert nach C. H. F. Peters - E. B. Knobel: *Ptolemy's Catalogue of Stars,* Washington 1915, S. 14, Anm. ⟨3⟩.

18 Vgl. Kunitzsch (s. Anm. 10), S. 85f.

annes Auendehut[19] sowie Michael Scotus bei der Übersetzung von al-Biṭrūǧīs *De motibus celorum* einen Juden namens *Abuteus*[20]. Auendehut beschreibt die dabei angewandte Übersetzungsmethode sogar noch näher: er habe den Text wortweise ins romanische Umgangsidiom übersetzt und Dominicus Gundisalvi danach weiter ins Lateinische (wobei also Dominicus selbst mit dem Arabischen überhaupt nicht in Berührung kam!). Der Einfluß der arabischen Umgangssprache, d. h. des spanisch-arabischen Dialekts, wird auf jeden Fall deutlich sichtbar bei den Transkriptionen arabischer Namen und Wörter in den lateinischen Übersetzungen. Hier tritt die starke *imāla* in Erscheinung (die Wiedergabe von arab. *ā* durch lat. *e*), ebenso unklassische Vokalisierungen (wie *mirac* statt *marāqq*) oder die Aussprache von schriftarab. *ǧ* als *y* (*alieuze* für *al-ǧawzā'*, etc.[21]). Auch bei der Übersetzung von Vokabeln macht sich spanisch-arabischer Sondergebrauch bemerkbar (Gerhard: *lupus* für arab. *sabuʿ*, „Raubtier", für das ptolemäische Sternbild *Thēríon*[22]).

Warum übrigens die lateinischen Übersetzer häufig oder gelegentlich arabische Wörter in Transkription in ihre Texte übernehmen, ist ebenfalls ungeklärt. In der Frühzeit (*Sententie astrolabii*) sind derartige Arabismen besonders häufig; in ihnen drückt sich einerseits die Unbeholfenheit des Übersetzers aus, der noch kein passendes lateinisches Äquivalent parat hat, andererseits werden sie bewußt eingesetzt, um dem Text den Nimbus arabischer Herkunft zu erhalten. Bei späteren Übersetzern, wie Gerhard, vermag man aber meist den Grund für die Beibehaltung solcher Arabismen nicht einzusehen, da es sich dabei häufig um simple Appellative handelt, deren Übersetzung keinerlei Schwierigkeit bereitet hätte und die sogar gelegentlich an anderen Stellen im selben Text glatt übersetzt sind. Interessant ist ebenfalls, daß die Arabismen manchmal sogar die arabischen Flexionsendungen bewahren.

Das Latein der Übersetzer in Spanien seinerseits ist auch nicht mehr das reine Latein klassischer Tradition. Abgesehen von der vielfach gezwungenen, bei Gerhard besonders streng dem Arabischen nachgeformten Satzbildung findet man auch manche Vokabeln, die nicht dem klassischen Latein angehören, sondern die aus dem romanischen Umgangsidiom der Zeit und Gegend abgeleitet sind (z. B.

19 Vgl. O. Bardenhewer: *Die pseudo-aristotelische Schrift Über das reine Gute,* Freiburg 1882, S. 132f.

20 Vgl. F. J. Carmody: *Arabic Astonomical and Astrological Sciences in Latin Translation. A Critical Bibliography*, Berkeley/Los Angeles 1956, S. 165f. (Nr. 35, a).

21 Vgl. Kunitzsch (s. Anm. 10), S. 108ff.

22 S. *Sternkatalog* (s. Anm. 11), Bd. 2, S. 161, Anm. 1.

bei Gerhard im Sternkatalog: *anc(h)a, grunnum, titillicus, treca/trica*[23]).

So ranken sich um die Entstehung der einzelnen arabisch-lateinischen Übersetzungen noch viele Fragen, für deren Beantwortung es jetzt noch zu früh ist. Es müssen erst noch viel mehr Texte in philologisch gut erschlossenen Ausgaben zugänglich gemacht werden.

Eine dieser Fragen wäre ferner, wie weit bei den einzelnen Übersetzern die Bildung reichte, sowohl sprachlich wie auch allgemein literarisch und spezifisch fachlich hinsichtlich der Fachgebiete, aus denen sie Texte übersetzten. Vielfach wurden, besonders in Einleitungen und Vorworten, bereits Zitate aus der klassischen lateinischen Literatur erkannt und nachgewiesen. Noch gar nicht hat man sich dagegen mit der Frage beschäftigt, wie weit die fachliche Spezialterminologie, die die Übersetzer in den verschiedenen Fachgebieten verwenden, klassisch-lateinische termini weiterbenutzt oder wiederbelebt bzw. weiterentwickelt und wie weit jene selbständig neue termini prägen und nach welchen Gesichtspunkten, mit welchen sprachlichen Mitteln. Zu all diesem fehlt es noch an sämtlichen Voraussetzungen, nicht nur, weil eben erst viel zu wenige Übersetzungstexte ediert sind, sondern auch, weil das klassische Erbe selbst auf diese Thematik hin noch nicht genügend durchgearbeitet und erschlossen ist (von Le Bœuffles oben erwähnter Studie über die klassisch-lateinische Terminologie zu Astronomie und Astrologie einmal abgesehen, die aber wohl auch noch nicht das letzte Wort darstellen kann, wie die Rezension von W. Hübner gezeigt hat[24]).

Ebenso ungeklärt ist die Frage nach den Fachkenntnissen der Übersetzer in den vielen Sachbereichen, aus denen sie Texte übertrugen (wie ganz besonders Gerhard von Cremona mit seinen über 80 Übersetzungen[25]). Man trifft in ihren Übersetzungen auf manches Absurde; ich habe auf solche Fälle bei Gerhard im *Almagest* hingewiesen[26]. Doch bleibt dabei offen, ob es sich hier wirklich um fehlende Sachkenntnis handelt oder nicht eher um die Befolgung eines bestimmten Übersetzerethos: es galt, den vorgegebenen Text nach bestem Vermögen getreu wiederzugeben, aber keinesfalls eigenmächtige Veränderungen an seiner Substanz vorzunehmen. Auch

23 Vgl. die Stellen in Kunitzsch (s. Anm. 10), gemäß dem lat. Index dort S. 364ff.

24 In: *Gnomon* 60 (1988), 509–516.

25 Vgl. zuletzt die Werkliste bei R. Lemay, Art. „Gerard of Cremona“, in: *Dictionary of Scientific Biography,* vol. XV (New York, 1978), 173ff., spez. S. 176–188 (Nr. 1–82).

26 Vgl. Kunitzsch (s. Anm. 10), S. 92f., 107.

hier wäre es wohl überheblich von uns, in den Fehler der Antiarabisten der Renaissance zu verfallen und vorschnelle oberflächliche Urteile zu fällen.

Zusammenfassend bleibt festzuhalten, daß die Erschließung und Erforschung der arabischen und der lateinischen Übersetzungsliteratur der Naturwissenschaften noch im Anfangsstadium stehen. Bevor es möglich wird, auf diesem Gebiet zu allgemeineren Folgerungen und Urteilen zu gelangen, müssen erst noch viel mehr Texte ediert werden. Das bleibt auf lange Jahre hinaus die vordringlichste Aufgabe der historischen Wissenschaft. Dabei wird es für die lateinischen Übersetzungstexte von grundsätzlicher Bedeutung sein, daß sie stets in durchgehender Abstimmung mit den jeweiligen arabischen Vorlagen behandelt werden, da nur so ihre authentische Darstellung möglich ist. Die Philologie wird also noch lange ihren Teil zur Erforschung der Geschichte der Naturwissenschaften beizutragen haben.

XI

THE CHAPTER ON THE FIXED STARS IN ZARĀDUSHT'S *KITĀB AL-MAWĀLĪD*

Since late Antiquity, and in the Middle Ages, it had become habitual to include among the constituent elements for astrological judgements - beside the Sun, Moon and the planets - the fixed stars also. They were assigned, either by constellations or individually, the "temperament" (κρᾶσις, Arabic *mizāj*, Latin *complexio*) of a planet (or sometimes of two) and were considered in the casting of horoscopes or other astrological operations. For this purpose astrological writings often contained a chapter on the fixed stars[1] or a related star table[2]. Together with the complex of Greek astrology this knowledge was handed on to the Orient - Middle Persian, Indian, and later Islamic - and then to Medieval Europe[3].

Among the texts thus transmitted there is an astrological compilation ascribed - already by the Greeks - to Zoroaster (and hence, to Zarādusht or Zardusht by the Arabs)[4]. It made its way from the Greeks into Middle Persian (Pahlavi) and then into Arabic[5]. As demonstrated by D. Pingree (following the book's own introduction; see note 5), the Pahlavi text was reworked and put into 'Newer Persian' by Māhānkard in about 637, and translated into Arabic by Sa'īd ibn

[1] E.g., Ptolemy's *Tetrabiblos* I,9; Abū Ma'shar's *Introductorium maius* II,1.

[2] For example, cf. "Type III" in my article "Das Fixsternverzeichnis...", in: *Byzantinische Zeitschrift* 57 (1964), 382-411, esp. 406-411 (reprinted in: P. K., *The Arabs and the Stars*, Variorum Reprints, Northampton 1989, as item II); see also the three successive articles on the "Stelle beibenie", in: *Zeitschrift der Deutschen Morgenländischen Gesellschaft* 118, 1968, 62-74; 120, 1970, 126-130; 131, 1981, 263-267 (repr. *ib.* as items XII-XIV) and the article "Abū Ma'šar..." in the same journal, 120 (1970), 103-125 (repr. *ib.* as item XVII).

[3] The papers cited in note 2 present Greek, Arabic and medieval Latin material.

[4] Cf. F. Sezgin, *Geschichte des arabischen Schrifttums*, VII (Leiden 1979), 81ff., esp. 85f.; M. Ullmann, *Die Natur- und Geheimwissenschaften im Islam*, Leiden 1972, 294f.

[5] Sezgin, *l.c.*; D. Pingree, Classical and Byzantine Astrology in Sassanian Persia, in: *Dumbarton Oaks Papers* 43 (1989), 227-239, esp. 234f.

Khurāsānkhurra in the time of the Abbasid propagandist Abū Muslim, that is in the years between 747 and 754. This book, entitled in Arabic *Kitāb al-mawālīd* (on genethlialogy), seems to form the fifth book of a major compilation in five books, an astrological 'Pentateuch'[6].

It is the aim of this article to edit and discuss the section on the fixed stars from the *Kitāb al-mawālīd*[7].

For the edition two manuscripts were available: Istanbul, Nuruosmaniye 2800 (II f, 14 foll., dated 658 H = 1260[8]; marked in this paper by N, and by ن in the Arabic edition) and Escorial 939 (foll. 18v ff., presumably copied in 511 H = 1117-18, in Maghribi script[9]; marked in this paper by S or س respectively). SEZGIN, l.c. p. 86, mentions five more, in Teheran.

In both manuscripts we find the text in the same version, of course with a number of defects and scribal errors, but these do not go beyond what is normal in the transmission of Arabic texts. For the edition I have chosen the best, or most appropriate, readings from both manuscripts, none of the two being generally "better" than the other.

The text carries elements from various stages in the course of its long history, Greek, Pahlavi and Arabic, each stage, as it seems, having added its own contributions[10]. The planetary temperaments added to stars nos. 9-13[11], the names of stars nos. 14 and 16, the fragmentary indications of the stars' latitudes, the discussion of the

[6] SEZGIN, *l.c.*

[7] I am grateful to Profs. D. PINGREE and G. J. TOOMER of Brown University, Providence, R. I., who started work on this text earlier, for their kind permission to edit this section. They also kindly provided me with photographs of the relevant part of the book from two manuscripts.

[8] Cf. M. KRAUSE, Stambuler Handschriften islamischer Mathematiker, in: *Quellen und Studien zur Geschichte der Mathematik, Astronomie und Physik*, Abt. B, Bd. 3, Heft 4 (Berlin, 1936), 471 (*sub* no. 16).

[9] Cf. H. DERENBOURG/H.-P.-J. RENAUD, *Les manuscrits arabes de l'Escurial*, II,3 (Paris 1941), 56 (*sub* 4°).

[10] D. PINGREE (in the paper cited above in note 5) edited and discussed four horoscopes from the *Kitāb al-mawālīd*. They are referred to dates in the years A.D. 549 (p. 234 and note 43), 629 (*ib.*, with note 45), 487 (*ib.*, with note 46) and 232 (p. 235 and note 48).

[11] There is partial coincidence with the temperaments listed in Ptolemy's *Tetrabiblos* I,9 (nos. 9 and 10) and those mentioned in Hermes' *Stelle beibenie* (nos. 11 and 13).

risings of nos. 4-5, the description of α Ori as "red" - all point to a Greek background. The vague indication of temperaments in the other stars as "lucky" or "unlucky" could betray pre-Ptolemaic or Sassanian influence. The Persian names mentioned for most of the stars demonstrate Persian influx; those names that could be recovered from the corrupted Arabic spellings (nos. 4, 5, 6, 10) show relatively "new" Persian forms[12]; this would mean that they were introduced, or recast, at a very late state, shortly before the Arabic version was made. Elements from the Arabic stage of development are: the Arabic names for most of the stars (they belong to the set of old, indigenous Arabic star names[13]), the allusion to the adoration of Sirius in pre-Islamic times[14], the mention of Yemen and the Bedouins in no. 15[15] and last but not least the citation of the Arabic translator Sa'īd himself, by name, in no. 4.

On the last page of the text in N[16] some comments are made on the phases of the fixed stars and their astrological evaluation. *Mediatio coeli* is mentioned, for the "present time" (*i.e.*, the time of the translator Sa'īd, around A.D. 750), for α CMa, as Gem 24°, and for α Aql, as Cap 20°. Further the precession rate is indicated as 1° in 100 years[17]

[12] In this they are different from the names in the Hermetic *De stellis beibeniis* (cf. above, note 2) which appear to be of a more archaic character. Incidentally, in his pursuit of Greek-Pahlavi-Arabic astrological traditions D. PINGREE has found that the Hermetic chapter on the fixed stars was also used by Abū Ma'shar in his *Kitāb aḥkām al-mawālīd.* Similarly, a version of the same was incorporated into a *Liber Aristotelis...* translated from Arabic by Hugo of Santalla. Cf. PINGREE's article cited above in note 5, p. 233.

[13] Cf. P. KUNITZSCH, *Untersuchungen zur Sternnomenklatur der Araber*, Wiesbaden 1961, 20f.

[14] A reference to this is also found in the *Qur'ān* 53, 49. Cf. J. HENNINGER, Über Sternkunde und Sternkult ..., in: *Zeitschrift für Ethnologie* 79 (1954), 82-117, esp. 94-96 (reprinted in J. H., *Arabica sacra*, Freiburg (Switzerland)/Göttingen 1981, see pp. 66-69); also Ibn Qutayba, *Kitāb al-anwā'*, Hyderabad 1956, 46.

[15] The statement that α Car passes through the zenith of the inhabitants of Yemen (and the Arabian deserts) appears rather arbitrary. Its δ in Sa'īd's time, A.D. 750, would have been ca. -51°. Thus, in a place in southern Yemen, at φ = 15°, its maximum altitude above the horizon would have been ca. 24° which is still far away from the zenith. Perhaps the phrase *yusāmitu ru'ūsahum* is not to be taken too strictly in the technical sense of "passing through the zenith", but rather as something like "(the star) is right in front of their view" or so.

[16] Fol. 14v, lines 19ff. Of this portion no copy from S was to hand.

[17] *Ib.*, line 23.

which is the value given by Ptolemy in the *Almagest*. Of α CMa it is also said that its place (in longitude) is "near the end of Gem", *fī 'l-jawzā' fī awākhirihi* [sic][18] (Ptolemy, in the star catalogue of his *Almagest*, epoch A.D. 137, had it in Gem 17°40′ so that in Sa'īd's time, *i.e.* about 615 years later, and using Ptolemy's rate of precession, it should have been in Gem 23°49′, or roughly 24°). The context makes it probable that these elements which distinctly point to Ptolemy were only added in the latest stage of development, by Sa'īd.

The text aims at mentioning the bright stars; this would imply mostly those which were classified by Ptolemy as of the 1st magnitude. But of the 16 stars mentioned only 10 are among Ptolemy's fifteen first magnitude stars, four are of a smaller magnitude (nos. 2, 8, 12 and 16). On the other hand, five of Ptolemy's first magnitude stars are absent from our text (β Leo, α PsA, ϑ Eri, α CMi and α Cen). Stars nos. 4 and 5 of the text cannot be identified[19].

The fixed stars are called in our text *al-kawākib al-biyābānīya*, with a Persian name whose explanation I reported in the article of 1981 (cf. above, note 2), p. 265, citing the Iranist W. B. Henning: it derives from a Pahlavi form *a-wiyābān-īg* (found in the *Bundahišn*, ch. 2) which literally renders Greek ἀπλανής, the current term for the fixed stars. The popular and erroneous explanation brought forward

[18] *Ib*., line 22.

[19] It is striking that both stars are given the same Persian name, *chashm-i shīr*, "the Lion's Eye", whereas neither of them is assigned an Arabic name. (In S, in no. 4, there is written - in the same hand - above the Persian name: *ya'nī fam al-ḥūt*, "that is, the Mouth of the Fish" [cf. the apparatus to the edition, note 8] which would designate α PsA. But this is not confirmed by the coordinates.) Singularly among the stars of the text, there are mentioned coordinates for these two: no. 4, long. Leo 10° , lat. 0°, and no. 5, long. Leo 10° , lat. 5° North, the longitudes being fixed for the epoch of the translator Sa'īd's time, *i e.* around 750. But in Ptolemy's star catalogue there are no stars, in the constellation of Leo, corresponding to these coordinates. One might think, at the first glance, of the two stars in the Lion's head, μ and ε Leo, which could be understood as the animal's "eyes", but their coordinates are too different. Also the two stars of the Arabic lunar mansion no. 9, *al-ṭarf*, generally interpreted as "the Lion's Eyes" (κ Cnc and λ Leo), cannot be considered, for the same reason. Further, all these stars are of the third or fourth magnitudes and would not fit among the "bright stars". There seems to have crept in a major confusion into the description of these two stars. Another question is how these coordinates (and those quoted above, from N) were obtained, obviously in a time before the translation of the *Almagest* into Arabic.

by al-Bīrūnī, from (new) Persian *biyābān*, “desert“ (*i.e.* "desert stars")[20], is therefore no longer valid[21].

The *Kitāb al-mawālīd*, of which an extract is here edited, is an interesting document from the early period of the Arabic reception of Greek science and a valuable piece of evidence for the important role played by Sassanian Persia in the transmission of Greek science to the Arabs[22]. Before embarking on the translation of Greek works directly, they had, and used, the advantage of the presence of such Persian material. The edition and study of such documents, of which PINGREE (in his article cited in note 5) has presented a rich choice, will shed more light on the early stage of the Arabic reception of Greek science and the forming of the technical vocabulary in Arabic.

Now follows the edited Arabic text extracted from the *Kitāb al-mawālīd*, accompanied by an English translation. The serial numbers added to the 16 stars are intended to facilitate the orientation between the two. For the Persian star names which could not be identified only transliterations of the corrupted spellings in the manuscripts are cited in the translation. Explanatory matter, in the translation, is included in square brackets. A *lacuna* (in no. 14) is indicated by pointed brackets, both in the text and translation.

[20] Cf. my article of 1968 (as in note 2, above), p. 64, with reference to al-Bīrūnī, *Kitāb al-tafhīm*, ed. R.R. WRIGHT, London 1934, 46 (§ 125).

[21] PINGREE has not taken notice of this and speaks once more of the "desert stars" (in the paper cited above in note 5, p. 233).

[22] In an article "Über das Frühstadium ...", in: *Saeculum* 26 (1975), 268-282, I have stressed in many places the importance of this aspect in the early period of the development of Arabic science.

نعت عدة من الكواكب البيابانية[١]

اعلم أن [1] الروزاهنك[٢] أجلّ كوكب ثابت فى الفلك وهو جنوبى وهو معبود العرب وإن البهائم تسرّ[٣] به إذا رأته وهو سعد صرف [٤]وهو الشعرى اليمانية[٤]. [2] [٥]النابح وهو النسر الطائر شمالى سعد صرف بيبانى[٥]. فأما[٦] [3] رجل الجوزاء[٧] سعد صرف وهو جنوبى. [4] جشم شير[٨] سعد صرف لا عرض له وكان[٩] فى عشر من الأسد[١٠] فى زمان سعيد . [5] جشم شير[١١] عرضه شمالى خمس درج وهو فى عشر من الأسد[١٢] ويسبق طلوعُه طلوعَ الأول من أجل أن عرض هذا فى الشمال وهو نحس صرف. [6] الجوهر النسر[١٣] الواقع سعد صرف شمالى عالٍ[١٤] وهوأجلّ من النسر الطائر و أسعد. [7] شعان[١٥] وهو الدبران قليل العرض فى الجنوب وهو[١٦] نحس صرف. ومقابله [8] قلب العقرب[١٧] وهو[١٨] نحس صرف. [9] كيل[١٩] سهر منكب الجوزاء[٢٠] الأحمر جنوبى ، قال[٢١] زرادشت[٢٢] ليس فى الفلك كوكب ثابت[٢٣] أنحس منه قوته مثل[٢٤] قوة المريخ سواء وفيه زحلية قليلة فإذا[٢٥] قارنه سعد[٢٦] أبطل قوته وإذا قارنه نحس[٢٧] زاد فى شرّه. [10] السماك الرامح يسمى جشم زرين[٢٨] شمالى عالٍ[٢٩] قوته مثل[٣٠] قوة المريخ. [11] سروس وهو[٣١] العيوق نصف قوته للمشترى ونصفها[٣٢] لزحل وهو شمالى. [12] الردف شمالى ويسمى بالفارسية سلار[٣٣] طبعه[٣٤] طبع المشترى وهو عالٍ[٣٥] فى الشمال. [13] السماك الأعزل قليل العرض فى الجنوب

[١] العنوان: ناقص في س || [٢] س: ورامند || [٣] ن: لتسر || [٤-٤] ناقص في س || [٥-٥] زيادة في س؛ ن على الهامش فقط: النسر الطائر، كأنه تصحيح بدلاً من: الشعرى اليمانية || [٦] ناقص في س || [٧] س: رجل الدلو || [٨] س: حشم سيد، وكُتب فوقه: يعنى فم الحوت || [٩] س: وهو || [١٠] س: عشر درج من الحمل || [١١] س: حسم سيد || [١٢] س: عشر درج من الحمل || [١٣] س يزيد: وهو || [١٤] ن: عالى || [١٥] س: شعبان || [١٦] ناقص في س || [١٧] س: قلب الرحان (كتابة الكلمة الثانية غير واضحة) || [١٨] ناقص في س || [١٩] ن: كل || [٢٠] س: منكب الدلو || [٢١] ن: وقال || [٢٢] ن: زردشت || [٢٣] ناقص في س || [٢٤] ناقص في س || [٢٥] ن: وإذا || [٢٦] س: قارنه سعداً ؛ ن: قارن سعداً || [٢٧] ن: قارن نحسًا || [٢٨] ن: حشم ارىن، والتصحيح في الهامش: زرىن؛ س: جسم ررين || [٢٩] ن: عالى || [٣٠] ناقص في س || [٣١] ن: هو || [٣٢] ن: ونصف || [٣٣] س: سلا || [٣٤] ن: وطبعه || [٣٥] ن: عالى ||

DESCRIPTION OF A NUMBER OF FIXED STARS

Know that [1] *al-rwz'hnk* (S: *wr'mnd*) is the most excellent fixed star in the sphere. It is southern. It is the object of adoration of the Arabs. Quadruped animals rejoice when seeing it. It is purely lucky. It is *al-shi'rā al-yamāniya* ["the Southern *Shi'rā*", α CMa, Sirius]. [2] *al-n'.ḥ*, it is *al-nasr al-ṭā'ir* ["the Flying Eagle", α Aql, Altair], northern, purely lucky, a fixed star. As for [3] *rijl al-jawzā'* ["Orion's Foot", β Ori, Rigel], it is purely lucky and it is southern. [4] *Chashm-i shīr* ["the Lion's Eye"] is purely lucky. It has no latitude [*i.e.*, its position is directly on the ecliptic]. In the time of Sa'īd it was in [longitude] Leo 10°. [5] *Chashm-i shīr* ["the Lion's Eye"], its latitude is 5° North and it is in [longitude] Leo 10°. Its rising precedes the rising of the first one, because the latitude of this [second] one is northern. It is purely unlucky. [6] *Al-jawhar* ["the Jewel, or Pearl"], *al-nasr al-wāqi'* ["the Falling Eagle", α Lyr, Vega], purely lucky, high up in the North. It is more excellent and luckier than *al-nasr al-ṭā'ir* [= no. 2]. [7] *Sh''n* (S: *sh'b'n*), it is *al-dabarān* [α Tau, Aldebaran]. It has a small southern latitude. It is purely unlucky. Opposite it [*i. e.*, 180° away in longitude] is [8] *qalb al-'aqrab* ["the Heart of Scorpius", α Sco, Antares]. It is purely unlucky. [9] *Kyl s'hr*, *mankib al-jawzā'* ["Orion's Shoulder", α Ori, Betelgeuse], the red one, southern. Zoroaster said, there is no fixed star in the sphere more unlucky than it; its virtue is exactly like the virtue of Mars and it has some [influence] of Saturn. When a lucky [planet] passes near it, it annihilates its [*i. .e.*, the planet's] virtue, and when an unlucky [planet] passes near it, it increases its [*i e.*, the planet's] evil. [10] *Al-simāk al-rāmiḥ* ["the *Simāk* Armed with a lance", α Boo, Arcturus], [in Persian] it is called *chashm-i zarrīn* ["the Golden Eye"], high up in the North. Its virtue is like the virtue of Mars. [11] *Srws*, it is *al-'ayyūq* [α Aur, Capella], half of its virtue belongs to Jupiter and half to Saturn. It is northern. [12] *Al-ridf* ["the One Riding on the same animal, behind the rider", α Cyg, Deneb], northern, in Persian it is called *sl'r* (S: *sl'*). Its nature is that of Jupiter. It is high up in the North. [13] *Al-simāk al-a'zal* ["the Unarmed *Simāk*", α Vir, Spica],

طبعه طبع المشترى وعطارد. [14] ‹...› وهو قلب[٣٦] الأسد[٣٧] الذى لا عرض له أحسن هذه السعود البيابانية[٣٨] تأثيراً لقربها[٣٩] من الطريقة ولأن[٤٠] سائر هذه[٤١] الكواكب لها عروض شمالية وجنوبية[٤٢]. [15] خرّه بازى[٤٣] سهيل وهو منخفض فى الجنوب وتأثيره فى أرض اليمن وبوادى[٤٤] العرب أكثر من تأثيره فى البلاد الشمالية لأنه يسامت رؤوسهم. [16] روحنه[٤٥] وسط منطقة الجوزاء[٤٦] سعد صرف[٤٧] عرضه فى الجنوب[٤٨]. فهذه الكواكب المشرقة الزاهرة التى[٤٩] إذا طلعت أشرقت وزهرت[٥٠].

[٣٦] س ن: وقلب || [٣٧] س: الحمل || [٣٨] س: الثابتة || [٣٩] س: لقربهما || [٤٠] س: لأن || [٤١] ناقص في س، وفي ن أضيف فوق السطر || [٤٢] س: عرض شمالى وجنوبى || [٤٣] س: حره نهاري || [٤٤] س: وفى بوادى || [٤٥] س: وحبه وحمسه (وفوق الاخيرة: ظ) || [٤٦] س: وسط سما منطقة الدلو || [٤٧] ن(في السطر): ضعيف، وفي الهامش التصحيح: صرف || [٤٨] س: جنوبى || [٤٩] ناقص في س || [٥٠] ن : شرقت وبوهت

of a small southern latitude. Its nature is that of Jupiter and Mercury. [14] < ... > and it is *qalb al-asad* ["the Lion's Heart", α Leo, Regulus], which has no latitude [*i.e.*, its position is directly on the ecliptic]. It has the best influence of these lucky fixed stars, because it is so close to the "Path" [*i. e.*, of the Sun: the ecliptic] and because all the others of these stars have northern or southern latitudes. [15] *Khrrh .'zy* (S: *ḥrh nh'ry*), *suhayl* [α Car, Canopus]. It is deep in the South. Its influence on the Yemen and the deserts of the Bedouins is greater than its influence on the northern regions, because it passes through their zenith [*i.e.*, of the people in those southern regions]. [16] *Rwḥ..h* (S: *wḥbh*, or *wḥms'h*), *wasaṭ minṭaqat al-jawzā'* ["the Middle of Orion's Girdle", ε Ori, *Alnilam*], purely lucky. Its latitude is southern. These are the brilliant and bright stars which, when they are visible, beam and shine.

XII

THE ASTRONOMER AL-ṢŪFĪ AS A SOURCE FOR ULUĠ BEG'S STAR CATALOGUE (1437)

Tamerlane's grandson Uluġ Beg (born 1394), who reigned over the empire during the last two years of his life before he was assassinated by his own son in 1449, is well-known for his scientific interests. In the dynasty's residence, Samarkand, he founded a *madrasa* (inaugurated in 1420) and afterwards also an observatory (construction perhaps begun in the same year, 1420).[1] While Uluġ Beg seems to have been an able mathematician and astronomer himself[2] he assembled a greater number of scientific collaborators for work in the *madrasa* and the observatory, the most famous among whom were Ġiyāṯ al-dīn Jamšīd al-Kāšī (d. 832/1429)[3], Qāḍīzāda al-Rūmī (d. after 844/1440, but before Uluġ Beg)[4] and 'Alā' al-dīn 'Alī al-Qūščī (d. 879/1474).

As a result of the work conducted in the observatory, a *zīj* was published which became known under various titles, such as *Zīj-i Sulṭānī-i Gūrkānī* or *Zīj-i Jadīd-i Sulṭānī* and others.[5] Like several other major *zīj*es it also contains a star catalogue which was established for the epoch beginning 841 H = July 1437.[6]

It has been debated whether this work was originally written in Persian or in Arabic – besides a greater number of manuscripts in Persian there exist also some in Arabic.[7] But it is now generally assumed that the original text was in Persian and that the Arabic versions are translations thereof.[8] This is also corroborated by our present study.

The complete text of Uluġ Beg's *zīj* has not yet been published, there exist only editions, translations and discussions of selected items from it.[9] For my research, I relied on the edition of the Persian text of the star catalogue, accompanied by a Latin translation and an ample commentary, by Thomas Hyde, Oxford 1665.[10]

---

1. *Cf.* Sayili, 260ff.; Samsó, 601.
2. *Cf.* the characteristic extract from a letter of Jamšīd al-Kāšī, in Sayili, 262.
3. See Kennedy (1960), 7.
4. See Sellheim, 160ff.
5. See Storey, 67f.; Kennedy (1956), 125f. (no. 12), abstract pp. 166f.
6. Ed. Hyde, 4f.
7. See Storey, 67ff., 71f.
8. *Cf.* Knobel, 5f.
9. See the survey in Storey, 69f.; *cf.* also Kennedy (1956), 166f.
10. See Bibliography.

Uluġ Beg's star catalogue, which basically repeats Ptolemy's star catalogue in the *Almagest* (epoch A.D. 137), has been the object of several studies, mainly with the objective to find out whether the data in the catalogue are the result of his own independent observations carried out in Samarkand or whether they were merely obtained by calculation and derived from existing older catalogues. The main work in this field was the great collation of Uluġ Beg's star coordinates by E. B. Knobel (1917) based upon 24 manuscripts, 21 of which were in Persian and 3 in Arabic (the collation of three manuscripts stems from C.H.F. Peters).[11] In recent times the problem of Uluġ Beg's star coordinates was discussed anew in a series of highly critical articles[12] which, however, I am afraid, also did not bring the question to a definite end.

Our aim in the present paper will not be so much the numerical values in Uluġ Beg's catalogue, but rather the textual descriptions of the individual 1,018 stars registered there.

It was declared by Uluġ Beg himself, in the short introduction prefixed to his star catalogue[13], that the work of reference from which he departed and to which he had recourse in making his star observations and in establishing his catalogue was the 'Book on the Constellations of the Fixed Stars' by al-Ṣūfī. For 27 deeply southern stars which could not be observed at the latitude of Samarkand he directly took over al-Ṣūfī's values and processed them, with a precession rate of 1° in 70 years, for his epoch 1437. Similarly, he omitted from his catalogue eight stars from Ptolemy's basic catalogue which al-Ṣūfī had declared to be wrongly established by Ptolemy and which he had not been able to identify in the sky.

Abū al-Ḥusayn ʿAbd al-Raḥmān ibn ʿUmar al-Ṣūfī (903-986), from Rayy and active in Iran in close relationship to the Buyid rulers there, composed a detailed book, in Arabic, on the fixed stars in which he laid down the results of his own observations and a thorough critique of Ptolemy's star catalogue. However, in the catalogue of the stars, arranged by constellations, which he added to his book and which was set up for the epoch 964, he faithfully rendered Ptolemy's data (the longitudes being augmented, for precession, by 12°42'), relying on the existent Arabic translations of the *Almagest*. The verbal descriptions of the positions of the stars within each constellation were taken over from the Arabic *Almagest* version by Isḥāq ibn Ḥunayn.[14]

For many centuries, al-Ṣūfī's book became the standard work in Islamic astronomy on the matter of the fixed stars and even radiated into medieval

11. See the survey in Knobel, 85.
12. Evans, 162-165; Shevchenko; Krisciunas.
13. Hyde, 2-5; Knobel, 8f.
14. Kunitzsch (1974), 51f. For our study, we relied on the Hyderabad edition of al-Ṣūfī's book and on Sezgin's facsimile edition of the oldest extant manuscript, Oxford, Bodleian Library, Marsh 144, copied in 400/1009-10, perhaps by the author's son.

Europe, although no formal Latin translation was made of it.[15] In the Orient, it was translated into Persian and Turkish several times, in various periods.[16] One Persian translation of al-Ṣūfī's book was made by the famous Iranian scholar and scientist Naṣīr al-dīn al-Ṭūsī (1201-1274).[17] A manuscript of al-Ṭūsī's translation is in Istanbul, Aya Sofya 2595; it is dated 647/1250, and some modern scholars are inclined to regard it as the autograph of al-Ṭūsī[18], a problem which we shall not discuss here further. The manuscript carries the mark of Uluġ Beg's library, which means that it would have been within the reach of Uluġ Beg and his collaborators.

The Istanbul manuscript of al-Ṭūsī's Persian translation of al-Ṣūfī's book (which itself was in Arabic) has been published in facsimile in 1969 and in a critical edition in 1972.[19]

The availability of all these texts, by al-Ṣūfī, al-Ṭūsī and Uluġ Beg, encouraged and enabled us to carry out comparisons the results of which are presented here.[20]

The verbal text of the star descriptions in Uluġ Beg's catalogue is generally Persian with some interspersed Arabic words.[21] Since he himself named al-Ṣūfī as a source of reference for problems concerning the star catalogue, one has to investigate whether there exists a relationship between the two texts. And if so, whether Uluġ Beg used al-Ṣūfī's text in the original Arabic version (and translated the Arabic into Persian himself) or whether he used an existing Persian version of it.

Any way, for research of this type one has to consult the original texts directly. Knobel's study of 1917 is not helpful for this purpose in so far as he simply reproduces, in the column of the verbal descriptions of the stars, the French translation of al-Ṣūfī's text by H.C.F.C. Schjellerup ('with some amendments').[22]

The comparison between Uluġ Beg's original text of the verbal star descriptions and that of al-Ṭūsī's Persian translation of al-Ṣūfī's book proves clearly that Uluġ Beg indeed relied upon al-Ṭūsī's Persian version.

To cut matters short, I shall only quote two examples here. The fifth star of Pegasus (τ Peg; Baily no. 319) is called by al-Ṣūfī: *amyal al-iṯnayn allaḏayn*

15. See Kunitzsch (1989), item XI, 64ff.
16. See Storey, 41.
17. Storey, 41, no. (1).
18. *Cf.* Kunitzsch (1989), item XI, 62 note 22.
19. See Bibliography, al-Ṭūsī (1969) and (1972).
20. A first report on it was given in Kunitzsch (1989), item XI, 61-64.
21. The characteristic terms in Persian from Uluġ Beg's catalogue have been included in the list of terms compiled from the Arabic versions of the *Almagest* and some other Arabic star catalogues, in Kunitzsch (1974), 217-348; see also the index of the Persian words, *ibid.* 362f. The word list compiled by C.H.F. Peters and rendered, in revised form, in Knobel, 95-109 (*cf. ibid.*, 10), is often erroneous and contains many words or forms that did not originally appear in Uluġ Beg's text.
22. *Cf.* Knobel, 25.

fi'l-badan taḥta'l-janāḥ ila'l-šamāl ('the one more inclined towards north of the two in the body, below the wing'). This was translated by al-Ṭūsī as: *šamālītarīn-i ān du kawkab ki bar tan and dar zīr-i janāḥ*, and the same wording is found in Uluġ Beg (only with the omission of the word *kawkab*).[23]

A second example from the end of the catalogue, in the constellation of the Southern Fish (Piscis Austrinus); al-Ṣūfī calls the sixth star in it (μ PsA; Baily no. 1016): *allaḏī 'ala'l-šawka al-janūbīya allatī 'ala'l-ẓahr* ('the one on the southern spine that is on the back'), which al-Ṭūsī renders as: *an-ki bar šawka-i janūbī ast ki dar pušt ast*, and which is similarly found in Uluġ Beg.[24]

It is thus evident that Uluġ Beg used al-Ṣūfī's book in al-Ṭūsī's Persian version and that he repeated *verbatim* al-Ṭūsī's rendering of al-Ṣūfī's text. We observe the typical style of al-Ṭūsī's translation: he transformed the formulae of the star descriptions basically into Persian, but retained here and there some of the Arabic specific nouns indicating the position of the stars, such as – in our examples – the 'wing' (*janāḥ*) and the 'spine' (*šawka*).

In examining the complete catalogue one finds a small number of minor differences in the wording between the texts of al-Ṭūsī and Uluġ Beg: one of them using an Arabic term and the other a Persian one instead, or *vice versa*.[25] These slight differences would appear natural in the course of a longer transmission of a text. About three hundred years passed between the writing of al-Ṣūfī's book and al-Ṭūsī's translation and another two hundred years between the translation and its use by Uluġ Beg. In the course of the intervening centuries each text must have suffered its own alterations at the hands of copyists, readers and well-meaning commentators and 'correctors'.

In pushing the analysis one step further we can even say that Uluġ Beg's verbal descriptions of the star positions in his catalogue are a Persian version of Ptolemy. The line of transmission runs from Ptolemy to the Arabic translation of the *Almagest* by Isḥāq ibn Ḥunayn (around A.D. 890); Isḥāq's text was used by al-Ṣūfī; al-Ṣūfī's text was translated into Persian by al-Ṭūsī, and al-Ṭūsī's text was finally reproduced by Uluġ Beg.

In this context it may be noted that al-Ṭūsī was quite familiar with the text of the *Almagest* and with the Arabic version of the star catalogue as well. Among his numerous writings, he also produced a recension of the *Almagest*, the so-called *Taḥrīr al-Majasṭī*, completed in 1247, that is three years before the possible date of his translation of al-Ṣūfī's book (i.e. 1250, *cf.* above). The *Taḥrīr* was written in Arabic and it was based – like al-Ṣūfī's text of the star descriptions – on the Arabic translation of the *Almagest* by Isḥāq ibn Ḥunayn.[26]

---

23. See Hyde, 54f.
24. Hyde, 150f.
25. For some details, see Kunitzsch (1989), item XI, 63.
26. *Cf.* Kunitzsch (1974), 47f.

The straightforward use of the Persian version of al-Ṣūfī's text by Uluġ Beg seems to support the view that the language in which his *zīj* was originally composed was Persian. If he had written it in Arabic, he most certainly would have used of al-Ṣūfī's original Arabic text, and a possible later Persian translator of the *zīj* would have hardly had recourse to an earlier Persian translation of al-Ṣūfī, but rather would have translated the star catalogue, together with the complete *zīj*, into Persian anew with the result that a completely different Persian text would have been produced.

At the end of this demonstration I should like to discuss the case of a single word in Uluġ Beg's star catalogue which has remained incomprehensible and which I am unable to identify.

The 11th star of Orion (χ^1 Ori; Baily no. 744) was placed by Ptolemy on the Κολλόροβος, a club carried by Orion in his right hand.

The Greek word can mean a 'club' or a 'shepherd's staff or crook'. Isḥāq ibn Ḥunayn rendered the term as *al-ʿaṣā ḏāt al-kullāb*, 'the staff with a hook', which was retained identically by al-Ṣūfī. In al-Ṭūsī's Persian translation it was rendered as عصاء اهنها, and this appears in Uluġ Beg as عصای ائنها [27]. The first word is the Arabic *ʿaṣā*, 'staff', taken over from al-Ṣūfī (and lastly from Isḥāq ibn Ḥunayn). The correct reading and meaning of the second word have remained hidden from me. S. M. Mahdavī, the Persian editor of al-Ṭūsī's translation[28], omitted the word completely and simply remarks *ġayr maqrū'*, illegible[29], although in the facsimile the spelling اهنها is clearly visible.[30] I inspected several manuscripts of Uluġ Beg's *zīj* and found, in addition to the two spellings mentioned above, further the forms آهكهای and انتها. [31] The ending *-hā* in the word may point to its quality as the plural of a noun. In this context we can mention that the medieval translator of the *Almagest* from Arabic into Latin, Gerard of Cremona, mistook the draft of the word *kullāb* ('hook') as *kilāb* (plural of *kalb*, 'dog') and translated it (elsewhere in the catalogue) as *hastile habens canes*.[32] But this cannot have happened with al-Ṭūsī and Uluġ Beg, because 'dog' in Persian is *sag* and its plural *saghā* would appear quite different in writing from the odd forms mentioned above.

The same term, *kollórhobos*, occurs a second time in Ptolemy's star catalogue, in the constellation of Bootes. In the 15th star of it (ω Boo; Baily no. 102) the term chosen by Isḥāq ibn Ḥunayn, *al-ʿaṣā ḏāt al-kullāb*, is found unchanged in all our texts, al-Ṣūfī, al-Ṭūsī and Uluġ Beg.[33] In the 8th star (μ Boo; Baily no. 95) we have the same situation; the Arabic *al-ʿaṣā ḏāt al-*

27. Hyde, 114f.
28. *Cf.* note 19, above, and the Bibliography.
29. In the edition, p. 244 with note 4.
30. Facsimile edition, p. 158.
31. Kunitzsch (1974), pp. 311 (no. 496) and 372.
32. Kunitzsch (1974), pp. 227 (no. 57), 229 (no. 63), 311 (no. 496).
33. Kunitzsch (1974), p. 229 (no. 63); Hyde, 20f.

kullāb appears in all the sources.[34] But here the Istanbul manuscript of al-Ṭūsī's version shows an addition appended to the end of the star's text: *ya'nī* اهنها *bar zada*, 'i.e. placed on the ...', with the same unknown word as in Orion.[35] This gloss was not taken over by Uluġ Beg. And the modern editor of al-Ṭūsī, M. S. Mahdavī, has likewise omitted it from his edition, without a notice in the apparatus.[36]

BIBLIOGRAPHY

EVANS, J. (1987), 'The Origin of the Ptolemaic Star Catalogue: Part 1', in: *Journal for the History of Astronomy* 18, pp. 155-172.

HYDE, Th. (1665), *Tabulae long(itudinis) ac lat(itudinis) stellarum fixarum, ex observatione Ulugh Beighi...*, Oxford.

KENNEDY, E. S.
— (1956), 'A Survey of Islamic Astronomical Tables', in: *Transactions of the American Philosophical Society*, N.S. 46, pp. 123-177.
— (1960), *The Planetary Equatorium of Jamshid... al-Kashi*, Princeton.

KNOBEL, E. B., (1917), *Ulugh Beg's Catalogue of Stars*, Washington.

KRISCIUNAS, K., (1993), 'A more complete analysis of the errors in Ulugh Beg's star catalogue', in: *Journal for the History of Astronomy* 24, pp. 269-280.

KUNITZSCH, P.
— (1974), *Der Almagest. Die Syntaxis Mathematica des Claudius Ptolemäus in arabisch-lateinischer Überlieferung*, Wiesbaden.
— (1989), *The Arabs and the Stars*, Northampton.

SAMSÓ, J., (1991), article 'Marṣad', in: *Encyclopaedia of Islam*, new ed., vol. VI, Leiden, pp. 599-602.

SAYILI, A., (1960), *The Observatory in Islam*, Ankara; repr. New York 1981.

SCHJELLERUP, H.C.F.C., (1874), *Description des étoiles fixes*, St.-Pétersbourg; repr. Frankfurt am Main 1986 (ed. F. Sezgin).

SELLHEIM, R. (1976), *Verzeichnis der orientalischen Handschriften in Deutschland*, XVII A 1: *Materialien zur arabischen Literaturgeschichte*, I, Wiesbaden.

SHEVCHENKO, M., (1990), 'An Analysis of Errors in the Star Catalogues of Ptolemy and Ulugh Beg', in: *Journal of the History of Astronomy* 21, pp. 187-201.

---

34. Kunitzsch (1974), p. 227 (no. 57); Hyde, 20f.
35. Al-Ṭūsī (1969), 34.
36. Al-Ṭūsī (1972), 50. In Knobel's vocabulary (*cf.* above, note 21) only the Arabic term is discussed, the enigmatic Persian word is not mentioned (Knobel, 96f., under Bootes).

STOREY, C.A., *Persian Literature. A Bio-bibliographical Survey*, II, 1, repr. London 1972 (1st ed., 1958).

al-ṢŪFĪ, Abū al-Ḥusayn

— (Hyderabad), *Kitāb ṣuwar al-kawākib al-ṯamāniya wa'l-arba'īn*, Hyderabad / Deccan 1373/1954; repr. Beirut: Dār al-Āfāq al-Jadīda 1401/1981.

— (Sezgin), *The Book of Constellations by 'Abd al-Raḥmān al-Ṣūfī* (facsimile ed. by F. Sezgin), Frankfurt am Main 1986.

al-ṬŪSĪ, Naṣīr al-dīn

— (1969), *Tarjama-i ṣuvar-i kavākib bi-qalam-i Ḫ'āja Naṣīr al-dīn al-Ṭūsī*, ed. P. N. Ḫānlarī, Tehran 1348š./1969.

— (1972), *Tarjama-i ṣuvar-i kavākib-i 'Abd al-Raḥmān Ṣūfī*, ed. Sayyid Mu'izz al-dīn Mahdavī, Tehran 1351š./1972.

ULUĠ BEG: see Hyde.

XIII

AL-ṢŪFĪ AND THE ASTROLABE STARS

The importance of the astronomer Abu 'l-Ḥusayn ʿAbd al-Raḥmān al-Ṣūfī (903–986), notably in the knowledge of the fixed stars and the constellations, and the influence exerted by his *Kitāb ṣuwar al-kawākib al-thābita* ("Book on the Constellations of the Fixed Stars", written around A. D. 964) both in the Orient and in Europe have been discussed in detail in an earlier volume of this journal[1]. In this present article we shall analyse what this prominent Islamic astronomer and specialist in matters of the fixed stars wrote about the stars which featured on astrolabes.

The astrolabe, inherited by the Arabs from late Antiquity and handed on by them to medieval Europe, was equipped on its front side with the *rete* ("spider"), a movable star map rotating over the underlying plate(s) and indicating the instantaneous position of the stellar sky relative to the local horizon. On the *rete* some fundamental fixed stars were marked serving to set the instrument according to the actual situation in the sky. In early times, the number of these stars ranged from 15 to 30 stars, mostly of the first and second magnitudes[2], in written treatises as well as on the instruments (cf. for example the oldest existing Arabic astrolabe, dated 315 H = 927/8, made by Nasṭūlus, now in the Kuwait National Museum: 17 stars; a table of astrolabe stars for the year 367 H = 978 by Maslama al-Majrīṭī: 21 stars; Maghrebi instruments of 420 to 605 H = 1029 to 1208/9: 26 to 29 stars; the oldest known Latin table of astrolabe stars, end of

[1] P. Kunitzsch, "The astronomer Abu 'l-Ḥusayn al-Ṣūfī and his Book on the Constellations", *ZGAIW* 3 (1986), 56–81.

[2] Of the 1,025 stars registered in Ptolemy's star catalogue in the *Almagest* 15 belong to the first magnitude, and 45 to the second. Most of the traditional astrolabe stars are found among these sixty stars, but there are also some of the third, and even fourth, magnitudes.

10th cent.: 27 stars). In the later development more and more stars were added; astrolabes from the 16th and 17th centuries, both in the Orient and in Europe, contain up to more than sixty stars among which, of course, several stars of magnitudes smaller than the second.

Al-Ṣūfī's contribution to the art of the astrolabe is documented in a number of written texts; there are no reports that he ever made such instruments himself[3].

With special regard to the stars to be used on the astrolabe, the following works of al-Ṣūfī must be consulted:

1. *Kitāb ṣuwar al-kawākib al-thābita* ("Book on the constellations of the fixed stars"; epoch for the longitudes in the star coordinates: A. D. 964)[4]. Here, in the course of the discussion of the individual stars of the 48 constellations, al-Ṣūfī describes 44 stars as being used on the astrolabe (five of them specifically on the "southern astrolabe")[5]. In other words, we do not have here

[3] On the other hand, one Ibn al-Sanbadī reports to have seen in the year 435 H = 1043/4 in the library (*khizānat al-kutub*) at Cairo a silver celestial globe made by Abu 'l-Ḥusayn al-Ṣūfī for the ruler 'Aḍud al-Dawla, weighing 3,000 dirhams and which had been purchased at a price of 3,000 dinars; see Ibn al-Qifṭī, *Ta'rīkh al-ḥukamā'*, Leipzig 1903, p. 440. Among al-Ṣūfī's writings there is also a treatise on the use of the celestial globe; cf. F. Sezgin, *Geschichte des arabischen Schrifttums*, VI (Leiden 1978), p. 215 (work no. 4); E. S. Kennedy, in *ZGAIW* 5 (1989), 48–93.

[4] For editions and translations, see 1986 (as in note 1, above), pp. 57–59.

[5] W. Hartner erroneously named only 37 out of these 44 stars: "The principle and use of the Astrolabe", in A. U. Pope, *A Survey of Persian Art*, vol. III (Oxford 1939), p. 2542, footnote 1 (reprinted as *Astrolabica*, N°. 1, Paris 1978). Afterwards the wrong number of 37 stars was transmitted by several other authors: H. J. J. Winter, p. 4 of his Introduction to the Hyderabad edition of al-Ṣūfī's Book on the Constellations; *idem*, p. 129 of an article on al-Ṣūfī in *Archives Internationales d'Histoire des Sciences* 8 (1955); J. M. Millás Vallicrosa, in *Isis* 48 (1957), 364; M. Destombes, in *Archives Internationales d'Histoire des Sciences* 15 (1962), 12 etc. In an addition to the reprint of his article, in *Oriens-Occidens*, I (Hildesheim 1968), p. 311, Hartner corrected the number to 44 [but still with a wrong epoch, 364 H = 974; al-Ṣūfī's epoch is actually A. D. 964]; from here, the number 44 was taken over by A. A. Al-Daffa' and J. J. Stroyls, "The art of the astrolabe", *Hamdard Islamicus* 6 (1983), 15–36, esp. p. 21. Cf. also P. Kunitzsch, *Arabische Sternnamen in Europa*, Wiesbaden 1959, p. 59 footnote 2, and *idem*, *Typen von Sternverzeichnissen* . . ., Wiesbaden 1966, p. 6 footnote 10.

an actual list of astrolabe stars with indication of coordinates; there are only the simple remarks about the use of the stars on the astrolabe, scattered throughout the book and added to the discussion of each of the 44 stars in passing. [This source will be called "A" afterwards.]

2. The Old-Spanish redaction of A, made under Alfonso X of Castile (reigned 1252–1284) and contained in his astronomical collection *Libros des saber*, Books I–IV[6], has assembled the astrolabe stars in a star table formed as a "rueda" (rosette) at the end of Book IV. Here 44 stars are registered with ecliptical coordinates (longitude = Ptolemy + 17°8′ for the epoch 1252), but it is surprising that eight of al-Ṣūfī's original stars were replaced by new stars (which are not among the 44 stars of A). This intervention seems to be due to Alfonso's astronomical translators and collaborators. [This source will be called "B".]

3. *Risāla fī 'l-'amal bi-l-asṭurlāb* ("Treatise on the use of the astrolabe"), in 170 chapters, edited in facsimile (as the second text) in: *Two Books on the Use of the Astrolabe* by 'Abd al-Raḥmān al-Ṣūfī, ed. F. Sezgin, Frankfurt am Main 1986, from Ms. Aya Sofya 2642 (dated 871 H = 1466/7). In the introduction al-Ṣūfī dedicates the work to the Buwayhid prince Abu 'l-Fawāris Shīrzīl [i. e., Shams al-Dawla], the son of his permanent patron 'Aḍud al-Dawla, and says that he has abbreviated it from a longer version in 1,760 chapters. (885 chapters of the longer version seem to be extant in Ms. Paris, B.N. ar. 5098, which however is defective at the beginning and the end.) The last (170th) chapter (pp. 118–129 of the facsimile ed.) is "On the well-known stars that are inscribed on the northern and southern astrolabes". It contains the description of 41 stars (all of them also found among the 44 stars of A except one star: beside the traditional ι Ceti, magnidude 3−, the brighter star β Ceti, magnitude 3+, was also added). No coordinates are given, the positions are only described relative to other known stars and asterisms. The descriptions are correct and mostly coincide with those in A. [This source will be called "C".]

4. *Kitāb al-'amal bi-l-asṭurlāb* ("Book on the use of the astrolabe"), in roughly 400 chapters. This seems to be an independent

[6] For details on the text, editions etc., see 1986 (as in note 1, above), pp. 64–66.

work, not related to, or abbreviated from, another treatise on the subject. For this study, two edited versions were available: the printed edition of Hyderabad, 1962 (in 386 chapters; the last [386th] chapter, pp. 340–350, is "On the figures of the fixed stars that are inscribed on the astrolabe"), based on the very late and faulty Ms. Paris, B.N. ar. 2493, dated Ṣafar 1283 H = June/July 1866; and the Frankfurt facsimile edition mentioned above (the first text; here in 402 chapters, reproduced from Ms. Topkapı Sarayı, Ahmet III, 3509, dated 676 H = 1277/8; the last [402nd] chapter, pp. 509–519, is on the stars). [This source will be called "D"; the Hyderabad edition will be cited as "h", the Frankfurt facsimile edition as "f".] This version lists 30 stars (three of which are "new" and not contained among the 44 stars of A; one of the three new stars is also among the eight new stars in B)[7]. Both sources, h and f, are faulty copies, but in many instances one, or the other, of the two has a better, or more complete, text so that a definite reading can be chosen with some certainty. The stars are described in their positions relative to other well-known stars, and to about half of them coordinates are added, *mediatio coeli* (which, however, is not named as such) and declination (*al-buʿd ʿan muʿaddil al-nahār*). In TABLE 2, below, I have edited the material in D in tabular form.

It is now clear that al-Ṣūfī, the well-known expert in matters of the fixed stars, has dealt with the astrolabe stars at least three times, in A, C and D (B can be left aside because it is a later development introduced by Alfonso's astronomers and influenced by their knowledge of contemporary astrolabes).

Of these three texts, A can be dated to around 964, the epoch of the star longitudes. At that time al-Ṣūfī, who was born in 903, was aged about sixty years.

C, the 170-chapter version abbreviated from the long 1,760-chapter work on the use of the astrolabe, contains the star descriptions in full congruence with the text of A. In one instance (p. 118,ult. - 119,1) al-Ṣūfī explicitly quotes his *Kitāb al-kawākib*, i. e. our A. Further, ʿAḍud al-Dawla's son Shams al-Dawla, to whom al-Ṣūfī dedicated text C, took his father's place in the gov-

[7] On C and D cf. SEZGIN (as in note 3, above), works nos. 3 and 2, respectively.

ernment of Kirman from about 972 and became ruler of Fars in 983. It is thus evident that C was written after 964, and more precisely – since it was dedicated to the *amīr* Shams al-Dawla – later in the 970s or near 980. Because of the perfect congruence of the star descriptions in the star chapter (which was extracted from the 1,760-chapter version together with the rest of text C) with the wording in A, it can be assumed that the long version was also written some time after 964.

The star chapter in D is essentially different from A and C in that the descriptions of the stars' locations are often confused or even incorrect. In A al-Ṣūfī had demonstrated his full mastery in the knowledge of the fixed stars, the constellations, their nomenclature and location in the sky. The same standard is also reflected in C. The information laid down in D, on the other hand, reflects a much inferior standard of knowledge. The descriptions of several stars are ambiguous or even contradictory to the statements in A. For example, of the star *mankib al-faras* (β Pegasi, no. 17 in TABLE 1, below) the following is said (C, h p. 349; f p. 518): "*mankib al-faras* is also placed on the astrolabe, it is the bright star above the northern one of *al-fargh al-muqaddam*[8]; it is also said that it is the northern star of *al-fargh al-muqaddam* itself, but I take it to be correct that it is the star above it; when you look up [to the sky], you see this star, which is *mankib al-faras*, and the northern star of *al-fargh al-awwal*[9] and another star together with the two, forming a triangle." This is obvious nonsense, because in the sky there is no "bright star" above β Pegasi. Furthermore, it becomes clear that when writing this treatise the author had not yet acquired a solid knowledge of the classical Ptolemaic constellations and the distribution of their stars[10], and was not yet able correctly to understand the indigenous Arabic star traditions and to identify stars in Arabic asterisms with the corresponding stars of the *Almagest*. It should also be mentioned that for three stars with substantial southern declinations al-Ṣūfī gives the maximum altitude above the hori-

[8] This asterism, forming the 26th Lunar Mansion, consists of αβ Pegasi, of which α is the southern, and β the northern one.

[9] *Al-fargh al-awwal* refers to the same stars as *al-fargh al-muqaddam*.

[10] The name *mankib al-faras* is not of indigenous Arabic origin, but is derived from the *Almagest* and designates β Pegasi.

zon at Rayy, his birthplace. It may therefore seem reasonable to regard D as a very early work of al-Ṣūfī, written at an early stage of his career, still at his birthplace Rayy, when he had not yet gained the mastery in the subject later shown in the "Book on the constellations" of 964 (our A). It is only natural to assume that in the roughly sixty years between his birth and the composition of A he wrote a number of works, the 400-chapter treatise (D) being an early one among them. Apart from the factual mistakes in the text, the transmission also seems to be weak. In both manuscripts there are omissions and many scribal errors, and in both h and f coordinates are only transmitted for about half of the thirty stars, but it seems very probable that they were originally added to all of them. It is interesting to see how this author, who later became the most important authority on matters of the fixed stars in Arabic-Islamic astronomy, began his career on a much lower level.

The total number of stars designed by al-Ṣūfī for use on the astrolabe, including the "new" stars in B, C and D, is 55. He started with a list of 30 stars in D and later named 44 stars in A and 41 in C (a survey of all the stars is given in TABLE 1, below). He thus contributed to an inflation in the use of astrolabe stars, which in later centuries involved more than sixty. It is noteworthy that among his astrolabe stars there are several rather faint and insignificant stars, e. g., no. 3 (in TABLE 1, below), mag. 3− (with no. 11, mag. 3, being quite near); no. 5, mag. 3− (with no. 6, mag. 2, being quite near); no. 14, mag. 4+ (with no. 13, mag. 2+, being quite near); no. 31, mag. 4− (with no conspicuous star nearby)[11]; no. 35, mag. 3− (again following the tradition; in C, at least, he has newly added the brighter star no. 53, mag. 3+).

In the two tables which conclude this paper the first surveys all the stars named by al-Ṣūfī for use on the astrolabe in the various sources described above, and the second shows the coordinates given in D. As a means of comparison, in TABLE 2 the corresponding coordinates are shown from the table of 25 astrolabe

[11] In A (and similarly C), al-Ṣūfī explains that he has retained β Sgr because in the *Almagest* it was assigned magnitude 2, although he himself had found it to be of magnitude 4−. It is evident that he prefers to follow the Ptolemaic tradition rather than his own observation.

stars of al-Farghānī calculated for the year 225 Yazdağird = A. D. 856/7[12]. The juxtaposition of the two groups of coordinates makes it sufficiently clear that al-Ṣūfī did not re-calculate the coordinates in D for his own epoch (probably in the 920s), but he rather retained older values found in some of his sources and which are closely related to al-Farghānī's coordinates of 856/7. This may be regarded as another interesting aspect of al-Ṣūfī's early work in 400 chapters (D).

There is yet another connection between D and al-Farghānī, also found in D, chapter 40 (f, p. 71f.; omitted in the Paris MS, and therefore not printed in h, p. 31). Here al-Ṣūfī has inserted a "Table of Distances", *Jadwal al-ab'ād*, which contains the same 25 stars as al-Farghānī's table of astrolabe stars. For each star there are given *al-ṭūl* and *al-'arḍ*, "the distance to the circle of the beginning of Aries" (which is the equator; this coordinate is declination), and the distances from the North and South Poles and from the tropics of Cancer and Capricornus. The coordinate here called *al-ṭūl* (i. e., "[ecliptical] longitude") proves to be the *mediatio* taken from al-Farghānī's table, without change. Similarly, the values for declination are al-Farghānī's. The four distances tabulated are derived from declination through addition and subtraction. The values in the column *al-'arḍ* (i. e., "[ecliptical] latidude") are indeed ecliptical latitudes, though here combined with *mediatio*. But the values are quite peculiar; they are not identical with the latitudes known from other important texts such as the *Almagest*, al-Battānī's star catalogue (which is based on the "old translation" of the *Almagest*, otherwise lost), the table of 24 stars in the *Zīj al-Mumtaḥan* and al-Farghānī's astrolabe star table (where the latitudes are identical with those in the *Zīj al-Mumtaḥan*). While the dependence of al-Ṣūfī's "Ta-

[12] The edition by M. Destombes, in *Actes du VIIIe Congrès International d'Histoire des Sciences* (*Florence 3–9 Septembre 1956*), p. 309 (= p. 1 of the offprint), contains only ecliptical longitudes and latitudes and has omitted *mediatio* and declination. Al-Farghānī's star table was derived from that of the *Zīj al-mumtaḥan* (of 214 H = 829/30) by adding 0°15′ to the longitudes, for precession, cf. P. Kunitzsch, in *ZDMG* 120 (1970), 281–287, esp. p. 283 (with footnote 7), repr. in P. Kunitzsch, *The Arabs and the Stars*, Northampton 1989 (Variorum Reprints, CS307), item III.

TABLE 1

Table of astrolabe stars according to various writings of al-Ṣūfī

Number	Names of the stars according to al-Ṣūfī	Modern astronomical identification	Magnitude acc. to al-Ṣūfī	A	H	B	C	D
1	*al-simāk al-rāmiḥ*	α Bootis	1	1	1	5	19	17
2	*al-munīr min al-fakka*	α Coronae Borealis	2	2	2	6	20	22
3	*ra'ʾs al-jāthī*	α Herculis	3 –	3	3	7	–	–
4	*al-nasr al-wāqiʿ*	α Lyrae	1	4	4	8	24	24
5	*minqār al-dajāja*	β Cygni	3 –	5	5	9	26	–
6	*dhanab al-dajāja* (also: *al-ridf*)	α Cygni	2	6	6	10	29	27
7	*al-kaff al-khaḍīb* (also: *sanām al-nāqa*)	β Cassiopeiae	3	7	–	11	34	30
8	*janb barshāwush*	α Persei	2	8	–	12	5	–
9	*raʾs al-ghūl*	β Persei	2 –	9	7	13	4	3
10	*al-ʿayyūq*	α Aurigae	1	10	8	14	9	5
11	*raʾs al-ḥawwāʾ*	α Ophiuchi	3	11	9	15	23	23
12	*ʿunuq al-ḥayya*	α Serpentis	3	12	10	–	21	–

13	*al-nasr al-ṭā'ir*	α Aquilae	2 +	13	11	17	25	25
14	*dhanab al-dulfīn*	ε Delphini	4 +	14	12	-	27	26
15	*surrat al-faras* (also: *ra's al-musalsala*)	α Andromedae	2 –	15	13	18	32	-
16	*janāḥ al-faras*	γ Pegasi	2 –	16	14	19	33	-
17	*mankib al-faras*	β Pegasi	2 –	17	15	20	30	28
18	*matn al-faras*	α Pegasi	2 –	18	16	21	31	-
19	*fam al-faras*	ε Pegasi	3	19	17	-	28	-
20	*janb al-musalsala* (also: *baṭn al-ḥūt*)	β Andromedae	2 –	20	18	22	-	-
21	*rijl al-musalsala* (also: *al-'anāq*)	γ Andromedae	3	21	19	23	-	-
22	*ra's al-muthallath*	α Trianguli	3	22	20	24	2	-
23	*al-nāṭiḥ*	α Arietis	3 +	23	21	25	1	-
24	*al-dabarān* (also: *'ayn al-thawr*)	α Tauri	1	24	22	26	6	4
25	*muqaddam al-dhirā'ayn* (also: *ra's al-taw'am*)	α Geminorum	2	25	23	27	11	10
26	*qalb al-asad*	α Leonis	1	26	24	28	14	14
27	*ẓahr al-asad*	δ Leonis	2	27	25	29	15	-
28	*dhanab al-asad* (also: *al-ṣarfa*)	β Leonis	1	28	26	30	16	16
29	*al-simāk al-a'zal*	α Virginis	1 –	29	27	31	18	18
30	*qalb al-'aqrab*	α Scorpii	2	30	28	32	22	20

Number	Names of the stars according to al-Ṣūfī	Modern astronomical identification	Magnitude acc. to al-Ṣūfī	A	H	B	C	D	
31	*ʿurqūb al-rāmī*	β Sagittarii	4 –	31	–	–	40	21	S
32	*dhanab al-jady*	δ Capricorni	3	32	29	33	–	–	
33	*fam al-ḥūt al-janūbī*	α Piscis Austrini	1	33	–	–	41	29	S
34	*al-kaff al-jadhmāʾ*	α Ceti	3	34	30	–	3	–	
35	*dhanab qayṭus*	ι Ceti	3 –	35	31	35	35	–	
36	*mankib al-jawzāʾ* (also: *yad al-jawzāʾ*)	α Orionis	1 –	36	32	36	8	7	
37	*rijl al-jawzāʾ*	β Orionis	1	37	33	38	7	6	
38	*ākhir al-nahr*	θ Eridani	1	38	–	–	37	12	S
39	(*al-shiʿrā*) *al-yamāniya* (auch: [*al-shiʿrā*] *al-ʿabūr*)	α Canis Maioris	1	39	34	39	10	8	
40	*al-shiʿrā al-shaʾāmiya* (auch: *al-shiʿrā al-ghumayṣāʾ*)	α Canis Minoris	1	40	35	40	12	9	
41	*suhayl*	α Carinae	1	41	–	–	38	13	S
42	*ʿunuq al-shujāʿ* (also: *al-fard*)	α Hydrae	2	42	36	41	13	15	
43	*janāḥ al-ghurāb al-ayman*	γ Corvi	3	43	37	43	17	–	
44	*rijl qanṭūris*	α Centauri	1	44	–	44	39	19	S

45	(espinazo de la ossa mayor)	α Ursae Maioris	2	1		
46	(cabo de la cola de la ossa mayor, *alcayd*)	η Ursae Maioris	2	2		
47	(oio de la serpiente)	β Draconis	3	3		
48	(somo de la cabeça de la serpiente)	γ Draconis	3	4		
49	(pescueço de la culuebra)	β Serpentis	3	16		
50	(uentre de caytoz)	ζ Ceti	3	34		2
51	(la mediana de las tres que son en la cinta [?] de Vrion)	ε Orionis	2	37		
52	(en la oriolla [?] septentrional de la ... [?])	ε Crateris	4	42		
53	*al-shuʿba al-janūbīya min dhanab qayṭus*	β Ceti	3 +		36	
54	*al-nāṭiḥ*	β Arietis	3			1
55	*ra's al-nahr*	?				11

TABLE 2

Table of the coordinates given in D

(In variants, h marks the reading in the Hyderabad edition, f the reading in the Frankfurt facsimile edition by F. SEZGIN. For comparison, *mediatio* and declination in al-Farghānī's astrolabe star table of 856/7 are included.)

Number in D	Number in TAB. 1	Modern identification	*mediatio*	Declination	Number in al-Farghānī	*mediatio*	Declination
1	54	β Ari	Ari 18°0	18°0	–	– –	–
2	50	ζ Cet	Ari 18°0	– 14°0	–	– –	–
3	9	β Per	Tau *	h 36°0 f 37°0	2	Tau 0°36	36°6
4	24	α Tau	Tau 24°0	13°35	4	Tau 24°30	13°35
5	10	α Aur	Gem *	43°40	8	Gem 0°12	43°40
6	37	β Ori					
7	36	α Ori					
8	39	α CMa	Gem **	– 15°50	9	Gem 28°37	– 15°48
9	40	α CMi	Cnc h 13°0 f 12°0	7°15	10	Cnc 8°13	7°16
10	25	α Gem	Cnc 5°0	32° . .	–	– –	–
11	55	?	Gem 4°0	[–] 14°0	–	– –	–
12	38	θ Eri		– 43°40	25		– 43°38
13	41	α Car					

14	26	α Leo							
15	42	α Hya							
16	28	β Leo							
17	1	α Boo	Lib	–	25°30	13	Lib	21°40	25°31
18	29	α Vir	Lib	–	– 4°45	12	Lib	6°20	– 4°44
									var. – 5°44
19	44	α Cen			[–] 45°0	23			– 44°43
20	30	α Sco	Sco	22°0	[–] 22°30	15	Sco	22°15	– 22°54
									var. – 23°14
21	31	β Sgr	Cap	*	– 45°0	22	Sgr	28°35	– 46°38
22	2	α CrB							
23	11	α Oph							
24	4	α Lyr			38°0	17			38°12
25	13	α Aql							
26	14	ε Del	Cap	24°0		–	–	–	–
27	6	α Cyg							
28	17	β Peg							
29	33	α PsA	Aqr	27°0	– 36°0	21	Aqr	27°0	– 37°7
30	7	β Cas							

ble of Distances" from al-Farghānī is fully clear, it remains impossible for the moment to explain whence or how al-Ṣūfī obtained the peculiar latitude values in this table.[13]

Notes to TABLE 1:

Column A: numbers according to the sequence of the stars in *Kitāb ṣuwar al-kawākib* (where the 48 constellations are described in the sequence of the *Almagest*); 44 stars.

Column H: numbers according to HARTNER, *Principle*, p. 2542, footnote 1; 37 stars (with seven omissions).

Column B: numbers in *Libros del saber*; 44 stars (including eight new stars not mentioned in A as astrolabe stars). I used a photograph of the "rueda" of the astrolabe stars from the oldest and most authoritative "Complutense" manuscript, fol. 23ᵛ, in TALLGREN, *Nombres*, p. 647 (a few words in nos. 51 and 52 remained illegible); the "rueda" was also redrawn in ed. RICO Y SINOBAS, vol. I, opposite p. 142; the same stars appear, printed in form of a list, in the edition of the Italian version from 1341 by P. KNECHT, pp. 234–239. (Cf. the bibliographical references in 1986 [as in note 1, above], pp. 64–66.)

Column C: numbers according to the sequence in al-Ṣūfī's astrolabe treatise, 170-chapter version, facsimile edition Frankfurt 1986, pp. 118–129 (ch. 170); 41 stars (including one new star not mentioned in A as an astrolabe star).

[13] Of our text D a remarkably old manuscript (dated end of Shaʿbān 372 / February 983, i. e. three years before al-Ṣūfī's death) in 128 folios is being offered for auction at Christie's, London, on April 24, 1990 (lot no. 114). The script of the text apparently belongs to the same Muẓaffar ibn Hibatallāh al-Aṣṭurlābī who signed the diagram on fol. 110ʳ. The MS contains text D in 402 chapters, that is, in the version represented in the Frankfurt facsimile edition (our f). Through the kind help of Prof. D. King, Frankfurt, I was able to inspect xerox copies of four pages from the MS (fols. 109ᵛ–110ʳ, the end of chapter 360 and the diagram as in f, pp. 463 and 465; and fols. 124ᵛ–125ʳ, a section from chapter 402 corresponding to f, pp. 512,1–514,5). The text in the MS is in some places more complete than in the Istanbul MS reproduced in facsimile, but on the other hand contains a few grammatical mistakes. As far as can be judged from the specimens mentioned, nothing speaks against the authenticity of the Christie's manuscript.

Column D: numbers according to the sequence in al-Ṣūfī's astrolabe treatise, 386/402-chapter version, as in ed. Hyderabad, pp. 340–350 (ch. 386), and in the facsimile edition Frankfurt 1986, pp. 509–519 (ch. 402); 30 stars (including three new stars not mentioned in A as astrolabe stars; one of the latter, our no. 50, is also found among the new stars in B). The sequence of the stars in C and D is roughly according to increasing longitude, or rather *mediatio coeli*, as usual in astrolabe star tables.

Stars nos. 31, 33, 38, 41 and 44 are explicitly stated in A to be used on the "southern astrolabe" (marked "S" on the right side of the table).

Notes to TABLE 2:

This table includes some modifications to an older edition in P. Kunitzsch, *Typen von Sternverzeichnissen* . . ., Wiesbaden 1966, p. 7 f. (footnote).

Al-Farghānī's values are taken from the following four Mss.: London, Brit. Lib. or. 5479; Berlin, Ahlwardt 5790, 5791 and 5792. Variants are only mentioned where two of the four Mss. stand against the other two.

For the three deeply southern stars nos. 13, 19 and 29 D mentions the maximum altitude above the horizon at Rayy, α Car: 4°0; α Cen: 10°0; α PsA: 19°0. For the first star no declination is given. For the second and third, the altitudes are consistent with 35° for the latitude of Rayy (too small by about 1°) using al-Farghānī's declinations. For al-Ṣūfī's declinations they are inconsistent, yielding latitudes almost $1\frac{1}{2}$° apart.

* Text : *fī awwal*. . ., "Near the beginning of . . ."

** Text: *fī ākhir al-jawzā'*, "Near the end of Gem".

No. 11: *ra's al-nahr*, "the Beginning of the River". The star is described as "a bright star below the right foot of Orion [according to the *Almagest* this is κ Orionis; but following the *Almagest*, the beginning of Eridanus is near β Ori, the left foot; perhaps a corruption has crept in, *al-yusrā* "left" was changed to *al-yumnā* "right"] towards the south" (then follow the coordinates). The coordinates roughly point to the star α Leporis which, however, cannot be involved here. In the *Almagest* the "Beginning of the River" is at λ Eridani, a small star (mag. 4 or 4+) which is never reported for use on the astrolabe.

XIV

AN ARABIC CELESTIAL GLOBE FROM THE SCHMIDT COLLECTION, VIENNA

Fig. 15 – 19

For some years, R. Schmidt's private collection in Vienna has possessed a bronze Arabic celestial globe (diameter 139 mm according to the unpublished expert opinion of E. SAVAGE-SMITH). It is not dated, but bears the inscription "Made by (sana'ahu or san'at) Muhammad ibn Mu'aiyad al-'Urdhi".

In respect of wording, position (left of the Great Bear constellation) and form of the letters, this inscription is identical with that on the Arabic celestial globe in the Mathematics-Physics Salon in Dresden. This provides decisive evidence for identification. In the following account, a detailed comparison of the Vienna globe (referred to as V) with the Dresden example (D) will be undertaken, which may provide some clues to the historical origins of V.

The following facts about D are known (1). In 1562 (certainly before 1564), the Elector Augustus of Saxony bought the instrument from the Coburg mathematician Nikolaus Valerius (A. DRECHSLER); it must therefore already have been in Europe for some time. The diameter is 144 mm according to DRECHSLER, but 146 mm according to SAVAGE-SMITH. It is a Type A globe according to the latter, that is, it contains all the 48 Ptolemaic constellations (which were introduced to Arabic-Islamic astronomers through translations from Greek, and which had been basically in use by them since the ninth century), and most of the 1025 individual stars recorded by Ptolemy. The writing is done in careful kufic script. The manufacturer is named as Muhammad ibn (i. e. the son of) Mu'aiyad al-'Urdhi.

No date is given for the star positions; however, there are two clues that allow one to be suggested. Mu'aiyad (al-Din) al-'Urdhi was one of the astronomers who were active in the Maragha observatory (Iran); he died in 1266. The manufacturer of D, Muhammad, was his son who worked in his father's group and may have survived him by several decades. Secondly, one can attempt to fix the date of D from the positions of the stars. G. W. S. BEIGEL had already attempted to do this in 1805 and arrived at the year 1289 (2). Less satisfactorily, DRECHSLER (1922, p.7) on the basis of a mere assumption assigned a round date of 200 in the Djalali era, or 1279 in our calendar. Another investigator, DESTOMBES, carefully measured the positions of 40 stars, showing an increase in longitude of about 17 degrees 51 minutes compared with the time of Ptolemy, and deduced from that a date of 1304.

I feel unable to accept such a precise positioning of the stars near the ecliptic, but likewise observed an increase in longitude of just under 18 degrees as against Ptolemy. If the star list in the Ilkhani Tables of al-Tusi, the leading astronomer of the Maragha observatory who died in 1274, is taken as a basis, which for 1232 uses the longitude value of Ptolemy plus 16 degrees 45 minutes, and if one adopts the customary Arabic precession constant of 1 degree in 66 years, then with a longitude difference from Ptolemy of 17 degrees 45 minutes we arrive at the year 1298. In another calculation, using the same constant beginning from Ptolemy's epoch, 137 A. D., the result is 1308.5 A. D. Should the change in longitude be nearer to 18 degrees, the corresponding date becomes still later. Without attributing greater accuracy to the instrument than it warrants (the thickness of the silver pins which mark the stars and which vary in size according to star magnitude, can amount to 1 degree on the ecliptic scale), a date of "ca 1305" can be assigned to D. This date is also in accordance with the possible lifespan of the manufacturer, namely, Muhammad, the son of Mu'aiyad al-'Urdhi.

For the comparison of the globes D and V there exist excellent detailed photos of the complete surface areas (photos by R. Schmidt). From these one can anticipate one conclusion: that V is a copy of, or was made after, D. The manufacturer of V has adopted all the characteristic features of D - the form of the constellation figures, the nomenclature, the general appearance of the writing (the Arabic characters are undotted on both instruments); he has even literally copied the manufacturer's note on D. At the same time, however, V is not so perfectly made and lacks astronomical expertise, with various omissions and many mistakes.

While the stars on D have been meticulously marked with silver pins of different sizes, showing the intention of exact astronomical location, those on V are represented rather crudely by circles of the same size, undifferentiated in terms of star magnitude. The stars are distributed only approximately by eye across the constellations without regard to the astronomical data, and the distribution is incomplete - even stars of the first magnitude are often missing.

On D the ecliptic and the equator are proportionately divided, degree by degree. Six great circles pass through the poles of the ecliptic, cutting the latter perpendicularly so as to subdivide the twelve signs of the zodiac at every 30 degrees. As a result of the obliquity of the ecliptic to the equator, the sizes of the corresponding segments on the equator vary by about 2.5 degrees on either side of the 30 degrees mean. Just as the signs of the zodiac on the ecliptic are divided into thirty degrees each, the manufacturer of V has also mechanically divided each of the corresponding twelve segments on the equator into 30 parts; as a result, some of the graduation marks on the equator no longer correspond to 1 degree.

Further evidence that V was copied from - or was made after - D is given by a series of spelling mistakes in star names, which are transcribed unchanged from D (3):

20 (al-'awâ'idh): on D, there is an r in place of w; the same mistake appears on V.

45 (second word, al-halya): on D the y of the Arabic word is omitted; similarly on V.

61 ('anâq al-ardh): on D the letters r and dh in the second word are not written; similarly on V.

105 (mankib al-djauzâ): on D the last letter of the second word has slipped into the graduation marks of the ecliptic and into the edge of the figure of Orion and is only recognisable with difficulty. The manufacturer of V, to whom this nomenclature was foreign, therefore only copied "al-djauz" (4). The form of the z used in this is, incidentally, remarkably similar to the forms of r and z to be found on a Moorish astrolabe in the Technical Museum in Vienna (5).

118 (hadâri wa al-wazn): D has here a wrong and meaningless m in place of w (for wa, "and"); the same applies to V.

Furthermore, a whole series of names on D have not been reproduced on V, partly because poor spacing has left insufficient room for them, and partly out of carelessness, for example the names D 8, 9, 10, 19, 24, 25, 30, 33, 42, 48, 69, 86, 87, 97 (only first word) and so on.

The repetition of the mistakes of D on V, together with the generally poor quality of V compared with D and numerous new errors on V (6), make the direction of the borrowing quite clear: V must have been derived from D - the opposite is inconceivable.

So much can be deduced from the comparison, but if one then asks about the possible dating of V, problems arise. In the first place, we do not know whether V has been directly copied from the globe D, or whether perhaps further specimens of D, constituting a set, were produced from the workshop of the same manufacturer Muhammad ibn Mu'aiyad al-'Urdhi. There are examples where several globes from the same manufacturer have been preserved (see the catalogue of SAVAGE-SMITH, 1985), just as occasionally several instruments from astrolabe manufacturers have been preserved (MAYER, 1956). Anyway, we must work on the assumption that the specimens passed down to us today from Arabic-Islamic times represent only the tip of an iceberg and that formerly a great many more of these instruments existed. In addition it can be assumed that D was already copied in ancient times and that such an imitation formed the model for V.

Secondly, in regard to V itself, this globe offers no evidence of age from its graphical and pictorial design. The copying of the figures and inscriptions could have taken place at any time after D was made.

Closer delimitation of age of V may be achieved by considering historical data. Let us first assume that V is a direct copy of D. D orignated around 1305, and was bought from a previous owner in Coburg in 1562, by which time it had already been in Europe for a while. V, on the other hand, must have been manufactured by an oriental workman, as the general style of lettering shows. The possibility that V could have been copied by a European sometime after D appeared in the West can be ruled out. Just as unacceptable is the notion that someone at one time allowed globe D in Dresden to be copied by an oriental person. If we take D to have been the only source, we are forced to conclude that V, at sometime between the manufacture of D (about 1305) and its importation into the west (which was certainly before 1562), was manufactured in the east by an oriental artisan. (The resemblance, mentioned above, of a single z on V to the often similarly written r and z on Moorish astrolabes of about 1300 and later may well be pure chance and will be disregarded here).

If, however, imitations or copies of D provided the model for V, the matter becomes unclear and practically any time after 1305 becomes possible.

The stand on which V is mounted poses special problems. The horizontal ring normally corresponds to the horizon, but in this case, unusually, it comprises the ecliptic. The twelve signs of the Zodiac are plotted on it, each comprising 30 degrees with numbers added for every five degrees. It is equally strange that the zodiac is arranged anticlockwise (just as in the case of the celestial image - concave), whilst globe V itself (and of course D also) shows the heavens in accordance with a global image (convex), with the zodiac arranged clockwise. The purpose of this unusal horizontal ecliptic that, compared with the globe, runs in the opposite direction is not apparent. Nevertheless, the characters on the ecliptic show the same characteristics as those on globe V (compare, especially, the d and the angular, as opposed to round, form of the n).

Whereas on the one hand it is tempting to accept that the globe and its mounting did not originally belong to each other and were arbitrarily put together at a later date, after the original stand had been lost, there appear to be, on the other hand, indications of an original relationship between the writing on the globe and the stand. It is conceivable that the same manufacturer who so obviously lacked specialised knowledge in making globe V, in a similar amateurish way manufactured the stand for it.

Concerning the arrangement of constellations on Arabic-Islamic celestial globes, as has already been indicated above, the Moslem astronomers followed the Greek-Ptolemaic tradition quite closely. In the late eight and ninth centuries the specialised Greek literature on the subject, in which the most important was Ptolemy's Almagest with its star catalogue, was translated into Arabic. Moreover, it seems that specimens of (late) classical instruments, including celestial globes, still existed in the first centuries of the Islamic era. An important document in this context ist the ceiling fresko in a dome of the desert palace of Qusair 'Amra (near Amman, Jordan) dating from around 715, which, evidently based on the model of a Greek celestial globe, portrays the heavens with the most important great circles. Still in the twelfth century, the learned Ibn al-Salah cited a text with a description of a Greek celestial globe; this globe must have been constructed, according to the specific coordinate examples that it contains, for an epoch around 738 (7).

So it can be seen how it was possible in the tenth century for the Persian astronomer 'Abd al-Rahman al-Sufi (903-986) to compile his "Constellation Book" in which, with the help of the Almagest translated into Arabic, he describes in detail the Ptolemaic constellations and their stars with Ptolemaic coordinates; in this work each constellation is portrayed twice, once as seen on the celestial globe (convex) and once as seen in the sky (concave). Almost certainly there may have been a classical celestial globe among his sources. In any case the illustrations of the 48 constellations introduced by al-Sufi became the model for the whole subsequent development of Arabic-Islamic astronomy. All graphic representations of the constellations, primarily on celestial globes and therefore also on our D and V, stem from the tradition of al-Sufi.

In one respect, however, the Arabs were inconsistent and incorrect: in Ptolemy's Almagest the constellations were described from an interior (concave) viewpoint. Thus the centrally standing observer sees the human figures depicting the constellations from the front, and their corresponding body parts are correctly arranged left and right. In a global, exterior (convex) view, the figures would be shown to the observer from the rear, as is the case with globes of classic and pure western tradition (compare for example the globe of the Atlas Farnese, or the globe of 1502 in the Cluny Museum, Paris (8)). The Islamic celestial globes of al-Sufi and his followers, and so also D and V, on the contrary allow the human figures in the global (exterior) image to present their front view to the observer. In order to make clear the difference from the interior (concave) view, they appear simply in mirror image - in other words, their left and right sides are exchanged.

Orion provides a concise example: the bright star Alpha (Betelgeuse) stands according to Ptolemy on his right shoulder; in the interior (concave) image, viewed from the front, this star appears to the observer on the left, and in a correct exterior (global) image, in rear view, it would appear on the right. In the misrepresented global image of al-Sufi and on Islamic globes in general, the figure is shown to the observer also from the front but now the star Alpha stands on the left shoulder, i.e. on the right, as the observer would see it. Why the Arabs chose this mistaken solution for the global image is not known. Perhaps they followed an already existing trend in late classical times; perhaps for some reason they wanted to avoid an arrangement where the figures presented their rear view to the observer; or perhaps a false analogy was made with constellations that appear in profile and in which, in changing from interior to exterior view, there only seems to be a change of side (for example, in the case of Leo: in the interior view the head appears on the right, but in the exterior view on the left).

Ptolemy showed the stars in his star catalogue with ecliptical coordinates, longitude and latitude. This was quite practical since the latitudes could continue to remain unchanged; only the longitudes would need to be converted according to the precession for other periods of time. The Arabs and subsequently Europeans in the Middle Ages adopted this system. Celestial globes could accordingly be constructed in a relatively uncomplicated manner. Perhaps this is also the reason why our specimen V is only oriented by the ecliptic - it has holes only for the poles of the ecliptic and so can only rotate around that axis, which certainly precludes its usefulness for practical astronomical purposes. Moreover, V has only marks, not axial holes, for the celestial north and south poles. On the other hand, D was fully equipped in the specialist sense and can be turned around both celestial and ecliptic poles, so that it is fully prepared for astronomical use.

This also shows, as has already been noted several times, the amateurish design of V.

We do not know whether V, copied from D directly or indirectly, was ordered by a customer, nor what purpose this instrument would serve. If the customer was a real astronomer, he could hardly have been content with the work and might have declined to accept it. Equally possible, it could have been that an artisan, wanting to swim with the business tide, made the copy on his own initiative, without a special order from a customer and without specialist guidance, in order to offer it on the market and to earn good money.

For posterity, V with all its defects is still an interesting object, first because of its evident relationship to the famous globe D, and secondly on its own merits as a product of the meeting ground between scientific and everyday knowledge.

NOTES

(1) References for this part are: A. DRECHSLER, Der Arabische Himmelsglobus des Mohammad ben Muyíd el-'Ordhi vom Jahre 1279 im Mathematisch-physikalischen Salon zu Dresden, 2nd edition (Dresden 1922); L. A. MAYER, Islamic Astrolabists and Their Works (Geneva 1956), esp. 72 et sequ.; M. DESTOMBES, Globes célestes et catalogues d'étoiles orientaux du Moyen-Age. In: Actes du VIIIème Congrès International d'Histoire des Sciences (1956) (Florence 1957), 313-24 (esp. Globe no. 6); E. SAVAGE-SMITH, Islamicate celestial globes, their history, construction, and use (Washington 1985), esp. 220, no. 5.

(2) Cited by DRECHSLER and DESTOMBES: BEIGEL's "Nachricht . . . " has now

been reprinted in F. SEZGIN (ed.), Arabische Instrumente in orientalistischen Studien, vol. 1 (Frankfurt a.M. 1990), 1-14.

(3) The numbers used here refer to DRECHSLER (1922), Table VII, where the inscriptions on D are fully quoted (130 numbers listed). The Arabic characters in DRECHSLER were not photomechanically reproduced but were hand copied by him (cf. p. 8); they do not give quite accurately all the finer points of the original, and some mistakes have crept in. The author for his part always used the photos of the globes for his comparisons.

(4) DRECHSLER incidentally made the same mistake in his copying of the name.

(5) Cf. M. G. FIRNEIS, A Moorish Astrolabe from Granada. In: History of Oriental Astronomy, IAU Colloquium 91 (New Delhi 1985) (Cambridge University Press 1987), 227-32. The instrument came from Turkish booty, probably abandoned in 1683 just outside Vienna; it then came into the possession of the Jesuit College in Kalksburg and from there passed to the Technical Museum in Vienna in 1937. There appear to be connections between this (anonymous, undated) instrument and those of Ibn Basso of Granada (died 1316/17). The instruments of Ibn Basso (cf. MAYER (1956), Plate VII, for example in the names al-thaur and al-djauzâ) also show such forms of r and z (cf. the illustration of FIRNEIS (1987) p. 231).

(6) For example, in the name D 62 the last letter th was read as an r, so that a meaningless word is written; in D 2 the third word (al-sughra) is spoilt, giving a meaningless al-mtry; in D 39 the first word is corrupted from mmsk to an incorrect msk; and in D 105 a meaningless mnk is written in place of the first word mnkb. It is plain that the copyist of V not only did not know the traditional established nomenclature for the constellations, but also engraved meaningless combinations of letters without caring what he was doing.

(7) Cf. P. KUNITZSCH (ed.), Ibn as-Salah, Zur Kritik der Koordinatenüberlieferung im Sternkatalog des Almagest (Göttingen 1975), 18, 72 et seq.

(8) Cf. M. DESTOMBES, Un globe céleste inédit de l'année 1502. In: Actes du XIe Congrès International d'Histoire des Sciences, vol. 3 (Warsaw etc. 1968), 73-81; see the illustrations of Hercules, Ophiuchus and Virgo. The question of the orientation of the human constellation figures of Ptolemy is also discussed by O. J. TALLGREN, Survivance arabo-romane du Catalogue d'étoiles de Ptolémée. In: Studia Orientalia 2 (Helsinki 1928), 202-83, esp, 208 et seq., § 7-8; he arrived at similar results.

(Translated by C. Embleton)

Abb./Fig. 15: Muhammad ibn Mu'aiyad al-'Urdhi, arabischer Himmelsglobus, Ø 13,7 cm. Sammlung R. Schmidt, Wien.

Photo: R. Schmidt

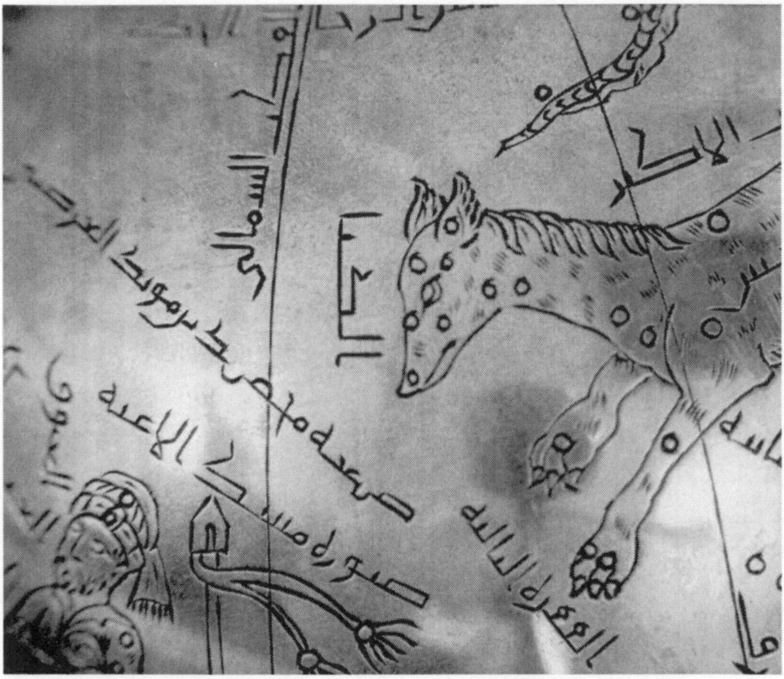

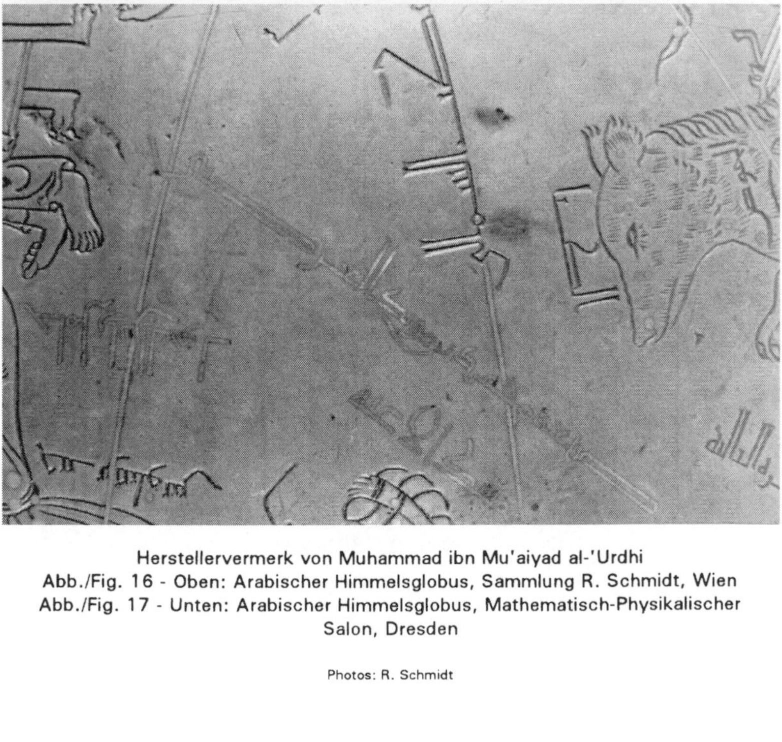

Herstellervermerk von Muhammad ibn Mu'aiyad al-'Urdhi
Abb./Fig. 16 - Oben: Arabischer Himmelsglobus, Sammlung R. Schmidt, Wien
Abb./Fig. 17 - Unten: Arabischer Himmelsglobus, Mathematisch-Physikalischer Salon, Dresden

Photos: R. Schmidt

Sternbild Orion
Abb./Fig. 18 - Oben:Arabischer Himmelsglobus, Sammlung R. Schmidt, Wien
Abb./Fig. 19 - Unten: Arabischer Himmelsglobus, Mathematisch-Physikalischer Salon, Dresden

Photos: R. Schmidt

XV

LES RELATIONS SCIENTIFIQUES ENTRE L'OCCIDENT ET LE MONDE ARABE A L'ÉPOQUE DE GERBERT

Les sciences dans l'Occident latin – nous ne parlerons pas de Byzance, bien que là-bas aussi aient eu lieu des évolutions similaires – subirent du Moyen Age jusqu'au XVIIe siècle une forte influence arabe. A côté des noms les plus connus de l'Antiquité nous rencontrons dans les écrits et traités du *quadrivium* les noms d'auteurs arabes ainsi qu'une terminologie scientifique elle aussi fréquemment d'origine arabe et dont une part importante subsiste de nos jours. Il s'agit d'un phénomène bien connu et nous savons que ce fonds arabe gagna l'Europe par « vagues » successives : traités de médecine à Salerne dans la deuxième moitié du XIe siècle ; un grand nombre de textes dans différents domaines scientifiques en Espagne au XIIe siècle ainsi que quelques autres en Sicile ; des éléments tardifs arrivèrent également au XIIIe siècle et, de manière plus disparate, au XIVe siècle. Les Etats latins créés en Orient par les croisades n'ont que très peu contribué à ce phénomène. Cependant la première vague d'assimilation du fonds arabe remonte à l'époque de Gerbert et eut également pour point de départ l'Espagne, plus précisément la Catalogne.

Le processus en tant que tel de la reprise par l'Europe de la science arabe est connu depuis longtemps et de nombreuses données concernant des auteurs, des textes, des traducteurs ou bien la tradition etc., sont devenues entre temps objets d'études, bien que nécessitant encore dans le détail de nombreux approfondissements. Du reste, une question fondamentale qui s'impose dans ce contexte n'a toujours pas trouvé de réponse décisive ou pouvant se résumer en une simple formule : quelle fut l'origine de cet intérêt que les Européens portèrent à la *doctrina Arabum*[1] ? Comment les Européens purent-ils concilier cet intérêt avec leur haine farouche des envahisseurs musulmans, de ces ennemis héréditaires de la foi chrétienne

---

Les notes ci-dessous sont abrégées : elles renvoient à une bibliographie placée à la suite du texte de la contribution.

1. Daniel de Morley, § 1 (p. 212).

Extrait de "*Gerbert l'Européen*", publié dans la collection "Mémoires, n° 3", par la société de la Haute-Auvergne en 1997.
Aurillac – France

qu'étaient les Sarrasins, comment purent-ils considérer ces ennemis comme leurs maîtres dans le domaine scientifique et en faire l'objet d'études ? Les textes eux-mêmes ne nous éclairent en rien sur ce sujet. Certes, de tout temps, les gens d'Eglise ont mis en garde contre le danger des sciences « païennes » et menacé tous ceux qui s'y intéresseraient – une attitude en tous points analogue pouvant du reste être observée du côté arabo-musulman[2] – mais ils ne purent cependant empêcher la pénétration de la *doctrina Arabum* dans toute l'Europe.

Une autre question se pose : comment « les Arabes » purent-ils faire dans le domaine scientifique des progrès, sources d'une telle renommée, d'un tel rayonnement ? A l'époque de Mahomet, au moment de la naissance de l'Islam, les Arabes étaient, dans la presqu'île arabique, en majeure partie de simples bédouins dont le mode de vie se limitait à affronter les problèmes quotidiens posés par le désert, restant largement étrangers aux développements culturels du nord et du nord-est, dans les empires byzantin et perse. Mais avec l'extension de la domination islamique après la mort de Mahomet (632), ils devinrent très rapidement les maîtres de vastes régions de l'empire byzantin ainsi que de l'empire perse et de zones avoisinantes. Après la consolidation de la nouvelle situation politique ils commencèrent bientôt à s'intéresser au patrimoine culturel des populations soumises. Ici aussi – comme plus tard en Occident – débuta une ample vague de traductions. Les principales œuvres de l'Antiquité tardive dans tous les domaines des sciences exactes furent alors traduites en arabe. Les sciences « arabes » apparurent. A leur pratique et à leur développement participèrent des membres de tous les peuples réunis dans le califat. Pour plusieurs siècles, l'arabe devint la langue commune à ces sciences. Les sciences de la culture islamique parvinrent donc aux Européens du Moyen Age en langue arabe et entrèrent dans la conscience collective comme « sciences arabes » tout simplement – terminologie qui est restée jusqu'à nos jours, même si elle n'est pas totalement correcte, ni sur le plan historique ni au regard de son contenu.

Après ces remarques préliminaires sur le panorama général des relations scientifiques arabo-européennes, nous nous intéresserons maintenant plus spécialement à l'époque de Gerbert, au moment des premiers contacts de l'Europe latine avec la science arabe qui, comme on le sait, s'effectuèrent dans le nord de l'Espagne, en Catalogne. A cet effet, nous évoquerons dans l'ordre les questions suivantes : quelle place tenaient alors les sciences arabes dans l'Espagne musulmane ? Qu'est-ce-que les Latins ont emprunté à ce fonds arabe ? Quelle part Gerbert prit-il dans ce processus ? Quelles furent les répercussions de ces premiers emprunts ?

*
* *

2. *Cf.* Goldziher.

Al-Andalus, l'Espagne musulmane, se trouvait très loin du centre culturel de l'empire islamique situé à l'est. Les premiers siècles qui suivirent la conquête de la péninsule ibérique (711) virent principalement la mise en place des structures de l'Etat musulman, son extension et sa défense, et finalement l'établissement de la dynastie des Omeyades. L'intérêt porté aux activités culturelles et plus spécialement scientifiques ne se manifesta que plus tard. Il est clair, de ce fait, qu'une science arabe dans l'Espagne musulmane n'est pas apparue d'elle-même, par une sorte de génération spontanée. Bien au contraire, la science hispano-arabe dépendit totalement des sciences de l'Orient arabe. Des œuvres scientifiques durent donc arriver d'Orient en *Al-Andalus* pour contribuer à la naissance et servir de base au développement ultérieur de la science hispano-arabe.

Des savants et historiens d'*Al-Andalus* comme Ṣāʿid al-Andalusī (vers 1068) et Ibn Ǧulǧul (vers 987) traitent dans leurs œuvres biographiques[3] du stade premier des sciences dans cet espace, notant les premiers rudiments dignes d'intérêt dans la seconde moitié du IXe siècle et parlant de leur évolution jusqu'à leur propre époque. Nous apprenons par eux les noms de savants, les domaines dans lesquels ils s'étaient spécialisés et leur connaissance de certains textes venus d'Orient. Beaucoup voyagèrent même en Orient pour compléter leurs connaissances et ils en rapportèrent des manuscrits importants. Y dominaient surtout les mathématiques, l'astrologie et l'astronomie, la médecine et la pharmacologie. Les principaux représentants du Xe siècle sont le mathématicien et astronome Maslama al-Maǧrīṭī (mort en 1007/08) et Abū l-Qāsim az-Zahrāwī (mort vers 1013), spécialisé en médecine. Maslama semble avoir suscité et influencé de manière indirecte les emprunts européens faits à l'époque de Gerbert, tandis que az-Zahrāwī ne fut « découvert » par les Latins qu'au XIIe siècle.

De tout ce que la science hispano-arabe avait à offrir dans la seconde moitié du Xe siècle, seules deux ou trois choses semblent avoir été connues des savants chrétiens dans les cloîtres de la région de Barcelone, en particulier celui de Santa Maria de Ripoll : l'astrolabe, quelques principes fondamentaux de l'astronomie et de l'astrologie arabes, de même que – peut-être en liaison avec l'abaque – les chiffres indo-arabes dans leur forme arabe-occidentale.

Maslama déploya une activité intense en mathématiques et en astronomie et forma un grand nombre d'élèves[4] qui furent à leur tour très productifs et donnèrent naissance à des œuvres dont certaines furent plus tard, au XIIe siècle, traduites en latin. Parmi les sujets abordés par Maslama se trouvait l'œuvre de Ptolémée, *Planisphaerium,* dont il disposa d'après la traduction arabe parue en Orient et pour laquelle il rédigea des remarques et des notes

3. Ṣāʿid al-Andalusī, p. 155 et suiv. = trad. Blachère, p. 120 et suiv. ; Ibn Ǧulǧul, p. 92 et suiv.

4. *Cf.* le résumé chez Samsó, p. 83 et suiv. (d'après Ṣāʿid al-Andalusī).

complémentaires[5], lesquelles furent traduites en latin au XIIe siècle également – en partie groupées avec le *Planisphaerium,* en partie séparément[6].

Dans ce contexte, il traita aussi des problèmes de l'astrolabe ; une table d'étoiles établie de sa main date de 978[7]. On peut penser que c'est bien le travail de Maslama et de son école sur l'astrolabe (dont le rayonnement se fit sentir jusqu'en Catalogne chrétienne) qui suscita l'intérêt des clercs latins de cette région pour cet instrument.

*

* *

Un groupe de manuscrits latins datant du XIe siècle comprend, sous une forme souvent étrangement disparate et éclatée, un *corpus* de textes contenant la description de l'astrolabe et des indications sur sa construction et son utilisation[8]. Ces textes foisonnent d'expressions scientifiques arabes[9]. Ils semblent avoir été partiellement traduits de textes arabes et conçus à partir d'astrolabes arabes, et partiellement rédigés librement, en s'appuyant sur des textes préexistants. Il a pu être démontré récemment qu'un paragraphe des *Sententie astrolabii* a été traduit du traité sur l'utilisation de l'astrolabe d'al-Ḫwārizmī (Bagdad, IXe siècle)[10], et un bref passage inséré dans le manuscrit parisien, B.N.F., lat. 7412, a pu être identifié comme étant une traduction de la version arabe du *Planisphaerium* de Ptolémée[11]. Ceci aussi corrobore l'hypothèse déjà exprimée précédemment d'un lien entre ces recherches latines sur l'astrolabe et les travaux de Maslama et son école[12]. Le centre des études latines sur l'astrolabe semble avoir été le cloître de Ripoll près de Barcelone. Lupitus de Barcelone – dont l'identification n'est toujours pas établie – aurait pu participer à ces travaux dans une mesure qui reste encore à préciser[13] ; il a dû participer à cette thématique comme traducteur de la langue arabe, car en 984 Gerbert lui demanda dans une lettre un *librum de astrologia translatum a te*[14]. Selon J.M. Millàs[15], le manuscrit

5. Voir à ce sujet Kunitzsch 1995/1996.

6. Les annotations et rajouts de Maslama sont en partie édités chez Vernet-Catalá ; le reste des annotations ainsi que l'ensemble des traductions latines chez Kunitzsch-Lorch.

7. Edité chez Vernet-Catalá, p. 45 ; Kunitzsch 1966, p. 17 et suiv. (« type IA »).

8. Voir les études et éditions chez Van de Vyver, Millàs, Bergmann.

9. Ces termes spécialisés sont édités et analysés chez Kunitzsch 1982.

10. Analysé et édité chez Kunitzsch 1987.

11. Edité et analysé chez Kunitzsch 1993.

12. Le *Planisphaerium* a été étudié de façon approfondie par Maslama qui y a ajouté des annotations et des compléments (voir plus haut). Il connaissait aussi les tables astronomiques d'al-Ḫwārizmī dont il fournit un remaniement (uniquement conservé dans la traduction latine d'Adélard de Bath ; édité par Suter, traduction anglaise de Neugebauer).

13. A propos de Lupitus, voir Lattin.

14. Fréquemment cité ; *cf.* Borst, p. 53.

15. Millàs, comme dans la note 8 et dans des études ultérieures.

225 de Ripoll aurait joué un rôle central dans tout cet ensemble ; il daterait du Xe siècle, aurait vu le jour à Ripoll et serait en quelque sorte le témoin principal de la date et du lieu de naissance du *corpus* sur l'astrolabe latin. Ceci fut mis en doute par la suite, en particulier parce qu'une date postérieure a été attribuée à ce manuscrit. Dans la dernière expertise, celle de Jean Vezin, on peut lire : « une écriture, peut-être catalane, de la première moitié ou du milieu du XIe siècle »[16]. Le débat est aujourd'hui dépassé depuis la découverte des « Fragments de Constance » par A. Borst – deux feuilles d'une compilation sur l'astrolabe composée essentiellement des textes qui apparaissent également dans le manuscrit de Ripoll, feuilles écrites vers ou peu après 1008 à Reichenau d'après un modèle apparu vers 995 à Fleury, vraisemblablement dans l'entourage de Constantin de Fleury[17]. Fleury a donc joué un rôle important pour la diffusion dans le nord de la France et le sud de l'Allemagne du *corpus* constitué à Ripoll[18]. L'existence de relations directes entre Fleury et Ripoll est attestée un peu plus tard par le fait que l'abbé Gauzlin fit venir vers 1020 un moine espagnol à Fleury, lequel apporta aussi des manuscrits d'Espagne[19]. La présence d'astrolabes arabes dans ces manuscrits est attestée – outre par certains détails des textes eux-mêmes – par le fait que plusieurs d'entre eux contiennent des dessins de parties d'astrolabes sur lesquels les copistes ont imité la graphie en caractères arabes. Le manuscrit parisien 7412 est particulièrement remarquable car il contient sur neuf pages la reproduction de tous les tympans ou disques d'un astrolabe arabe en caractères arabes, et même le nom de son créateur, Ḫalaf ibn al-Muᶜāḏ, fut copié avec le reste[20].

Un témoignage complémentaire des contacts arabo-occidentaux à l'époque de Gerbert est constitué par le texte astrologique *Mathematica Alhandrei summi astrologi* (manuscrit le plus ancien : Paris, B.N.F., lat. 17868 du Xe siècle) qui contient comme signes évidents de l'influence arabe les noms arabes des planètes, des signes zodiacaux et des mansions lunaires[21]. L'auteur, Alhandreus, également appelé Alchandrinus, Arcandam etc. dans des variantes, n'a toujours pas été identifié. Les mêmes éléments

---

16. Cité par Borst, p. 43, remarque 60.

17. *Cf.* Borst, p. 69.

18. Borst, p. 68.

19. Borst, p. 75.

20. Voir à ce sujet Kunitzsch, *Traces*. 19v, 20r, 23r et 23v sont reproduits chez Poulle 1964, planches I-IV ; 23v également chez Van de Vyver 1931, planche III. Autres exemples : ms Chartres 214 (XIIe siècle), 30r (tympan d'un astrolabe avec graduation du limbe ; reproduit chez Van de Vyver 1931, planche II) ; ms Londres, Brit. Lib., Old Royal 15 B IX (XIe-XIIe siècle), 71r (reproduit chez Bergmann, p. 233) ; ms Berne, Burgerbibliothek 196 (XIe siècle), 3r et 7v.

21. Les noms sont repris chez Millàs, p. 251, 256.

arabes apparaissent aussi dans le texte *De ratione sphere* copié dans les manuscrits du XIIe siècle[22].

Enfin : les chiffres indo-arabes. Comme on le sait, ils apparaissent pour la première fois en Occident sous leur forme arabe-occidentale en 976 (*codex Vigilanus*) et en 992 (*codex Emilianus*)[23] insérés dans les *Etymologiae* d'Isidore de Séville. En ce qui concerne leur emploi sur les pièces de l'abaque, ils sont cités textuellement tout d'abord par le Pseudo-Boèce et chez Bernelinus (après l'an 1000) comme une alternative aux caractères grecs de A à Θ utilisés jusqu'alors pour les chiffres. Ce n'est que chez Gerland de Besançon (2e moitié du XIe siècle) et Radolphe de Laon (1ère moitié du XIIe siècle) qu'ils remplacent totalement les caractères grecs[24]. A côté d'eux apparaissent depuis le Pseudo-Boèce, également dans les abaques, certains noms pour les chiffres de 1 à 9, parmi lesquels quelques-uns au moins laissent transparaître une origine orientale (un : berbère, quatre et huit : arabes, cinq : également arabe ou éventuellement berbère)[25].

Dans ces trois domaines, nous avons tracé les contours des éléments de la science arabe qui, à l'époque de Gerbert, dans la première période des contacts scientifiques arabo-occidentaux, gagnèrent l'Europe.

*
* *

Gerbert participa-t-il lui-même directement à ces emprunts ? Cette question trouve une réponse satisfaisante dans les propos de G. Beaujouan qui déclarait lors d'un colloque sur Gerbert à Bobbio en 1983 : « Dans le cas de Gerbert, il y a un déconcertant contraste entre sa réputation d'introducteur de la science arabe en Occident et la quasi-absence de traces d'influence arabe dans ses écrits indubitablement authentiques »[26].

Pour ce qui concerne l'abaque, les dernières recherches de W. Bergmann ont prouvé de façon détaillée que, par rapport à l'abaque du Moyen Age certaines innovations étaient déjà en partie présentes avant même la période des travaux de Gerbert, d'autres n'étant apparues que plus tard. Le texte de Gerbert sur l'abaque ne permet à aucun moment de dire quelles innovations étaient parvenues à sa connaissance, à supposer qu'il y en ait eu, et encore moins d'affirmer qu'il les aurait lui-même introduites en Occident. Pour les chiffres, là aussi l'évolution semble avoir commencé avant Gerbert :

---

22. Ces noms se trouvent chez Millàs, p. 262, 264 et suiv. (d'après ms Oxford, Bodleian Library, Digby 83 (XIIe siècle)). Sur ces textes, voir aussi Van de Vyver 1936.

23. Les deux *codices* se trouvent à la bibliothèque de l'Escurial. Reproductions chez Van der Waerden-Folkerts, p. 54, 55 ; *cf.* également Bergmann, p. 209.

24. Voir les détails chez Bergmann, p. 175 et suiv.

25. 1 : *igin* (berbère *yun* ou *igin*) ; 4 : *arbas* (arabe *arbaʿa*) ; 8 : *temenias* (arabe *ṯamāniya*) ; 5 : *quimas* (arabe *ḫamsa* ou berbère *səmmus*).

26. Beaujouan, p. 646.

on ne trouve dans ses écrits aucune indication allant dans le sens d'une innovation ; leur utilisation dans le cadre de l'abaque n'est visible qu'après Gerbert[27]. Pour l'astrolabe non plus il n'a pas participé à l'introduction ou à la reprise d'éléments arabes. Le texte *De utilitatibus astrolabii* (qui fut rédigé en complément aux textes les plus anciens sur l'astrolabe et dont il reprend la terminologie) pourrait, certes, avoir été écrit par Gerbert (l'éditeur Boubnov le rangea, il y a cent ans, parmi les œuvres ne pouvant être attribuées avec certitude à Gerbert ; récemment, au cours du débat qui se poursuit, A. Borst parla prudemment d'un « élève de Gerbert » comme étant son auteur). Mais l'absence de tout élément arabe dans les véritables écrits de Gerbert fait apparaître ici comme plutôt invraisemblable sa qualité d'auteur pour un tel traité[28]. Il est certes connu que Gerbert séjourna entre 967 à 970 à Ripoll, en Catalogne, pour ses études sur le *quadrivium*. Cependant, il est vraisemblable qu'à cette époque les études d'astrolabes arabes n'y avaient pas encore commencé (la table d'étoiles de Maslama date de 978 !). Gerbert n'est donc sans doute pas entré en contact avec la science arabe à Ripoll. Plus tard, il a manifestement essayé de s'informer sur elle (nous pensons à sa lettre de 984 à Lupitus), mais aucune répercussion ne peut en être décelée dans ses écrits authentiques.

*
* *

Jetons en conclusion un regard rapide sur les conséquences de ce premier emprunt européen d'éléments arabes vers la fin du Xe siècle. L'étude du contenu et de la terminologie des textes sur l'astrolabe est reprise dans *De utilitatibus astrolabii* dont l'auteur est donc peut-être un élève de Gerbert, puis chez Ascelinus d'Augsbourg[29], Hermann de Reichenau (1017-1054)[30],

---

27. M. Folkerts, Munich, m'indique cependant trois anciens manuscrits comprenant – en partie en relation avec Gerbert – également les chiffres et / ou leurs noms ainsi que le vers se rapportant à Gerbert « *Gerbertus Latio numeros abacique figuras* » : Paris, B.N.F, lat. 8663 (début du XIe siècle), f. 49v (reproduction de l'abaque à 27 colonnes ; les chiffres arabes sont joints ; la reproduction fait partie d'un ensemble de documents se rapportant aux différentes rédactions des *Regulae* de Gerbert, f. 47r – 49r ; *cf.* M.-Th. Vernet, p. 43) ; Vat. lat. 644 (cette partie : XIe siècle), f. 77v – 78r (reproduction de l'abaque à 27 colonnes, avec le vers, les chiffres et les noms) ; Berne, Burgerbibliothek 250 (Xe siècle), f. 1r (reproduction de l'abaque à 27 colonnes, avec le vers, mais sans les chiffres ni les noms ; la reproduction se trouve ici de manière isolée, suivie au f. 1v – 11r du *Calculus Victorii*). Un article du prof. Folkerts sur ces manuscrits sera publié bientôt dans un volume en l'honneur de Gino Arrighi.

28. Dans *Glossar* (1982), p. 479, j'avais également indiqué qu'aucune objection importante, ni sur le fonds ni sur la chronologie, ne permettait de mettre en doute la paternité de Gerbert. A la lumière des faits décrits ci-dessus, il va de soi que cette affirmation ne peut être maintenue sous cette forme.

29. *Cf.* Bergmann, p. 99 et suiv. (le texte de Ascelinus est édité *ibid.*, p. 223 et suiv.) ; Borst, p. 75 et suiv.

30. Le texte de Hermann a été dernièrement édité par Drecker.

et Fulbert de Chartres (mort en 1028)[31]. A. Borst parle d'astrolabes (le plus souvent apparemment arabes) présents ici et là dans des écoles monastiques[32]. Les noms arabes cités dans les textes d'Alhandreus sur l'astrologie réapparaissent chez Adémar de Chabannes (mort en 1034)[33] ; un manuscrit d'Adémar (avec des parties autographes) contient des caractères imités de l'arabe en calligraphie coufique[34]. Ces textes de la première moitié du XIe siècle confirment que le point de départ généralement admis aujourd'hui de ces premiers emprunts occidentaux d'éléments arabes se situe plus ou moins vers 980-1000.

La lourdeur du latin et le caractère étranger du sujet firent que toutes ces notions et connaissances disparurent à nouveau peu à peu de la conscience des savants. Ce n'est que la grande vague de traductions du XIIe siècle qui permit à l'astrolabe et aux autres connaissances arabes – ou du moins véhiculées par les Arabes – de connaître un succès définitif sur la base la plus large et pour plusieurs siècles.

*
* *

BIBLIOGRAPHIE

Beaujouan : G. Beaujouan, « Les apocryphes mathématiques de Gerbert », dans *Gerberto : scienza, storia e mito. Atti del Gerberti symposium (25-27 luglio 1983),* Bobbio, 1985, p. 645-658.

Bergmann : W. Bergmann, *Innovationen im Quadrivium des 10. und 11. Jahrhunderts,* Stuttgart, 1985 (Sudhoffs Archiv, Beihefte, 26).

Bischoff : B. Bischoff, *Anecdota novissima, Texte des vierten bis sechzehnten Jahrhunderts,* Stuttgart, 1984 (Quellen und Untersuchungen zur lateinischen Philologie des Mittelalters, 7).

Borst : A. Borst, *Astrolab und Klosterreform an der Jahrtausendwende,* Heidelberg, 1989 (Sitzungsberichte der Heidelberger Akademie der Wissenschaften, Phil.-hist. Kl., 1989, 1).

Daniel von Morley, *Philosophia,* éd. G. Maurach, dans *Mittellateinisches Jahrbuch,* 14 (1979), p. 204-255.

Drecker : J. Drecker, « Hermannus Contractus 'Über das Astrolab' », dans *Isis,* 16 (1931), p. 200-219.

31. Les extraits de Fulbert on été édités par McVaugh-Behrends.
32. Borst, p. 72 et suiv.
33. Son traité astrologique est édité chez Bischoff, p. 183-191.
34. Voir reproduction 182 chez Sievernich – Budde, p. 169.

Goldziher : I. Goldziher, *Stellung der alten islamischen Orthodoxie zu den antiken Wissenschaften,* Berlin, 1916 (Abhandlungen der Kgl. Preussischen Akademie der Wissenschaften, Phil.– hist. Kl., 1915, 5).

al-Ḫwārizmī : voir *Neugebauer* et *Suter*

Ibn Ǧulǧul : Ibn Ǧulǧul, Ṭabaqāt al-aṭibbāʾ wa-l-ḥukamāʾ, éd. F. Sayyid, Beyrouth, 1985.

Kunitzsch 1966 : P. Kunitzsch, *Typen von Sternverzeichnissen in astronomischen Handschriften des zehnten bis vierzehnten Jahrhunderts,* Wiesbaden, 1966.

Kunitzsch 1982 : P. Kunitzsch, *Glossar der arabischen Fachausdrücke in der mittelalterlichen europäischen Astrolabliteratur,* Göttingen, 1983 (Nachrichten der Akademie der Wissenschaften in Göttingen, Phil.-hist. Kl., 1982, 11).

Kunitzsch 1987 : P. Kunitzsch, « Al-Khwārizmī as a source for the 'Sententie astrolabii' », dans *From deferent to equant : A volume of studies... in honor of E.S. Kennedy,* New York, 1987 (*Annals* of the New York Academy of Sciences, vol. 500), p. 227-236.

Kunitzsch 1993 : P. Kunitzsch, « Fragments of Ptolemy's *Planisphaerium* in an early Latin translation », dans *Centaurus,* 36 (1993), p. 97-101.

Kunitzsch 1995-1996 : P. Kunitzsch, « The role of Al-Andalus in the transmission of Ptolemy's *Planisphaerieum* and *Almagest* », dans *Zeitschrift für Geschichte der Arabisch-Islamischen Wissenschaften,* 10 (1995-1996), p. 147-155.

Kunitzsch, *Traces* : P. Kunitzsch, « Traces of a tenth-century Spanish-Arabic astrolabe » (à paraître).

Lattin : H. P. Lattin, « Lupitus Barchinonensis », dans *Speculum,* 7 (1932), p. 58-64.

Mc Vaugh-Behrends : M. McVaugh – F. Behrends, « Fulbert of Chartres' notes on Arabic astronomy, dans *Manuscripta »,* 15 (1971), p. 172-177.

Millàs : J. M. Millàs Vallicrosa, *Assaig d'història de les idees físiques i matemàtiques a la Catalunya medieval,* Barcelone, 1931.

Neugebauer : O. Neugebauer, *The astronomical tables of Al-Khwārizmī,* Copenhague, 1962 (Kongel. Danske Videnskabernes Selskab, Hist.-filos. Skrifter, 4, 2).

Poulle 1964 : E. Poulle, « Le traité d'astrolabe de Raymond de Marseille », dans *Studi medievali,* 3e série, V, 2 (1964), p. 866-900.

Poulle 1985 : E. Poulle, « L'astronomie de Gerbert », dans *Gerberti symposium* (*cf.* ci-dessus, à Beaujouan), p. 597-617.

Ṣāᶜid al-Andalusī : Ṣāᶜid al-Andalusī, *Ṭabaqāt al-umam,* éd. Ḥ. Bū ᶜAlwān, Beyrouth, 1985 ; trad. française par R. Blachère, *Livre des catégories des nations,* Paris, 1935 (Publications de l'institut des hautes études marocaines, t. XXVIII).

Samsó : J. Samsó, *Las ciencias de los antiguos en al-Andalus,* Madrid, 1992.

Sievernich - Budde : G. Sievernich - H. Budde, *Europa und der Orient 800-1900,* Gütersloh - Munich, 1989.

Suter : H. Suter (éd.), *Die astronomischen Tafeln des Muḥammed ibn Mūsā Al-Khwārizmī...,* Copenhague, 1914 (Kongel. Danske Videnskabernes Selskab, Skrifter, 7. Raekke, Hist. og Filos. Afd. 3, 1).

Van de Vyver 1931 : A. van de Vyver, « Les premières traductions latines (Xᵉ-XIᵉ s.) de traités arabes sur l'astrolabe », dans *1ᵉʳ Congrès international de géographie historique,* t. II : *Mémoires,* Bruxelles, 1931, p. 266-290.

Van de Vyver 1936 : A. van de Vyver, « Les plus anciennes traductions latines médiévales (Xᵉ-XIᵉ siècles) de traités d'astronomie et d'astrologie », dans *Osiris,* 1 (1936), p. 658-691.

van der Waerden - Folkerts : B.L. van der Waerden - M. Folkerts, *Written numbers,* Milton Keynes, 1976 (The Open University. History of mathematics - counting, numerals and calculation, 3).

Vernet - Catalá : J. Vernet - M.A. Catalá, « Las obras matemáticas de Maslama de Madrid », dans *Al-Andalus,* 30 (1965), p. 15-45.

M.-Th. Vernet : M.-Th. Vernet, « Notes de dom André Wilmart † sur quelques manuscrits latins anciens de la Bibliothèque nationale de Paris (fin) », dans *Bulletin d'information de l'institut de recherche et d'histoire des textes,* 8 (1959), p. 7-45.

*
* *

RÉSUMÉ

Après quelques remarques préliminaires sur l'apparition des sciences dites « arabes » en Orient, nous traitons de l'introduction de ces sciences dans l'Espagne musulmane jusqu'au temps de Gerbert, puis de ce qui vint à la connaissance des savants européens. Le centre des premiers contacts avec la science « arabe » fut la Catalogne des années 980-1000. Parmi les sujets qui furent repris dans les manuscrits latins figurent l'astrolabe, les connaissances de base en astrologie et en astronomie ainsi que les chiffres arabes, ces derniers étant mis en relation avec le nouvel abaque selon des modalités obscures. Autant qu'on puisse en juger d'après les textes qui lui sont attribués avec certitude, Gerbert ne paraît pas avoir pris part à cette réception de la science arabe, mais il a manifestement tenté d'aborder ces nouvelles connaissances.

ABSTRACT

A survey is given of the development of « Arabic » science in the Muslim East and its introduction into Muslim Spain up to Gerbert's time. Then there is a description of what came to the knowledge of Latin scholars of these sciences. The first western European contacts with « Arabic » science developed in Catalonia around 980-1000 AD. The subjects treated were the astrolabe, the basic knowledge of Arabic astronomy and astrology and Arabic numerals ; the latter (after beginnings under unknown conditions) connected with the new, medieval, abacus. As can be judged from the authentic writings, Gerbert apparently was not himself involved in the process of translating Arabic material, but he took care to acquaint himself with the new scientific material.

ZUSAMMENFASSUNG

Nach einleitenden Bemerkungen über die Entstehung der « arabischen » Wissenschaften im Orient folgen Angaben über die Übernahme dieser Wissenschaften im muslimischen Spanien bis zu Gerberts Zeit. Sodann wird weiter besprochen, was davon zur Kenntnis europäischer Gelehrter kam. Zentrum der ersten westeuropäischen Kontakte mit « arabischer » Wissenschaft war Katalonien in der Zeit um 980-1000. Zu den Gegenständen, die übernommen und in lateinischen Texten behandelt werden, gehören das Astrolab und astronomisch-astrologische Grundkenntnisse sowie die arabischen Ziffern, letztere (nach ungewissen Anfängen) im Zusammenhang mit dem neuen Abakus. Gerbert scheint, soweit seine als echt geltenden Schriften erkennen lassen, nicht selbst an den Übernahmeprozessen beteiligt gewesen zu sein, aber er hat offenbar versucht, sich mit den neuen Kenntnissen vertraut zu machen.

XVI

TRACES OF A TENTH-CENTURY SPANISH-ARABIC ASTROLABE*

The astrolabe was handed down to the Arabs from late Antiquity. The earliest stages of its invention and development cannot yet be safely established. The oldest extant texts describing the planispheric astrolabe as we know it are by Philoponus (6th century, Alexandria), in Greek, and by Severus Sebokht (around AD 660, northern Syria), in Syriac[1]. Arabic texts on the astrolabe exist in greater number, from the ninth century on.

There has survived one Greek astrolabe, but it dates from a relatively late period, AD 1062[2].

The oldest extant dated Eastern Arabic astrolabe was made by Nasṭūlus and is dated 315 H = AD 927-28[3].

The oldest known surviving astrolabes from the Islamic West are by Muḥammad ibn al-Ṣaffār, one incomplete, made in Cordoba and dated

* Revised version of a paper read at the Sixth International Symposium on the History of Arabic Science, Ras Al Khaimah, UAR, December 16-20, 1996. The Symposium was jointly organized by the Institute for the History of Arabic Science, University of Aleppo, and the Documentaries and Studies Center, Ras Al Khaimah. The author wishes to express his gratitude for being invited to participate in the event and for the generous hospitality during his stay in Ras Al Khaimah.

---

1 For a short survey of the instrument's early history and the research literature cf. P. Kunitzsch, *Glossar der arabischen Fachausdrücke in der mittelalterlichen europäischen Astrolabliteratur*, Göttingen 1983, 461f.

2 Now in the Museo dell' Età Cristiana, Brescia. Cf. O.M. Dalton, The Byzantine Astrolabe at Brescia, *Proceedings of the British Academy* 12 (1926), 133-146.

3 Now in the Dar al-Athar al-Islamiyyah, Kuwait. See D.A. King, Early Islamic Astronomical Instruments in Kuwaiti Collections, in A. Fullerton and G. Fehérvári (eds.), *Kuwait, Arts and Architecture, A Collection of Essays*, Kuwait 1995, 76-83.

417 H = AD 1026-27[4], the other one complete, made in Toledo and dated 420 H = AD 1029[5].

The knowledge and practice of the astrolabe were then taken over from the Arabs by European scholars in the late tenth century in Catalonia, north-eastern Spain, and – more intensively – in the course of the great translation movement from Arabic into Latin in Spain in the twelfth century.

The oldest Latin texts on the astrolabe from the late tenth century show that those Christian scholars had at their disposal both Arabic texts and instruments and, as it seems, also the help of native speakers of Arabic. Portions of the *Sententie astrolabii*[6], one of those oldest texts, have been identified as being translated from a treatise on the use of the astrolabe by al-Khwārizmī (9th century, Baghdad)[7] and from the Arabic version of Ptolemy's *Planisphaerium*, the well-known text on the projection of the sphere onto a plane, which was generally understood – and mostly in the West – as the basic text for the construction of the planispheric astrolabe[8].

One of the manuscripts carrying scattered portions of the *Sententie astrolabii* and other related texts is MS Paris, BN lat. 7412. Fol. 1r-23v of the manuscript date from the middle of the 11th century and were written by two French hands, probably in Reims[9]. The last nine

[4] Now in the Royal Scottish Museum, Edinburgh. For these two instruments cf. the forthcoming catalogue of Islamic astrolabes by D.A. King (cf. his preliminary report in *Bulletin of the Scientific Instrument Society*, No. 31 (1991), 3-7).

[5] Kept in the Staatsbibliothek, Preussischer Kulturbesitz, Berlin. Cf. F. Woepcke, *Über ein ... arabisches Astrolabium*, Berlin 1858; repr. In F. Sezgin (ed.), *Arabische Instrumente in orientalistischen Studien*, vol. ii (Frankfurt/Main, 1991), 1-36.

[6] The *Sententie astrolabii* and several other related texts were edited by J.M. Millás Vallicrosa, *Assaig d'història de les idees físiques i matemàtiques a la Catalunya medieval*, Barcelona 1931.

[7] See P. Kunitzsch, Al-Khwārizmī as a Source for the Sententie astrolabii, *From Deferent to Equant: A Volume of Studies... in Honor of E.S. Kennedy*, New York 1987 (Annals of the New York Academy of Sciences, vol. 500), 227-236.

[8] See P. Kunitzsch, Fragments of Ptolemy's *Planisphaerium* in an Early Latin Translation, *Centaurus* 36 (1993), 97-101.

[9] I owe these indications to a detailed description of the manuscript kindly supplied to me by Prof. A. Borst, Konstanz, who is currently preparing a critical edition of Hermann the Lame's (d. 1054) scientific writings. A collation

pages of this section in the manuscript contain drawings of a complete astrolabe – the rete, three plates (with front and back) and the *mater* (with front and back). Most of the inscriptions of this astrolabe were copied by the Latin draftsman; but these inscriptions are *in Arabic*. He even copied (on the back of the mater) the maker's name, Khalaf ibn al-Muʿādh. No date is mentioned on the instrument, but from the general situation – date of the manuscript, state of transmission of the texts in it, the contents of the astrolabe itself – it can be concluded that the Arabic astrolabe here copied was older than the two oldest extant Andalusian astrolabes by Muḥammad ibn al-Ṣaffār and that it dates perhaps from a time around AD 1000 or earlier. For us, it is therefore the oldest specimen of a Spanish-Arabic astrolabe. It can be seen that the style of writing on the original instrument is the well-known Andalusian kufi found also on the surviving Spanish-Arabic astrolabes. Of course, the Latin draftsman who did not know Arabic himself and who simply copied the model as good as he could, committed many errors in the composition of the Arabic words. But what there is, fully suffices to recognize the contents of the inscriptions on Khalaf ibn al-Muʿādh's astrolabe. The maker's name, Khalaf ibn al-Muʿādh, is not mentioned by Ṣāʿid al-Andalusī in the lists of Spanish-Arabic astronomers in his *Ṭabaqāt al-umam* (AD 1068). But the elements of which the name is composed are well attested elsewhere in al-Andalus and therefore appear to be genuine.

The series of drawings begins on fol. 19v with the drawing of the rete. The rete has pointers for 27 stars. The Arabic names of the stars are added in more or less corrupted Latin transliteration. To each name is added a letter or another symbol, and at the bottom of the page variant spellings of each name are given according to these letters and symbols[10]. The same 27 stars were also assembled in a star table, with coordinates, in the text *De mensura astrolabii* (Inc. *Philosophi quorum sagaci studio...*) which belongs to the milieu of the *Sententie astrolabii*[11]. The signs of the zodiac are divided in steps of 6 degrees (cf.

of the contents of MS 7412 was also given by M. Destombes, *Archives Internationales d'Histoire des Sciences* 15 (1962), 41-43, and by E. Poulle (footnote 15), 900 (note 3).

[10] The names in both forms were edited by P. Kunitzsch, *Arabische Sternnamen in Europa*, Wiesbaden 1959, 90f.

[11] See the edition in Millás (footnote 6), 301f. It is the star table of "Type III" in P. Kunitzsch, *Typen von Sternverzeichnissen in astronomischen Handschriften des zehnten bis vierzehnten Jahrhunderts*, Wiesbaden 1966, 23ff.

also the almucantars), and to each sign is added its name in Arabic and in Latin.

Fol. 20r shows the plate for the 7th climate [Fig. 1]. In this astrolabe the plates were made for the latitudes of the seven climates and not for specific localities, which may be regarded as another sign pointing to the antiquity of the instrument. It shows the circles of Cancer, Aries-Libra and Capricorn, the almucantars (in steps of 6 degrees – the astrolabe therefore was of the *sudsī* type; cf. also the zodiac on the rete) and the curves of the seasonal hours. Inscribed, in Arabic, are the numbers of degrees of the almucantars (from 6° to 90°, in *abjad* notation), the names of East (*al-mashriq*) and West (*al-maghrib*), the name of the lower vertical line (*khaṭṭ al-zawāl*) and the numbers – in words – of the twelve (seasonal) hours, *al-ūlā* (the first) to *al-thāniya ʿashara* (sic; the twelfth). Below the center is written *al-iqlīm al-sābiʿ ʿarḍuhu mḥ lb, nahāruhu al-aṭwal yw sāʿa* ("the seventh climate, its latitude is 48°32', its longest day is 16 hours").

Fol. 20v shows the plate for the 6th climate. The circles, curves and lines are drawn as on fol. 20r, but most of the text is not repeated here. Below the center is written *al-iqlīm al-sādis ʿarḍuhu mh ʾ, nahāruhu al-aṭwal yh l* ("the sixth climate, its latitude is 45° 1', its longest day is 15,30 [hours]").

The plate for the 5th climate follows on fol. 21r. It is drawn as the two preceding plates, but with very few Arabic notations. Below the center we find *al-iqlīm al-khāmis ʿarḍuhu m yw, nahāruhu al-aṭwal yh* ("the fifth climate, its latitude is 40°16', its longest day is 15 [hours]").

Fol. 21v shows the plate for the 4th climate. The inscription reads *al-iqlīm al-rābiʿ ʿarḍuhu lw w, nahāruhu al-aṭwal yd l* ("the fourth climate, its latitude is 36°6', its longest day is 14,30 [hours]").

Next follows the plate for the 3rd climate, on fol. 22r. The inscription runs *al-iqlīm al-thālith ʿarḍuhu l kb, nahāruhu al-aṭwal* ... ("the third climate, its latitude is 30°22', its longest day is ...[12]").

Fol. 22v shows the plate for the 2nd climate. Below the center it has *al-iqlīm al-thānī ʿarḍuhu kj yw, sāʿāt nahārihi al-aṭwal yj l* ("the second climate, its latitude is 23°16', the hours of its longest day are 13,30").

The earliest form of this star table seems to appear in MS 7412, fol. 5v; cf. the remarks in Kunitzsch (footnote 1), 482f. (note 15[a]), on an inadequate edition by W. Bergmann.

[12] The number (14 hours) is not visible on the photocopy.

On fol. 23r there is the drawing of the mater with the limb (divided, anticlockwise, into 360°, marked in maghrebi *abjad* notation in steps of 5 degreees, from 5° to 360°). The interior of the mater carries the circles, curves and lines for the 1st climate. Below the center is written *al-iqlīm al-awwal ʿarḍuhu yw kz, nahāruhu al-aṭwal yj sāʿa* ("the first climate, its latitude is 16°27', its longest day is 13 hours").

In the drawing of each plate the number of the climate is also added in Roman numerals: VII, VI, V, IIII, III and I (II is not visible on the photocopy).

The values given here for the latitudes (of the middle of each) of the seven climates are found identically in a text from the milieu of the *Sententie astrolabii.* In this text the legends for each climate appear twice, in an upper line in Arabic (in Latin transliteration) and in the line below in Latin[13]. These values are basically identical with those given by Ptolemy in *Almagest* ii,13 (with three minor differences: in V Ptolemy has 40°56', in IV 36° and in II 23°51').

Finally, on fol. 23v we have the drawing of the back of the mater. In the outer rim it has the graduation of the four quadrants, with *abjad* notation (in the maghrebi style) in steps of 5 degrees, from 5° (at the horizontal line) to 90° (on top and at the bottom). Next follows the zodiac, with *abjad* notation for each sign, again in steps of 5 degrees, from 5° to 30° (written out only in Taurus and Gemini and left blanc in the other signs). The names of the signs are written down both in Arabic and Latin. The innermost system shows the calendar with the Julian month names both in Arabic and Latin (the Latin names are mostly abbreviated; the Arabic names are *ynyr*, *fbryr*, *mrs*, *ʾbryl*, *mʾyh*, *ywnyh*, *ywlyh*, *ʾghst*, *stnbr*, *ʾktwbr*, *nwnbr*, *djnbr* – almost fully dotted). The innermost ring indicates the months' days, adding *abjad* notation for each 5 days in a month. In the lower right quadrant there is a shadow square graduated from 1 to 12 in *abjad* notation along the horizontal and the vertical scales. Diagonally inside the shadow square, towards the center, runs the maker's signature, truly copied by our Latin draftsman: *ʿamal Khalaf ibn al-Muʿādh*, "The work of

[13] See the edition of the text in Millás (footnote 6), 290-292. It appears also in our MS 7412, fols. 8v (defective) and 10r (complete), In a derived text (*De utilitatibus astrolabii,* Inc. *Quicumque astronomice discere peritiam discipline* ...; usually ascribed to Gerbert or one of his disciples) the numbers of the latitude values were often corrupted; cf. the edition by N. Bubnov, *Gerberti postea Silvestri II papae Opera Mathematica,* Berlin 1899, 141f.; see also E. Honigmann, *Die sieben Klimata ...,* Heidelberg 1929, 189-191.

Khalaf ibn al-Muʿādh" [Fig. 2]. The throne is indicated at the bottom (the entire plate is drawn upside down on the page).

The drawings in MS 7412 had not remained unnoticed. In 1931, foll. 19v and 23v were reproduced by A. van de Vyver[14]. E. Poulle reproduced foll. 19v, 20r, 23r and 23v in 1964 and added a short comment[15]. In 1980 W. Bergmann called the drawings on foll. 20r-23v "Zeichnungen zur Astrolabkonstruktion" (drawings for constructing an astrolabe)[16]. But none of these authors realized the historical importance of the drawings as the oldest available representation of a Spanish-Arabic astrolabe nor did any of them recognize the maker's name on fol. 23v.

There are known several other Latin manuscripts of the same period containing drawings of astrolabe plates with poor imitations of *abjad* notation in Arabic and in Latin transliteration. But MS 7412 is unique in presenting the full set of plates of a complete Arabic astrolabe including the inscriptions in a nicely readable kufi script.

[14] A, van de Vyver, Les premières traductions latines (X^e^-XIe s.) de traités arabes sur l'astrolabe, *1er Congrès International de Géographie Historique*, t. ii: *Mémoires*, Bruxelles 1931, 266-290; see Pl. I and III.

[15] E. Poulle, Le traité de l'astrolabe de Raymond de Marseille, *Studi medievali*, 3^a^ Serie, V,2 (1964), 866-900; see Pl. I-IV. For the comment, see p. 900. Poulle rounded off the latitudes on the plates to full degrees.

[16] W. Bergmann, Der Traktat "De mensura astrolabii" des Hermann von Reichenau, *Francia* 8 (1980), 65-103; see p. 85f.

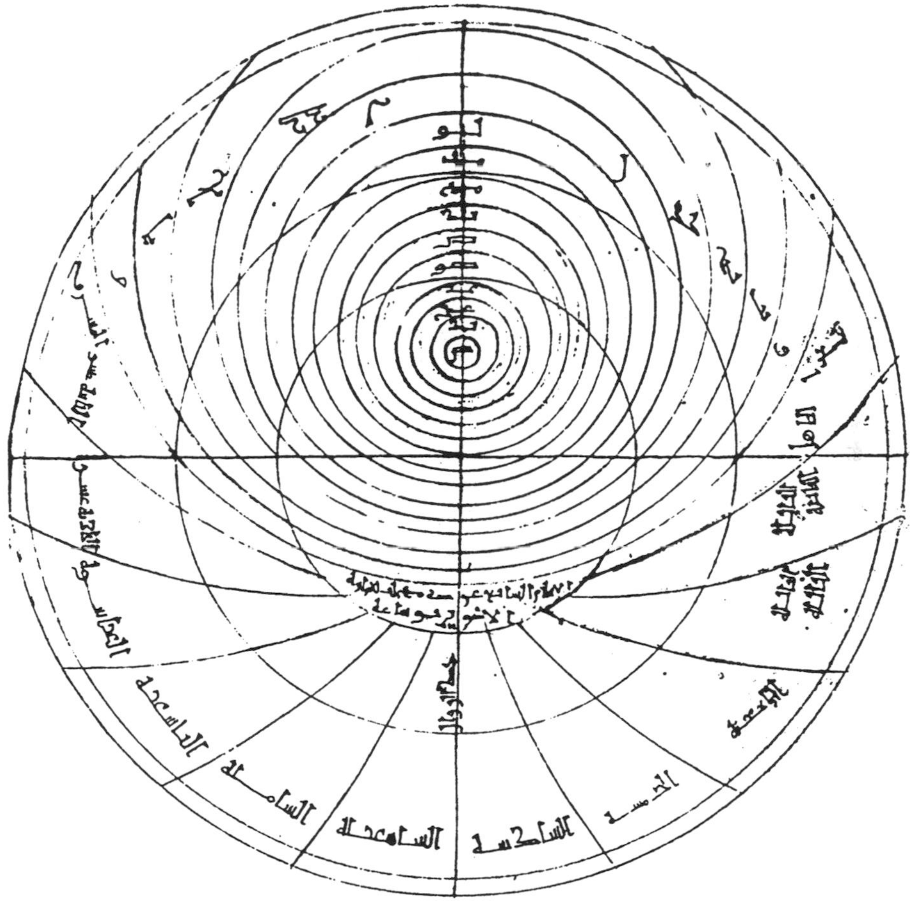

Figure 1: Plate for the seventh climate
MS Paris, BN lat. 7412, 20r

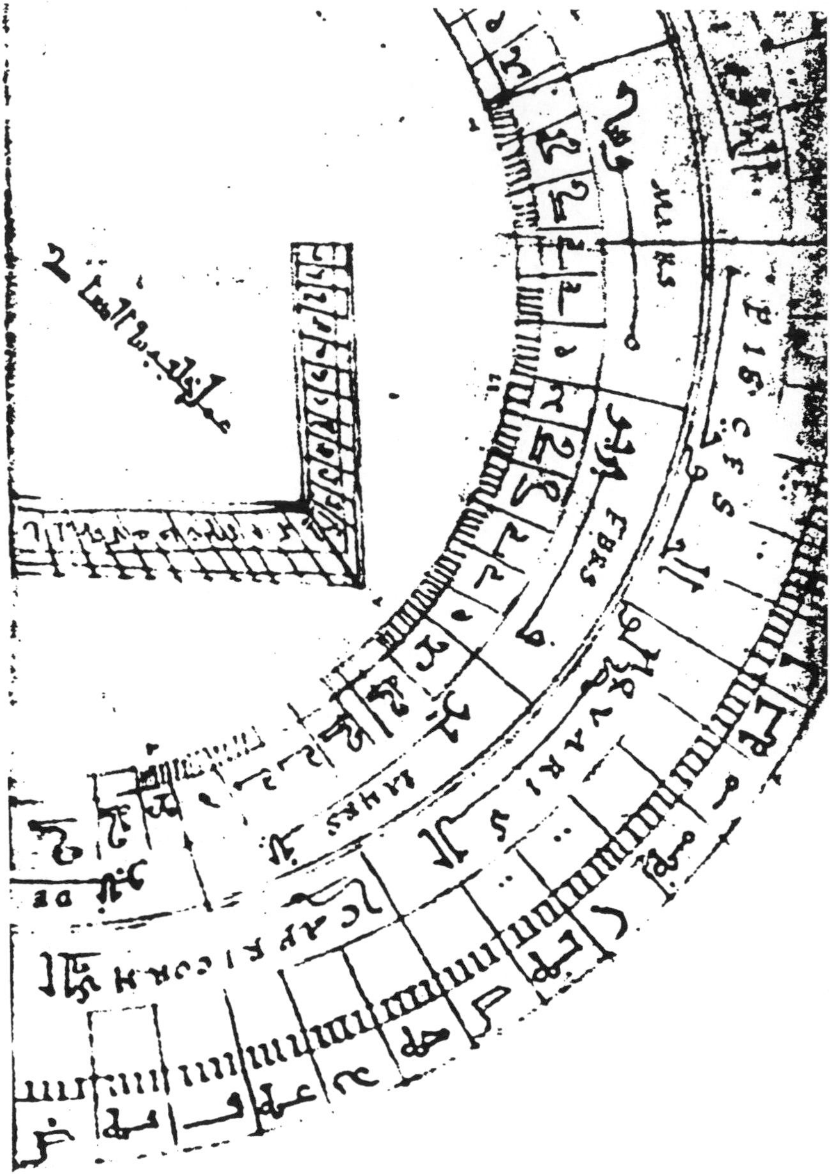

Figure 2: The back of the astrolabe.
The maker's name in the shadow square: *ʿamal Khalaf ibn al-Muʿādh*
MS Paris, BN lat. 7412, 23v

XVII

LA TABLE DES CLIMATS DANS LE CORPUS DES PLUS ANCIENS TEXTES LATINS SUR L'ASTROLABE

La connaissance de l'astrolabe est parvenue aux Européens du Moyen Age par les Arabes en deux périodes successives, la première dans les deux dernières décennies du dixième siècle, en Catalogne, et la deuxième dans l'âge d'or des traductions arabo-latines du douzième siècle, toujours en Espagne. Les textes de la première période étaient encore très lourds et ce sont les textes plus élaborés et plus précis du douzième siècle qui établirent finalement en Europe la connaissance de l'astrolabe et de son usage pour les siècles suivants.

Nous connaissons les textes de l'ancien corpus astrolabique du dixième siècle surtout depuis les études et éditions de Bubnov, Millás et autres.

L'astrolabe planisphérique se développa dans l'antiquité tardive. Ptolémée en livra le fondement théorique dans son traité sur la projection stéréographique, le *Planisphaerium*, qui fut traduit en arabe et qui était connu des astronomes arabes en Espagne musulmane au dixième siècle, comme par exemple de Maslama et de son école. Les plus anciennes descriptions de cet instrument, dans la forme connue depuis le Moyen Age chez les Arabes et en Europe, sont celles de Philopon (Alexandrie, VIe siècle)[1] et de Severus Sebokht (Syrie, vers 660, en syriaque)[2]. Ensuite les Arabes de l'Orient empruntèrent la connaissance de l'astrolabe, des textes sur sa construction et son usage furent élaborés et un grand nombre d'instruments furent construits. Les connaissances scientifiques des Arabes de l'Orient parvinrent à leurs compatriotes en Espagne musulmane, où, dans la deuxième moitié du dixième siècle, c'est surtout l'astronome et mathématicien Maslama (mort en 1007/08) et son école qui s'en occupèrent.

Il nous semble très probable que ce sont les activités de Maslama et

---

[1] Etude, réédition et traduction par SEGONDS 1981.
[2] Ed. par NAU 1899.

de son école qui suscitèrent dans les cercles érudits de la Catalogne cet intérêt pour un instrument pratique, qui pouvait servir à montrer les phénomènes célestes fondamentaux, à calculer le temps et à faire d'autres calculs.

D'après le texte le plus intéressant et plus complet de l'ancien corpus, les *Sententie astrolabii*, il est évident — surtout dans la section B de Millás[3], comportant une description de l'instrument, de ses parties et ses dessins avec les noms arabes[4] — que l'auteur l'écrivit avec un astrolabe arabe devant soi et à son côté un homme qui, sachant l'arabe et connaissant bien l'astrolabe, lui livrait l'information nécessaire, partie après partie. Mais ce n'est pas seulement un astrolabe qu'il tenait à la main, sans doute disposait-il aussi de textes arabes. J'ai pu établir, il y a quelques années, que la section C des *Sententie astrolabii*, sur l'usage de l'instrument, fut largement traduite du traité d'al-H̱wārizmī (Bagdad, IXe siècle)[5]. En outre, j'ai pu identifier quelques lignes de texte contenues dans le ms. Paris, BnF lat. 7412 — qui contient un mélange d'extraits de notre ancien corpus — comme étant une traduction latine de la version arabe du *Planisphaerium*[6] (lequel, comme nous le savons bien, était connu de Maslama, qui y a ajouté un certain nombre de commentaires[7]). Tout cela montre que ces premières études de l'astrolabe chez les Latins étaient fondées sur un choix de matériaux arabes, comprenant des astrolabes ainsi que différents textes.

Un phénomène étrange à observer concernant cet ancien corpus de textes astrolabiques est qu'ils sont transmis par plusieurs manuscrits dans une forme dispersée, qui varie d'un manuscrit à l'autre.

Notre manuscrit Avranches, BM 235 en est un bon exemple[8]. Nous n'entrerons pas ici plus avant dans cette question difficile. Nous nous occuperons plutôt de la table des sept climats, un de ces morceaux de texte insérés ici et là au cours des compilations successives du corpus ancien.

---

[3] MILLÁS, p. 276,22-279,110.

[4] Pour une étude de cette terminologie v. KUNITZSCH 1983.

[5] Voir KUNITZSCH 1987.

[6] Voir KUNITZSCH 1993.

[7] Voir VERNET - CATALÁ ; KUNITZSCH - LORCH.

[8] Voir la description détaillée du contenu de la section astrolabique chez BERGMANN, p. 228-231.

L'ancienne histoire de ces climats fut étudiée en détail par E. Honigmann dans son livre de 1929[9]. Les climats sont des bandes de la Terre s'étendant le long de certains parallèles de l'équateur. Peut-être est-ce Eratosthène qui les a utilisés le premier. Ptolémée donne dans *l'Almageste* II 6 une liste de plus de trente parallèles en mentionnant, pour chacun d'eux, la durée du jour le plus long et la distance par rapport à l'équateur, c'est-à-dire la latitude géographique. Les sept climats sont associés à sept parallèles toujours à la distance d'une demi-heure l'un de l'autre, allant d'une durée de jour de 13 à 16 heures. Les données astronomiques de ces sept parallèles — durée du jour le plus long et latitude — sont reprises par Ptolémée dans les titres des tables du livre 2, chapitre 13 de l'*Almageste*. Ces sept parallèles, avec leurs climats, sont alors un choix parmi les parallèles mentionnés dans le chapitre 6. Après Ptolémée, surtout chez les astronomes et géographes arabes et, à leur suite, au Moyen Age européen, ce système des sept climats servit comme moyen convenable pour la division du quart habité de la terre, de l'*oikoumenè*.

La table des sept climats dans l'ancien corpus de textes latins sur l'astrolabe est construite selon une forme stéréotypée. Pour chaque climat sont données sur la ligne supérieure les données astronomiques en arabe, translittérées en caractères latins, accompagnées sur la ligne inférieure d'une traduction latine. C'est toujours la même formule qui se répète pour chaque climat : [premier, etc.] climat, sa latitude est... [degrés] et... [minutes], son jour le plus long est... heures. (Dans la version arabe, il y a, pour la latitude, seulement les valeurs numériques, sans qu'on y ajoute les mots "degrés" et "minutes".)

De la transcription latine on peut reconstruire le texte de la formule arabe : al-iqlīm [al-awwal, etc.], ᶜarḍuhu ..., nahāruhu al-aṭwal ... sāᶜa. Dans la version latine, les valeurs numériques sont exprimées en chiffres romains. Pour l'arabe, la transcription latine donne, signe par signe, les noms des lettres de l'alphabet arabe désignant les nombres.

Il est connu que les Arabes ont utilisé dans leurs écrits scientifiques pour la notation des nombres soit des mots (écrivant par exemple "douze" ou "vingt-trois", etc.), soit les lettres de leur alphabet, chacune portant la valeur d'un nombre (comme, par exemple, *d* = 4, *y* = 10, *yd* = 14, *kb* = 22, etc.),

---

[9] HONIGMANN 1929.

c'est-à-dire le système appelé abǧad. Un système de notation similaire était utilisé par les Grecs.

Il va sans dire que cette forme de notation — de l'arabe ainsi que des chiffres romains — ouvrait le chemin à de nombreuses corruptions dans les valeurs numériques. Néanmoins il est encore possible de reconnaître nettement les valeurs des latitudes (pour les heures, il n'y a pas de grands problèmes, parce qu'elles se suivent dans une série régulière, de 13, 13 1/2, etc. jusqu'à 16).

Ce que nous trouvons pour les latitudes des sept climats dans notre table, ce sont exactement les valeurs données par Ptolémée dans l'*Almageste*, à l'exception des deuxième et quatrième climats et avec une erreur de transcription dans le cinquième climat.

Ainsi, la latitude du premier climat est de 16°27' (ou, en arabe translittéré, *ie uuau kef zein* = *yw kz*). Celle du deuxième est donnée comme étant de 23°16' (en arabe, *kef gim ie uuau* = *kǧ yw*), tandis que chez Ptolémée elle est de 23°51' (la confusion dans les minutes entre 50 et 10 s'explique facilement dans l'écriture arabe où — en composition avec les unités — les lettres désignant ces deux valeurs se distinguent seulement par un point, qui est souvent omis, avec pour conséquence qu'au lieu de *n* = 50, avec un point dessus, on lisait *y* = 10, sans point) ; pour le changement de 1 (51') chez Ptolémée à 6 (16') dans notre table, je ne peux pour le moment présenter aucune explication. Pour le troisième climat, nous avons 30°22' (en arabe, *lem kef be* = *l kb*) ; pour le quatrième, 36°6' (en arabe, *lem uuau et uuau* = *lw w*), où l'on relève une erreur de transmission : le *uuau* indiquant six minutes ne semble être qu'une répétition erronée de la même lettre dans les degrés. Dans le cinquième climat, nous avons 40°16' (en arabe, *mim ie uuau* = *m yw*), contre 40°56' chez Ptolémée (avec la confusion dans la lecture de l'arabe entre 50 et 10, déjà mentionnée). Le sixième donne 45°1' (en arabe, *mim he elif* = *mh'*), et le septième 48°32' (en arabe, *mim ha lem be* = *mḥ lb*), comme chez Ptolémée.

La table des climats fut éditée par Millás d'après trois manuscrits comme section E des *Sententie astrolabii*[10], faisant partie de quelques

[10] MILLÁS, p. 290-292.

chapitres associés aux *Sententie*. Elle se trouve aussi dans le ms. Bern, Burgerbibliothek 196, f. 8ᵛ.

Quelque temps après, la table des climats fut copiée et incluse par d'autres auteurs dans leurs écrits. Ainsi nous la trouvons comme chapitre 18 du *De utilitatibus astrolabii*, peut-être de Gerbert ou d'un de ses disciples[11]. Dans ces adaptations, le texte fut écrit comme un texte ordinaire, à longues lignes, et non plus comme un texte avec traduction interlinéaire. Dans les manuscrits de ces adaptations, on trouve ou la version complète (comme dans l'édition du *De utilitatibus* de Bubnov, ou dans les manuscrits Paris, BnF lat. 7412, f. 10, et Avranches, BM 235, f. 29ᵛ), ou une version abrégée dans laquelle la traduction latine du texte des climats IV-VI et le texte complet du septième climat sont omis (par exemple Paris, BnF lat. 7412, f. 8ᵛ ; München, Clm 14689, d'après Bubnov, p. 141 et Honigman, p. 190). Hermann de Reichenau, encore plus tard, parle des climats quasi en passant au chapitre 2 de son traité sur la construction de l'astrolabe ; pour les latitudes, il donne seulement des valeurs arrondies, en degrés et sans minutes[12].

Nous ne touchons pas ici au problème de la transmission latine et des variantes dans la transcription des nombres arabes et dans les chiffres romains des latitudes. Ce que nous voulions montrer est que cette table des climats présente pour les latitudes des sept climats exactement les valeurs de Ptolémée.

A la fin, il reste à répondre à une question : où l'auteur latin du dixième siècle pourrait-il avoir puisé les mesures exactes des climats et de leurs latitudes ?

On peut estimer qu'il est quasi impossible qu'il les ait puisées dans l'*Almageste* même. Bien qu'il soit reconnu que l'*Almageste* arriva jusqu'en Espagne musulmane et qu'il y fut étudié par Maslama[13], il serait surprenant qu'un moine catalan en ait reçu un exemplaire arabe dans son cloître et qu'il ait su le lire et le traduire en latin. Nous savons bien qu'il faudra attendre jusqu'au douzième siècle pour que des textes de cette difficulté et de ce

---

[11] Ed. par BUBNOV, p. 109-147 ; ch. 18 : p. 138-142.
[12] Voir l'édition de DRECKER, p. 205.
[13] Voir Ṣāᶜid al-Andalusī, p. 169 = trad. Blachère, p. 129.

volume puissent être traduits.

Non — l'explication est beaucoup plus facile, et une bonne fortune nous aide à répondre à cette question.

En fait, l'auteur a trouvé ces informations sur les sept climats dans un astrolabe arabe. Nous avons déjà exprimé notre conviction que d'autres parties des *Sententie astrolabii* furent écrites avec un astrolabe arabe devant les yeux. Le même raisonnement vaut pour les climats.

Alors que presque tous les astrolabes arabes conservés comprennent des tympans, ou disques, construits pour les latitudes de diverses villes, il semble que la première génération d'astrolabes arabes ait été équipée de tympans pour les sept climats.

Cela nous est clairement montré par le manuscrit latin, Paris, BnF 7412, qui contient une compilation de textes astrolabiques appartenant à l'ancien corpus. A la fin des textes (aux f. 19^{v}-23^{v}) suivent neuf pages présentant des dessins de l'araignée, du dos et des tympans pour les sept climats d'un astrolabe arabe andalou.

Le bon copiste latin n'a pas seulement dessiné tous les détails gravés sur ces pièces ; il a copié encore, aussi bien qu'il le pouvait, les inscriptions arabes en coufique andalou. Le manuscrit datant du milieu du onzième siècle, l'instrument copié peut bien dater d'environ l'an mille ou même remonter au temps de Maslama. Ce serait, dans ce cas, le plus ancien astrolabe andalou connu à ce jour[14]. Même le nom de son constructeur, Ḫalaf ibn al-Mu ᶜāḏ, fut retranscrit par le copiste latin[15].

Quant aux tympans des climats, nous y trouvons les mêmes valeurs de latitude que dans la table des climats figurant dans les manuscrits. Même les formules verbales pour chaque climat sont identiques à celles que nous pouvions reconstituer d'après les transcriptions des manuscrits latins : al-iqlīm al-awwal, ᶜarḍuhu yw kz, nahāruhu al-aṭwal yǧ sāᶜa, "le premier climat, sa latitude est de 16°27', son jour le plus long est de 13 heures", etc.

---

[14] Le plus ancien astrolabe andalou conservé fut construit par Muḥammad ibn al-Ṣaffār à Tolède en 1029. Voir MAYER, p. 75.

[15] Voir, sur cet astrolabe, KUNITZSCH 1996-98.

pour les autres climats[16].

On peut alors supposer que la table des climats apparaissant dans le corpus des plus anciens textes latins sur l'astrolabe fut directement lue sur les tympans d'un astrolabe andalou qui se trouvait dans les mains de l'auteur latin. Ce ne devait pas être nécessairement l'astrolabe même de Ḫalaf ibn al-Muᶜāḏ reproduit dans la série des dessins du manuscrit parisien 7412, c'était sans doute un instrument du même style et de l'ancien type, c'est-à-dire avec aussi des tympans pour les sept climats[17].

Les traités arabes sur l'astrolabe ne contiennent pas ordinairement une table des sept climats avec leurs latitudes ptoléméennes. Il semble que ce fut encore le cas dans le corpus astrolabique latin du dixième siècle. Une fois encore, cette table ne semble pas avoir une place fixe dans un des traités, que ce soit le *Sententie astrolabii* ou un des textes associés. Il semble plutôt qu'un des érudits latins, voyant que les tympans de l'astrolabe andalou qui lui servait de modèle étaient construits pour les sept climats, crut important d'établir une table de ces climats et d'indiquer les valeurs des latitudes pour quiconque voudrait construire un astrolabe dans l'Ouest. Cela

---

[16] Les tympans des septième et premier climats sont reproduits chez POULLE, pl. II et III. Dans le texte descriptif accompagnant les planches, les valeurs des latitudes sont arrondies au degré supérieur (49° au lieu de 48°32' pour le septième, et 17° au lieu de 16°27' pour le premier climat).

[17] La liste des sept climats se trouve aussi, avec les mêmes valeurs numériques que dans l'*Almageste*, dans les *Tables manuelles* de Ptolémée, d'où elle est passée dans la version latine de l'an 535, le *Preceptum canonis Ptolomei* (v. PINGREE, p. 358 ss ; pour les valeurs des latitudes, v. p. 359, notes 22 et 23). Le *Preceptum* est transmis par neuf manuscrits, dont un est notre Avranches 235, f. 1-26 (v. PINGREE, p. 368-9). Pour le premier climat, le *Preceptum* donne 15°15' (au lieu de 16°27' de l'*Almageste*, des *Tables manuelles*, de l'astrolabe andalou et de la table des climats latine discutée ici), une valeur dont HONIGMANN (p. 106) ne reconnaît pas l'origine. Dans la transmission de notre table des climats, on trouve aussi selon quelques manuscrits 15° pour le premier climat (au lieu de 16°) ; mais cela s'explique comme une simple faute de copie (XV pour XVI) et ne doit pas être pris comme un emprunt au *Preceptum*. Bien plutôt, notre table est purement dérivée, comme je l'ai démontré ici, de la tradition arabo- andalouse qui, à son tour, suit les valeurs de l'*Almageste*. Les deux traductions arabes de l'*Almageste* connues en Espagne musulmane présentent nettement la valeur de 16°27' (version al-Ḥağğāğ, ms. Leiden or. 680, f. 30, et version Isḥāq remaniée par Ṯābit, ms. Tunis 07116, non folioté). (Entre-temps, l'édition du *Preceptum canonis Ptolomei* par D. PINGREE est parue à Louvain-la-Neuve, 1997 [Corpus des astronomes byzantins, VIII]. Les climats y sont traités dans le ch. 1, p. 24-25, avec commentaire à la p. 123.)

expliquerait pourquoi cette table paraît comme un élément détaché, non constitutif d'un des textes mêmes de l'ancien corpus astrolabique latin.

En somme, la discussion de la table des climats, s'ajoutant à nos autres observations concernant les textes de l'ancien corpus astrolabique, révèle les efforts des érudits latins vers l'an mille autour de l'astrolabe. Utilisant des textes et des astrolabes arabes et aidés par des médiateurs sachant l'arabe, ils ont essayé d'établir, pour la première fois en Europe, une sorte de manuel précisant la construction et l'usage de cet instrument qui leur paraissait si séduisant. Mais leurs efforts ne menèrent pas à un succès durable. Ce succès fut réservé aux traductions et textes sur l'astrolabe du douzième siècle.

BIBLIOGRAPHIE

BERGMANN W., *Innovationen im Quadrivium des 10. und 11. Jahrhunderts* (Sudhoffs Archiv, Beiheft 11), Stuttgart, 1985.

BUBNOV N. (éd.), *Gerberti postea Silvestri II papae Opera mathematica*, Berlin, 1899.

DRECKER J., "Hermannus Contractus Ueber das Astrolab", *Isis* 16, 1931, p. 200-219.

HONIGMANN E., *Die sieben Klimata und die ΠΟΛΕΙΣ ΕΠΙΣΗΜΟΙ*, Heidelberg, 1929.

KUNITZSCH P. (1983), *Glossar der arabischen Fachausdrücke in der mittelalterlichen europäischen Astrolabliteratur*, Göttingen, 1983 (Nachrichten der Akademie der Wissenschaften in Göttingen, Phil.-hist. Kl., 1982, 11).

KUNITZSCH P. (1987), "Al-Khwārizmī as a Source for the *Sententie astrolabii*", dans *From Deferent to Equant : A Volume of Studies... in Honor of E. S. Kennedy*, New York, 1987 (Annals of the New York Academy of Sciences, vol. 500), p. 227-236.

KUNITZSCH P. (1993), "Fragments of Ptolemy's Planisphaerium in an Early Latin Translation", *Centaurus* 36, 1993, p. 97-101.

KUNITZSCH P. (1996/98), "Traces of a Tenth-Century Spanish-Arabic Astrolabe", *Zeitschrift für Geschichte der Arabisch-Islamischen Wissen-*

schaften 12, 1998, p. 113-120 (communication au Sixième Symposium International d'Histoire de la Science Arabe, Ras Al Khaimah, Déc. 1996).

KUNITZSCH P. - LORCH R., *Maslama's Notes on Ptolemy's Planisphaerium and Related Texts*, München, 1994 (Sitzungsberichte der Bayerischen Akademie der Wissenschaften, Phil.-hist. Kl., 1994, 2).

MAYER L.A., *Islamic Astrolabists and Their Works*, Genève, 1956.

NAU F., "Severus Sebokht, Le traité sur l'astrolabe plan", *Journal Asiatique* 1899, p. 58-86 et 238-273 (texte), p. 87-101 et 274-303 (traduction française).

MILLÁS VALLICROSA J.M., *Assaig d'història de les idees fisiques i mathemàtiques a la Catalunya medieval*, Barcelona, 1931.

PINGREE D., "The Preceptum Canonis Ptolomei", dans *Rencontres des cultures dans la philosophie médiévale*, Actes du Colloque international de Cassino 15-17 juin 1989, Louvain-la-Neuve - Cassino, 1990, p. 355-375.

POULLE E., "Le traité d'astrolabe de Raymond de Marseille", *Studi medievali* 5, 1964, p. 866-900 (avec Planches I-IV).

ṢĀ[c]ID AL-ANDALUSĪ, Ṭabaqāt al-umam, éd. Ḥ Bū.-[c]Alwān, Beyrouth, 1985 ; trad. française par R. Blachère, *Livre des catégories de nations* (Publications de l'Institut des Hautes Etudes Marocaines, t. XXVIII), Paris, 1935.

SEGONDS A.P. (éd.), Jean Philopon, "Traité de l'astrolabe" (Astrolabica, 2), Paris, 1981.

VERNET J. - CATALÁ M.A, "Las obras matematicas de Maslama de Madrid", *Al-Andalus* 30, 1965, p. 15-45.

XVIII

THE STARS ON THE RETE OF THE SO-CALLED "CAROLINGIAN ASTROLABE"

In 1962 M. Destombes[1] described an astrolabe in his possession[2] which he thought to be the oldest extant Latin astrolabe and which he dated to c. AD 980. He called it the "Carolingian Astrolabe". It is now housed by the Institut du Monde Arabe in Paris[3]. The authenticity and the date of the instrument have been, since then, the object of heavy debate and no unanimity could be reached among scholars on these fundamental questions[4]. If authentic, the astrolabe may belong to the same period (late tenth century) and area (North-Eastern Spain, the region of Barcelona) as the oldest Latin texts written on the construction and use of the astrolabe which were completely based on Arabic material[5]. Direct Arabic influence is also visible on the Carolingian Astrolabe where the numbers on the back and on the plates are indicated by Latin letters truly rendering the corresponding Arabic *abjad*-letters (cf. Plates II and III in Destombes 1962)[6].

The problem of the instrument's authenticity will not be re-

---

[1] Marcel Destombes, "Un astrolabe carolingien et l'origine de nos chiffres arabes". *Archives Internationales d'Histoire des Sciences* 15 (1962), 3-45.

[2] Lately, it has become known that the instrument was purchased at a price of 500,000 ancients francs (of 1961).

[3] Cf. J. Mouliérac, "La Collection de Marcel Destombes". *Astrolabica* 5 (1989), 78.

[4] The latest event in this connection was the Table Ronde held by the Société Internationale de l'Astrolabe in Paris on February 26, 1990, in which the problems around the Carolingian Astrolabe were discussed by MM. G. Beaujouan, E. Poulle, P. Kunitzsch and D. King under the presidency of A.J. Turner.

[5] I.e., the *Sententie astrolabii* and the annexed texts as edited by J.M. Millàs Vallicrosa, *Assaig d'història de les idees físiques i matemàtiques a la Catalunya medieval*. Barcelona, 1931.

[6] For these letters cf. P. Kunitzsch, "Letters in Geometrical Diagrams, Greek-Arabic-Latin". *Zeitschrift für Geschichte der Arabisch-Islamischen Wissenschaften* 7 (1991/92), 1 ff., esp. Table 1 (and text p. 3).

examined in this paper. Rather, an attempt will be made to identify the 20 stars marked on the instrument's rete by unnamed pointers (cf. Destombes, Plate I, and Figure 1, below)[7].

For this purpose let us assume that the rete did originate, as proposed by Destombes, around 980 in North-Eastern Spain. A reasonable way to identify the unnamed stars on the rete would then be to compare their positions with those found in astrolabe star tables or on astrolabes of the same period and area, i.e. Muslim Spain and Latin Europe of the late tenth and the eleventh centuries.

In that sense, the following material should be considered for comparison:

A. Star tables

1. Table of 21 astrolabe stars established by Maslama for the end of the year 367 H = A.D. 978, with coordinates longitude/ latitude and *mediatio*/ declination, in MS Paris, B.N. ar. 4821 (dat. 544 H = 1149-50), 81^{v}; partial ed. by Destombes 1962, Table I; further ed. by J. Vernet and M.A. Catalá, in: *Al-Andalus* 30 (1965), 45 (with annotations by M.A. Catalá, pp. 46 f.); P. Kunitzsch, *Typen von Sternverzeichnissen in astronomischen Handschriften des zehnten bis vierzehnten Jahrhunderts*, Wiesbaden 1966, pp. 17 f. (Type I A).

2. Table of 26 astrolabe stars, based on Arabic material, in MS Paris, B.N. lat. 7412 (11th c.), 5^{v}, with coordinates *mediatio* (here called *altitudo*)/ some other parameter (called *latitudo*)[8]; this table seems to be a variation of the widely spread table of "Type III" (see here, no. 3). There is a very defective ed., followed by consequently wrong conclusions, by W. Bergmann, in: *Francia* 8 (1980), 84-88. The mediations alone were edited by Destombes 1962, Table II.

3. Table of 27 astrolabe stars with coordinates *mediatio* (called

---

[7] The star tables drawn up by Destombes (1962: Tables I and II) are incomplete, with several copying and other errors, and somehow confusing.

[8] The numbers are different from the "marginal longitude" in "Type III" (cf. below, no. 3). Through calculation we found that the coordinate *latitudo* in 7412 would best correspond to the maximum altitude of the stars at a geographical latitude of 39° (Valencia?). A historical explanation of this cannot be given here.

latitudo)/"marginal longitude" (called *altitudo* - cf. above, no. 2)[9], established in connection with the texts centered around the *Sententie astrolabii*, later also adopted by such authors as Ascelinus and Hermann the Lame, and still copied in the 14th/15th centuries. A recent ed. (citing also older editions) is: Kunitzsch 1966, pp. 28-30 ("Type III").

B. Astrolabes and astrolabe drawings

4. Drawing of a rete carrying the 27 stars of "Type III", in MS Paris, B.N. lat. 7412, 19^{v}, forming part of a series of drawings copied from a Spanish-Arabic astrolabe made by one Khalaf ibn al-Muᶜādh (even this name was copied, in Arabic script, on the drawing of the back of the instrument, fol. 23^{v}); reproduced by A. van de Vyver, "Les premières traductions latines (X^{e}-XIe s.) de traités arabes sur l'astrolabe", in *1er Congrès International de Géographie Historique*, t. II: *Mémoires*, Bruxelles 1931, pp. 266-290, Planche I; and by E. Poulle, "Le traité d'astrolabe de Raymond de Marseille", in: *Studi medievali*, 3^{a} Serie, V,2 (1964), 866-900, Planche I. See below, Figure 2.

5. Drawing of a rete carrying the 27 stars of "Type III", in MS Vat. Reg. lat. 598 (11th c.), 120^{r}; reproduced by Millàs 1931, Làmina X. See below, Figure 3.

6. Rete of an anonymous Spanish-Arabic astrolabe (according to Prof. D. King "probably the earliest Andalusian astrolabe which has been preserved for us"); the rete is in the early Eastern Islamic style. Front reproduced by R.T. Gunther, *The Astrolabes of the World*, Oxford 1932, I, p. 244, Fig. 117. The rete carries all the 27 stars of "Type III" plus the star α Ser (= no. 13 of the Carolingian Astrolabe) which was not contained in Type III. See below, Figure 4.

7. Rete of an astrolabe by Muḥammad ibn al-Ṣaffār, Toledo, 420 H = A.D. 1029; reproduced by F. Woepcke, *Über ein...arabisches Astrolabium*, Berlin 1858, Tafel II. The rete carries the 27 stars of "Type III" plus α Hya (separate from Type III, no. 22: *aldiraan* / *muqaddam al-dhirāᶜayn*) and α Ser. See below, Fig. 5.

---

[9] On the second value, which could only be obtained by reading off from an astrolabe (and which was never found in any Arabic text), see J.D. North, *Richard of Wallingford*, Oxford, 1976, III, pp. 159-161 (Appendix 26); W. Bergmann, in: *Francia* 8 (1980), 70-75; *idem*, *Innovationen im Quadrivium des 10. und 11. Jahrhunderts*, Stuttgart 1985, pp. 44-53.

8. Rete of an astrolabe by Aḥmad ibn Muḥammad al-Naqqāsh, Saragossa, 472 H = A.D. 1079-80. The rete carries 24 stars (one pointer is missing, but the star name is there), among which 18 of the Carolingian stars (Carolingian nos. 2 and 8 are not inscribed on this instrument). See the recent description, with photos, by D.A. King, in: *Focus Behaim Globus* (Catalogue of an exhibition at Germanisches Nationalmuseum, Nuremberg, Dec. 1992- Feb. 1993), vol. II, pp. 568-570 (no. 1.70).

In addition, there could be compared the 11th-century astrolabes of other Spanish-Arabic astrolabe makers such as Ibrāhīm ibn Saᶜīd al-Sahlī and Muḥammad ibn Saᶜīd al-Sahlī. Exactly the 27 stars of Type III still appear on the astrolabe of 605 H = A.D. 1208-09 by Abū Bakr ibn Yūsuf of Marrakush[10].

In a first step we compared the positions of the stars on the Carolingian Astrolabe with star positions on other astrolabes or drawings (sources no. 4-8, above). We arrived at the following list for the twenty stars of the Carolingian Astrolabe (they are all unnamed; in the list below, we add, in brackets, the corresponding names from the star table of Type III). The sequence is according to mediation.

1. zeta Ceti (pantangaitot)
2. beta Persei (algol)
3. alpha Tauri (aldevaran)
4. beta Orionis (rigel)
5. alpha Canis Maioris (alhabor)
6. alpha Canis Minoris (algoize)
7. alpha Leonis (calbalaze)
8. gamma Corvi (algurab)
9. alpha Virginis (alcimec)
10. alpha Bootis (alramech)
11. alpha Coronae Borealis (alfecat)
12. alpha Scorpii (calbagrab)
13. alpha Serpentis (not in Type III; Arabic *ᶜunuq al-ḥayya*, "the Serpent's Neck")
14. alpha Ophiuchi (alhawi)

---

[10] See F. Sarrus, *Description d'un astrolabe, construit à Maroc en l'an 1208*, Strasbourg 1850, Planche 4 (reproduction of the rete).

15. alpha Lyrae (wega)
16. alpha Aquilae (altair)
17. delta Capricorni (denebalix)
18. epsilon Delphini (delfin)
19. alpha Cygni (alrif)
20. beta Pegasi (alferat)

All these stars belong to the set of astrolabe stars commonly used in Arabic and Latin sources, texts and instruments, of that time. Star no. 13 is not mentioned or used in the Latin sources, but it occurs on several Andalusian astrolabes of the eleventh century and later.

It seems that several of the star pointers on the Carolingian Astrolabe in its present state are bent away from their original positions which accounts for major deviations in some positions. The equatorial bar, in the lower part of the rete, is not symmetrical (with regard to the vertical axis), its left side being extended too far to the left; consequently the position of alpha Tauri (no. 3), near the end of the bar, is also too far to the left and falls under a mediation more than ten degrees too small for the possible epoch of 980.

Whereas at the Table Ronde of 1990 (cf. note 4) Kunitzsch was of the opinion that the Carolingian instrument was fabricated in the Arabic part of Spain and was brought to the Christian part in the North as a "blank" instrument, i.e., without any inscriptions, with the purpose that Latin inscriptions be added there, he now tends to assume that the instrument was made in the Christian North in imitation of an Arabic model astrolabe. The art of making such instruments was still unknown here, and it seems that a Latin scholar or craftsman, unexperienced in the theory and praxis of the astrolabe, for the first time tried to make an astrolabe himself.

The second step in our research was to measure and / or calculate the positions of the assumed stars on the Carolingian Astrolabe and in the sources mentioned above and to compare them with "theoretical" values of mediation and declination calculated for the epoch A.D. 980. This epoch marks the time when Christian scholars in North Eastern Spain appear to have had first contact with Arabic astronomy, and especially the astrolabe, and when they started to compose their first writings on the instrument (*Sententie astrolabii*, etc.). Also from roughly this epoch (exactly: 978; see above, source no. 1) dates the table of astrolabe stars by Maslama. Further, it is the epoch for which Destombes thought the Carolingian Astrolabe might have been constructed. The values obtained through measurement and calculation confirm the identifications found by the first rough comparison of the star positions on the Carolingian instrument and in the other sources.

The results of the measurements and calculations of star positions for three star lists, two drawings and two astrolabes are shown in two tables below. The numbers between brackets refer to the list of sources described in the introduction.

The procedures by which the data for the various sources have been obtained depend on the available material. The above mentioned "theoretical" data were calculated from the Ptolemaic longitude and latitude by correcting them for precession by adding 12° 40' (the value used by Maslama in his star list (1)) to the Ptolemaic longitude[11]. In the calculation modern mathematical relations were employed: first the longitude and the latitude were converted into right ascension and declination, and then the mediation was obtained from the right ascension, using an inclination angle of 23°.5.

The data of the stars in the *star lists* stem directly from their sources, except the declinations of the Type III stars (3), which had to be derived indirectly, because declinations are not listed in this star table. Because declinations cannot be used directly in making an astrolabe, other types of coordinates have been introduced in some star lists. The values listed under the heading "altitudo" in Type III (3) stem from such an alternative coordinate. Declinations were derived from them by assuming that this "altitudo" represents the "chord angle" defined by two radial lines on the astrolabe: one going through the star concerned and the other through the point of intersection of a particular "chord" through that star with the outer rim of the astrolabe[12]. The particular "chord" concerned is perpendicular to the radius of the star. In mathematical terms the

[11] The Ptolemaic data were taken from P. Kunitzsch (ed.), *Der Sternkatalog des Almagest..., III: Gesamtkonkordanz der Sternkoordinaten*, Wiesbaden, 1991. That Maslama himself knew the *Almagest* is stated by Ṣāᶜid al-Andalusī, *Ṭabaqāt al-umam*, ed. H. Bū ᶜAlwān, Beirut 1985, p. 169, 2-3 (French transl. by R. Blachère, *Livre des catégories des nations*, Paris 1935, p. 129).

[12] Cf. above, note 9.

relation between this "chord angle" and the declination (Dec) of a star is given by

$$\cos(\text{"chord angle"}) = \tan((90^\circ - \text{Dec}) / 2) / \tan((90^\circ + 23^\circ.5) / 2).$$

The preferable procedure for obtaining mediations and declinations of the stars depicted in the *drawings* and on the *astrolabes* included in this comparison would have been the direct measurement of the mediations, using the division of the zodiac, and of the declinations, using the declination scale sometimes encountered on instruments. However, since the declination scale is often lacking, one can only guess which of the several methods known for plotting the star positions was employed in constructing the instrument. Under such circumstances only a "modern" method allows one to determine the declinations of the stars. One such "modern" method is based on measuring the star positions, depicted in a drawing or on the rete of an astrolabe, with respect to a Cartesian coordinate system[13]. By using an *X-Y* coordinate system, oriented such that the origin coincides with the north pole, the positive *X*-axis with the radius to the first point of Aries and the positive *Y*-axis with the radius to the first point of Cancer, one can quantify the positions of the stars on a drawing or the rete of an astrolabe in terms of an x and a y value, its coordinates with respect to the *X-Y* coordinate system. The measured data can subsequently be converted into mediations and declinations by using the usual mathematical relations for coordinate transformations. Only under certain conditions do the data obtained in this way represent a precise "historical record": the more accurate the division of the scales employed in making the astrolabe, the closer our data will be to the true "historical" record[14]. This does not imply that the *X-Y* method for determining the mediations and declinations of the stars is practically useless. A comparison between the mediations of the stars on the Carolingian Astrolabe obtained by this method and by direct measurement, respectively, shows that even in this case of an astrolabe with a very unevenly divided zodiac, the differences in most cases amount to only one

---

[13] A photocopy of the rete of the Carolingian Astrolabe was kindly provided by Professor Gerard Turner. For the other astrolabes photos and copies thereof were used.

[14] In a recent study of an astrolabe of the sixteenth century the *X-Y*-method turned out to be entirely reliable. See G. Turner and E. Dekker, "An Astrolabe attributed to Gerard Mercator, c. 1570", in *Annals of Science* 50 (1993), 403-443.

or two degrees. Such differences should be kept in mind, but they are small compared to other errors involved in our analysis and to be discussed below. Therefore, with respect to the analysis of such "other" errors the *X-Y* method may be considered adequate for our purpose.

Once the mediation and declination of the stars had been determined for the various lists, drawings and astrolabes, we compared them with the corresponding "theoretical" value Ptolemy + 12° 40' and calculated the various differences in mediation (DMed) and in declination (DDec) by subtracting the former from the latter (Pt - source). The results of these measurements and calculations have been summarized in our two tables, one for the mediations (Table I) and the other for the declinations (Table II).

In both tables the first four columns give the reference numbers of the stars for the Carolingian Astrolabe and the three star lists included. In the next two columns the astronomical identification and the modern designation of the stars is recorded by the number of each star according to Baily and Peters-Knobel and its modern name, respectively. In the next two columns of Table I the mediation of each star as measured on the Carolingian Astrolabe and the corresponding value after Ptolemy + 12° 40' are listed. In the last eight columns the differences with respect to Ptolemy + 12° 40' as derived for all lists, drawings and astrolabes are given[15]. In Table II the corresponding data are listed for the declinations.

The lower two lines of the tables list the mean values of the differences and their standard deviation. We may interpret these mean values and standard deviations statistically when a normal errors pattern can be ascertained. Error patterns in the statistical sense are caused by some random process, like plotting data, now a little off in one direction, then in another direction. However, the "human" errors which concern us most in the present analysis are of a different kind. In order to elucidate this we shall discuss here a few of such cases encountered in the present analysis.

The most common errors in star lists are the result of copying and manipulating data. Some such errors, like a change of LX into XL of the "altitudo" of α Tau and α Leo in Type III (3), stand out by causing deviations of 20° or more and would have completely disturbed the error

---

[15] As a result of rounding off, a difference between 8 and 8 may result in 1 (Table I, no. 9).

pattern if we had not corrected them (compare Table II).

When studying an instrument completely different kinds of errors are encountered in addition to copying errors, as is illustrated by the astrolabes included in our study. The majority of the deviations in mediation, derived for the astrolabe dated 1029, are negative. This may indicate a later epoch. However, the majority of the deviations in declination are positive and not significantly smaller. Here obviously an attempt has been made to correct for precession by bending the pointers of the astrolabe and in this process the pattern of errors is totally disturbed. Clearly, without a more detailed analysis of the error pattern of both coordinates of an astrolabe, it is delusive to interpret negative mean values as evidence for a different epoch. Damage to the pointers of an astrolabe also causes large deviations. The large values for the errors in mediation and declination of α CrB of the astrolabe dated 1079 reflects a broken pointer.

Although we cannot without further anlysis interpret the standard deviations calculated in Tables I and II according to their usual statistical meaning, we can employ them nevertheless as an overall indication for the inaccuracy of the positional data. Using the standard deviation in this more general sense, it is clear from the values given in Tables I and II that the overall accuracy of the Carolingian Astrolabe is about twice as bad as that of any of the other sources included in the present comparison[16]. Even the drawings surpass the Carolingian Astrolabe in positional accuracy. Certainly, when judged by eleventh-century standards, the Carolingian Astrolabe is not a very accurate piece of work.

More generally, the values displayed in the two tables below - apart from showing the low standard of accuracy in the Carolingian Astrolabe itself - demonstrate that the star positions in practically all the other contemporary sources (leaving aside Maslama's pure book-work) deviate more or less from the expected values. Deviations of 5 degrees or more are not exceptional. The star positions on the retes of other astrolabes stemming from the source material available at the time and discussed here can therefore be expected to be even less accurate because new errors may have been introduced while making the astrolabe.

---

[16] In case the excessive deviations of α Ser (in mediation) and of β Per (in declination) are ignored the standard deviation in mediation and declination reduce to 6.3 and 6.4, respectively.

TABLE I: ANALYSIS OF THE MEDIATIONS (units in degrees)

Car. Astr. Nr	Type IA Nr	Ms. 7412 Nr	Type III Nr	Baily Pet.-Kn. Nr	Ident. Mod. name	Sign	Car.● Astr. Med. Deg.	Ptol. +12° 40' Med. Deg.
1		10	16	725	ζ Cet	ARI	24	16
2	1	23	24	202	β Per	TAU	6	3
3	2	1	19	393	α Tau	TAU	13	27
4	3	14	15	768	β Ori	GEM	7	8
5	6	15	14	818	α CMa	CNC	4	0
6	7	3	21	848	α CMi	CNC	12	11
7	9	4	23	469	α Leo	LEO	14	15
8		17	13	931	γ Crv	VIR	18	20
9	12	5	12	510	α Vir	LIB	8	8
10	13	25	1	110	α Boo*	LIB	36	24
11	14	18	2	111	α CrB	SCO	16	15
12	15	6	11	553	α Sco*	SCO	31	24
13				271	α Ser*	SCO	35	15
14	18	19	3	234	α Oph	SGR	18	13
15	16	13	9	149	α Lyr	CAP	1	0
16	17	7	4	288	α Aql	CAP	13	13
17		8	18	624	δ Cap	AQR	1	10
18		20	5	301	ε Del*	CAP	34	25
19	19	26	8	163	α Cyg*	CAP	40	30
20	20	21	6	317	β Peg	PSC	8	2

* The mediations of these stars on the Carolingian Astrolabe are in fact in the next sign, which is expressed by a value above 30.

● The mediations of the Carolingian Astrolabe were measured directly; those of the drawings and the other two astrolabes were derived indirectly by measuring the star positions with respect to a Cartesian coordinate system.

Car. Astr. Nr	Star lists			Drawings		Astrolabes		
	T. IA (1)	T.III (3)	Ms B.N. lat. 7412 (2)	Ms B.N. lat. 7412 (4)	Ms Vat. lat. 598 (5)	Car. Astr.	1029 Astr. (7)	1079 Astr. (8)
	Dmed Pt-Ms	Dmed Pt-Ms	DMed Pt-Ms	DMed Pt-Dr	DMed Pt-Dr	DMed Pt-A	DMed Pt-A	Dmed Pt-A
1		-4	-2	0	0	-8	-3	-2
2		-7	-3	-1	-4	-3	-2	
3	0	0	0	4	3	14	-8	-1
4	0	0	1	2	2	1	-1	0
5	0	-1	-1	0	0	-4	-2	0
6	0	0	0	0	-6	-1	-2	1
7	1	-4	-3	-3	-8	1	-3	1
8		2	2	2	2	2	2	
9	1	0	1	2	-2	1	-1	-1
10	0	0	0	-2	-4	-12	-1	-2
11		-1	-1	-5	-6	-1	-12	-9
12	0	-1	-1	-3	-4	-7	0	-1
13						-20	-5	1
14	-11	-11	0	0	-11	-5	-5	-2
15	0	-1	-1	1	0	-1	6	0
16	0	-1	-1	2	1	0	-6	2
17		3	3	2	8	9	1	3
18		0	0	0	2	-9	-7	-1
19	0	1	1	7	5	-10	-2	1
20	0	2	2	4	5	-6	-3	0
Mean value		-1.2	-0.2	0.6	-0.9	-3.0	-2.7	-0.5
Standard deviation		3.2	1.4	2.9	4.9	7.3	3.9	2.6

TABLE II: ANALYSIS OF THE DECLINATIONS (units in degrees)

Car. Astr. Nr	Type IA Nr	Ms. 7412 Nr	Type III Nr	Baily Pet.-Kn. Nr	Ident. Mod. name	Car. Astr. Dec.	Ptol. +12° 40' Dec.
1		10	16	725	ζ Cet	-6	-15
2	1	23	24	202	β Per	59	37
3	2	1	19	393	α Tau*	14	14
4	3	14	15	768	β Ori	-9	-10
5	6	15	14	818	α CMa	-10	-16
6	7	3	21	848	α CMi	18	7
7	9	4	23	469	α Leo*	16	16
8		17	13	931	γ Crv	-7	-12
9	12	5	12	510	α Vir	-11	-6
10	13	25	1	110	α Boo	25	25
11	14	18	2	111	α CrB	18	31
12	15	6	11	553	α Sco	-22	-23
13				271	α Ser	8	10
14	18	19	3	234	α Oph	5	14
15	16	13	9	149	α Lyr	32	39
16	17	7	4	288	α Aql	6	6
17		8	18	624	δ Cap	-23	-20
18		20	5	301	ε Del	10	8
19	19	26	8	163	α Cyg	41	42
20	20	21	6	317	β Peg	35	23

* In the calculation of the declinations of these stars for Type III the "altitudo" has been increased by 20° (change of XL into LX).

Car. Astr.	Star lists●			Drawings●●		Astrolabes●●		
	T. IA (1)	T.III (3)	Ms B.N. lat. 7412 (2)**	Ms B.N. lat. 7412 (4)	Ms Vat. lat. 598 (5)	Car. Astr.	1029 Astr. (7)	1079 Astr. (8)
Nr	DDec Pt-Ms	DDec Pt-Ms	DDec Pt-Ms	DDec Pt-Dr	DDec Pt-Dr	DDec Pt-A	DDec Pt-A	DDec Pt-A
1		-3		4	-5	-10	3	0
2		0		-6	-10	-22	3	
3	0	-4		6	-1	0	7	1
4	0	-1		5	-1	-1	3	7
5	-5	-1		3	-3	-5	4	3
6	0	-2		2	-2	-11	4	4
7	0	-1		-1	4	1	4	5
8		-4		0	-4	-5	3	
9	0	2		0	7	6	4	6
10	0	1		-3	3	-1	4	7
11		-6		1	-3	12	-2	-13
12	0	-1		2	0	-1	2	4
13						2	-2	-1
14	5	3		1	9	8	3	6
15	0	-1		-4	2	6	1	-4
16	0	0		-4	1	1	5	6
17		1		6	2	3	6	8
18		-5		-1	-5	-1	-1	2
19	0	-1		0	-1	1	-2	-1
20	0	-2		3	-1	-13	2	4
Mean value		-1.3		0.7	-0.6	-1.4	2.6	2.4
Standard deviation		2.5		3.4	4.5	7.9	2.6	5.0

● The declinations of Type III were calculated through the relation cos (altitudo) = Tan ((90° - Dec) /2)/ Tan ((90° + 23°.5)/ 2).

●● The declinations of the drawings and the astrolabes were derived indirectly by measuring the star positions with respect to a Cartesian coordinate system.

** Not used here; the values in Paris B.N. lat. 7412, under the title "latitudo" seem to represent altitudes of the stars for a geographical latitude of c. 39° (Valencia?).

Addendum to p. 656, note 8, and p. 667, note **: In a letter of June 18, 2000, Elly Dekker informed me that the mean value (of "latitudo" in MS 7412) of ca. 39° must be replaced by 38.2° ± 0.2°. This new value would point to Cordoba.

Figure 1

Rete of the Carolingian Astrolabe with stars numbered from 1 to 20; adapted from Destombes, Planche I.

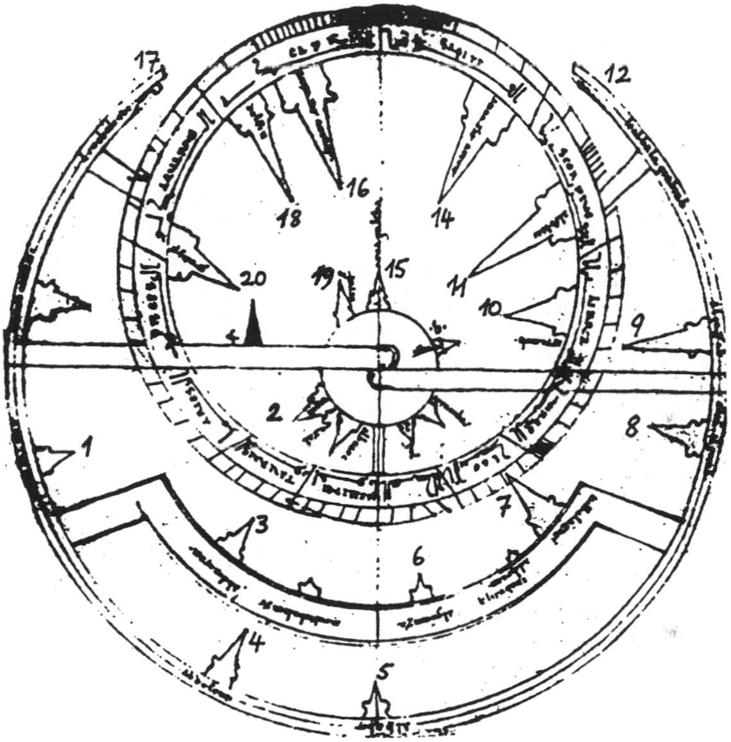

Figure 2

Rete copied from a Spanish-Arabic astrolabe, in MS Paris, BN lat. 7412, 19ᵛ; the Carolingian stars are numbered from 1 to 20 (reduced size). See above, source no. 4.

Figure 3

Drawing of a rete in MS Vat. Reg. lat. 598, 120r; the Carolingian stars are numbered from 1 to 20; adapted from Millàs, Làmina X (reduced size). See above, source no. 5.

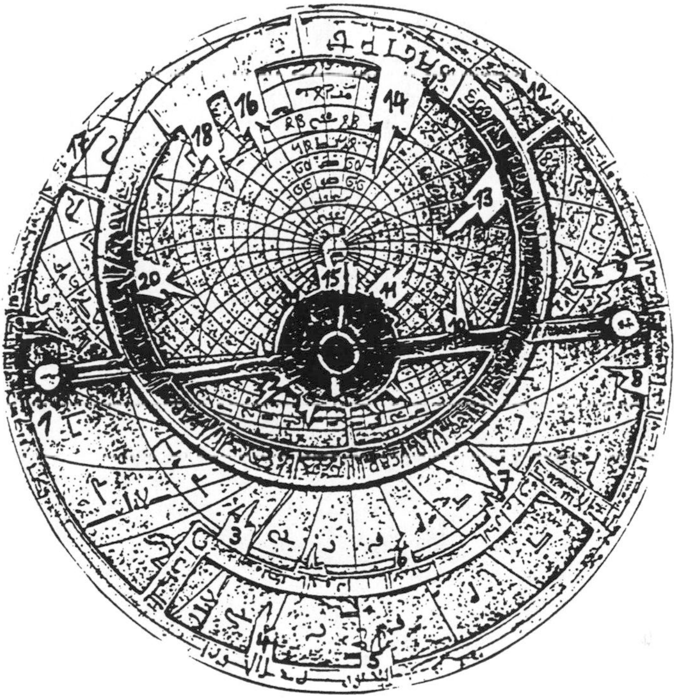

Figure 4

Rete of an anoymous early Spanish-Arabic astrolabe; the Carolingian stars are numbered from 1 to 20; adapted from Gunther I, Fig. 117 (reduced size). See above, source no. 6.

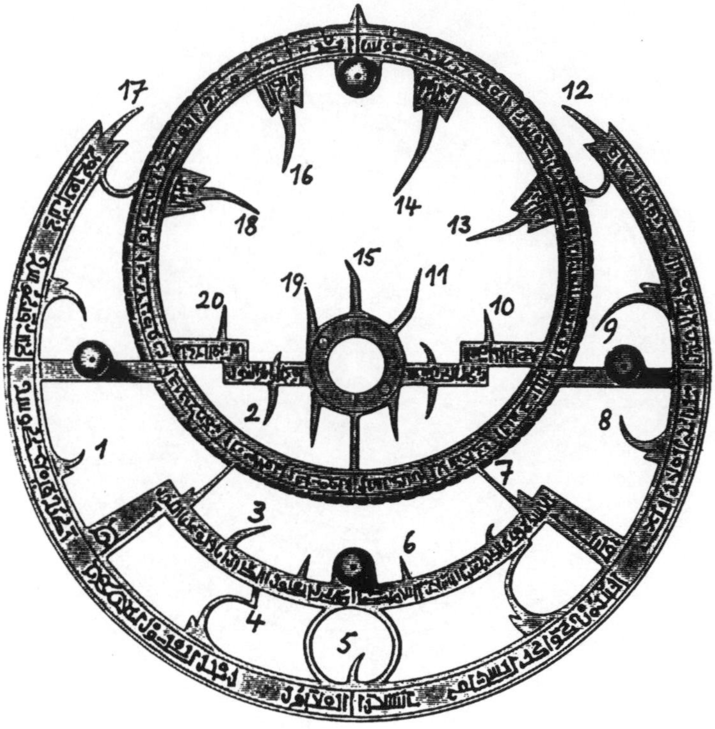

Figure 5

Rete of the astrolabe of Muḥammad ibn al-Ṣaffār; the Carolingian stars are numbered from 1 to 20; adapted from Woepcke, Tafel II (reduced size). See above, source no. 7.

XIX

THREE DUBIOUS STARS IN THE OLDEST EUROPEAN TABLE OF ASTROLABE STARS

The acquaintance of the Western, Latin, Europeans with the planispheric astrolabe began in the late 10th century in North-Eastern Spain, in the region of Barcelona. From the existing Latin testimonies it can be derived that scholars in that area came into contact both with Arabic texts on the instrument and with Arabic astrolabes. These Arabic materials came, of course, from the parts of Andalus conquered and dominated by the Arabs since 711. In the second half of the 10th century the leading astronomer there was Maslama al-Majrīṭī (d. 398 H = AD 1007-8). Of him it is well known that he had at his disposal a number of important astronomical texts from the Arabic East, as e.g. the *Almagest*, al-Battānī's *Zīj*, Ptolemy's *Planisphaerium*, etc. No treatise proper on the astrolabe of Maslama is so far known. But he made a revision of the Arabic translation of the *Planisphaerium*[1] and added to it several notes, an extra-chapter and a few chapters concerning the astrolabe[2] which are accompanied by a table of 21 astrolabe stars[3]. Certainly, what reached the Latin scholars in North-Eastern Spain of

[1] It survives in the Latin translation by Hermannus de Carinthia, 1143; edited by J.L. Heiberg, Ptolemaeus, *Opera*, II, Leipzig 1907, pp. 227-259; German translation by J. Drecker, in *Isis* 9 (1927), 255-278.

[2] The notes are in part included in the Latin translation (cf. note 1). The extra-chapter and the astrolabe chapters were edited by J. Vernet and M.A. Catalá, in *Al-Andalus* 30 (1965), 15-45. The notes, in several versions, were edited by P. Kunitzsch and R. Lorch, *Maslama's Notes on Ptolemy's* Planisphaerium *and Related Texts*, München 1994 (SB Bayer. Akad. d. Wiss., Phil.-hist. Kl., 1994, 2).

[3] The star table was edited, as "Type I A", by P. Kunitzsch, *Typen von Sternverzeichnissen in astronomischen Handschriften des zehnten bis vierzehnten Jahrhunderts*, Wiesbaden 1966, p. 17. (In the following, the types of star tables here edited will shortly be cited by their Roman numbers plus the individual stars in Arabic numbers; thus III, 27 will mean the 27th star in the table edited here as "Type III"). For the star coordinates used in this and other star tables the following symbols will be used: λ = ecliptical longitude, β = ecliptical latitude; μ = *mediatio caeli* (i.e. the degree of the ecliptic culminating – or passing at the meridian – together with a star; Ar. *tawassuṭ* or *mamarr*), δ = declination (Ar. *al-buʿd ʿan muʿaddil al-nahār*, distance from the equator).

the knowledge of the astrolabe was largely inspired by the activities of Maslama and his school[4].

The "Old Corpus" of Latin texts on the astrolabe comprises a number of treatises on the description, the construction and the use of the instrument[5]. One of these texts, *De mensura astrolabii*, Inc. *Philosophi quorum sagaci studio* (called h' by Millás), was accompanied by a table of 27 stars to be placed on the rete of the astrolabe; it was edited by P. Kunitzsch in 1966 as "Type III" (cf. note 3, above). In the following decades more material became known, so that we are now able to distinguish four stages in the development and use of this table.

Its oldest appearance seems to be in MS Paris, BNF lat. 7412 (first half 11th c.), 5v. Here the table is still organized in the Arabic way: it has to be read from right to left (but a direct model of the table could not be spotted in an Arabic source so far). It contains 26 stars (one star, η UMa, is omitted) with two coordinates: μ (here called *altitudo*) and another value (called *latitudo*) which may "best correspond to the maximum altitude of the stars at a geographical latitude of 39° (Valencia?)"[6]. The stars are mentioned with their Arabic names, in Latin transliteration; to several of the Arabic names tentative, often suitable, Latin translations are added. In a re-organized form and with 27 stars, then, the table appears in the text h' and in endless repetitions in manuscripts down to the 15th century. In this (main) form the table registers as *latitudo* the value of μ, and as *altitudo* a special value not found elsewhere in Arabic or later Latin texts; the value seems to be read off from an (Arabic) astrolabe present in the hands of the first author of that type of the table. The *mediatio* (here: *latitudo*) could be read off on the graduated ecliptic ring of the rete of an astrolabe by putting a ruler through the centre of the instrument and through the star. To find the second coordinate, here called *altitudo*, the ruler was put at right angles through the first line at the position of the star until

[4] Maslama's star table exists also in a Latin translation, of unknown provenance, though in a rather confused form ("Type I").

[5] The texts of the "Old Corpus" were edited by J.M. Millás Vallicrosa, *Assaig d'història de les idees físiques i matemàtiques a la Catalunya medieval*, Barcelona 1931.

[6] Proposed by E. Dekker; see P. Kunitzsch and E. Dekker, The stars on the rete of the so-called 'Carolingian Astrolabe', in: *From Baghdad to Barcelona. Studies in the Islamic exact sciences in honour of Prof. Juan Vernet*, Barcelona 1996, II, p. 656 note 8. The table was edited, in an unsatisfactory form, by W. Bergmann, in *Francia* 8 (1980), 84; a partial edition was given by M. Destombes, in *Archives internationales d'histoire des sciences* 15 (1962), 27 (Table II, columns 2-4).

the graduated outer rim of the astrolabe. Here, the number of degrees between the meeting point of the first line on the rim and each of the two meeting points of the second line were counted, and this (i.e., in one of the two directions) is the value called *altitudo*[7]. In a next stage the table was taken over, in this form, by Ascelinus of Augsburg (early 11th c.) in his treatise on the construction of the astrolabe[8]. Afterwards Hermannus Contractus (1013-1054) used it – apparently taken over from Ascelinus – also in a treatise on the construction of the instrument[9].

The values of the coordinates are basically the same in all the four sub-types of the star table, apart from scribal errors in the Roman numerals and with the exception of the *latitudo* in MS 7412, which stands alone by itself. The Arabic names of 26 of the 27 stars can be safely identified and, subsequently, the astronomical identification of 24 of them can be established without any doubt. There remain, however, three dubious stars in the list which will be discussed in detail, below. (I follow the numbering and spelling in the edition of "Type III"; it should be noted that the names appear in a great variety of spellings in thc manuscripts.)

1. No. 7, alhcadib

The identity of this name is easy to establish: it is (*al-kaff*) *al-khaḍīb*, "the (hand) tinted (with henna)". It originated in the old Arabic star lore, where, in a tradition, *al-thurayyā* (the Pleiades) was imagined as the head of a woman from which two arms emanate each ending in a hand, one hand toward the north, called *al-kaff al-khaḍīb* and formed by βαγδε Cas, and the other hand toward the south, called *al-kaff al-jadhmā'*, "the truncated hand", formed by λαγδνμ Cet[10]. Later, in

[7] In some later reproduction of the star table, this coordinate was aptly called *longitudo ex utraque parte* ("Type XI").

[8] Edited from one manuscript by W. Bergmann, *Innovationen im Quadrivium des 10. und 11. Jahrhunderts*, Stuttgart 1985, pp. 223-225; new edition, from five manuscripts (with additional notes from a sixth manuscript on a separate sheet added to the off-print), by C. Burnett, in *Annals of Science* 55 (1998), 343ff. (with Figs. 13-15).

[9] Edited, from MS Munich, Clm 14836 (written still during Hermannus' lifetime), by J. Drecker, in *Isis* 16 (1931), 200-219. In Ascelinus' version of the star table three of the Arabic star names were split up and re-composed in an extraordinary way: patangaitoz (ζ Cet), denebgaitoz (ι Cet) and denebaliedi (δ Cap) became gaitozpatan, gaitozderep and liedideneb. Hermannus took the list over with these peculiar spellings.

[10] See P. Kunitzsch, *Untersuchungen zur Sternnomenklatur der Araber*, Wiesbaden 1961, nos. 306 (*al-thurayyā* and the details of the two arms), 136a-b (*al-kaff al-khaḍīb*)

astronomical use, the star β Cas alone was also called *al-kaff al-khaḍīb*, and α Cet alone *al-kaff al-jadhmā'*. Both stars are listed, under these names, as astrolabe stars by al-Ṣūfī[11].

The coordinates of β Cas in the *Almagest* are λ Ari 7°50, β 51°40, and accordingly in al-Battānī's star catalogue in his *Zīj* λ Ari 19°, β 51°40[12]. The star table of the *Mumtaḥan zīj* (214 H = 829-30) gives values obtained by independent observation, λ Ari 18°24 [in the MS erroneously 13°24], β 51°45 [in the MS 11°45][13]. From the *Mumtaḥan* are derived the values of al-Farghānī's table of astrolabe stars (dated 225 Yazd. = 856-7), λ Ari 18°39, β 51°45, and the values in the Berlin MS of Ḥabash's *Zīj* (dated 304 H = 916-7), λ Ari 19°41, β 51°45 (the Istanbul MS of the *Zīj* repeats the *Mumtaḥan* values, unchanged; the same values are also mentioned in an example in Ḥabash's treatise on the melon-shaped astrolabe[14]).

For use on the astrolabe the ecliptical coordinates λ/β had to be converted to μ/δ. Some of the aforementioned texts add to their tables columns for μ and δ. (In al-Battānī's table of 75 fundamental stars [ed. Nallino, II, pp. 178-186] β Cas is not contained.) The *Mumtaḥan* family of tables shows the following values: *Mumtaḥan zīj* μ Psc 16°5 [in the minutes rather to be read 0 instead of 5], δ + 52°51; the same [i.e., with 16°0 for μ] in the Istanbul MS of Ḥabash's *Zīj* and in the calculated example in Ḥabash's treatise on the melon-shaped astrolabe; al-Farghānī μ Psc 16°13, δ + 53°2; Ḥabash, *Zīj*, Berlin MS μ Psc 16°17, δ + 53°26. The *mediatio* of β Cas is thus clearly in the sign of Psc, and its declination slightly above 50° north.

With approximately this position – close to the left of the polar ring around the centre, shortly above the horizontal bar, in Psc – the star *al-kaff al-khaḍīb*, or shortly *al-khaḍīb*, is to be seen on practically all Arabic astrolabes, Eastern, Andalusian and Maghrebi alike[15].

and 137 (*al-kaff al-jadhmā'*).

[11] Cf. P. Kunitzsch, *Al-Ṣūfī and the Astrolabe Stars*, in this journal, 6 (1990), 151-166; see Table 1, nos. 7 and 34.

[12] Al-Battānī, *Opus astronomicum*, ed. Nallino, II, p. 150.

[13] J. Vernet, Las "Tabulae Probatae", in: *Homenaje a Millás-Vallicrosa*, II (Barcelona 1956), pp. 501ff.; see p. 519.

[14] *The Melon-Shaped Astrolabe in Arabic Astronomy*, ed., transl. and comm. by E.S. Kennedy, P. Kunitzsch and R.P. Lorch, Stuttgart 1999, pp. 72-73 (with commentary on p. 127f.).

[15] Cf., when no better illustrations are available, the pictures in B. Stautz, *Untersuchungen von mathematisch-astronomischen Darstellungen auf mittelalterlichen Astrolabien islamischer und europäischer Herkunft*, Bassum 1997, pp. 178ff. The inscriptions are mostly illegible, but the pointer for β Cas can easily be recognized on

So far matters are fully clear and do not leave room for doubt or discussion. But when we then turn to al-Andalus and inspect Andalusian star tables we find that here against the name of *al-kaff al-khaḍīb* (β Cas) there are written coordinates quite different, pointing to no specific star or, as some scholars assume, to α And[16].

The oldest known Spanish-Arabic table of astrolabe stars, set up by Qāsim ibn Muṭarrif al-Qaṭṭān for 300 H = 912-3, has λ 341° [= Psc 11°], β 29°[17]. Maslama in his star table for 978 (and adding 12°40 to the λ of the *Almagest*) gives λ 347° [= Psc 17°], β 29° (Type I A, 21). The same values are also found in a star table travelling in a manuscript of al-Zarqāllu's treatise on the *azafea* and said to be calculated for 459 H = 1066-7[18]. (In Maslama's table, where usually μ/δ are also given, these values are omitted for this star.) It is out of question that the Andalusian tradition is erroneous for this star: against the name of β Cas it transmits coordinates which do not correspond exactly to any major star in Cas, Peg or And according to the *Almagest*.

When we consider α And – as some scholars do –, we find in the *Almagest* λ Psc 17°50, β 26°. For this star Type XIII, 13 (perhaps for 527 H [sic, instead of 577 in the manuscripts] = 1132-3) has λ Ari 2°25, β 26°, and μ 350°4 [= Psc 20°4], δ + 24°39. In Type XV, 30 (Ibn al-Kammād) we have, against the name of β Cas, the correct coordinates λ/β of α And, and the Latin *expositio* of the star names calls the star accordingly *caput mulieris id est andromede*[19]. Also some Hebrew star tables with Andalusian background show the same confusion: Abraham bar Ḥiyya (AD 1104), name of β Cas, but λ/β of α And, and similarly μ 349°40 [= Psc 19°40], δ + 24°24; and an anonymous table for AD 1392: name of β Cas, but λ/β of α And with μ Tau 23°1, δ + 26°4[20].

most of the retes here shown.

[16] Only Type IV, 28 has the corret μ Psc 21°, δ + 54°, and Type II, 17 has RA [or rather, μ?] 80°8 [starting the count at Sgr 1°; therefore = Psc 20°8], δ + 50°30.

[17] Cf. M. Comes, La primera tabla de estrellas documentada en Al-Andalus, in: *Actes de les I trobades d'història de la ciència i de la tècnica*, Trobades Científiques de la Mediterrània (Maó, 11-13 setembre 1991), Barcelona 1994, pp. 95-109; see p. 106.

[18] Cf. P. Kunitzsch, *The Arabs and the Stars*, Northampton 1989, item IV, p. 194, no. 18.

[19] Cf. also B.R. Goldstein and J. Chabás, Ibn al-Kammâd's Star List, in *Centaurus* 38 (1996), 317ff., esp. 320-323, Tables 1-4, no. 30 (here, some manuscripts have β 29°).

[20] Cf. B.R. Goldstein, Star Lists in Hebrew, in *Centaurus* 28 (1985), 185ff.; see the table p. 188f., no. 1, and the table p. 200f., no. 2.

And indeed, α And is found on some Arabic astrolabes shortly above the horizontal bar, in Psc, but much more distant from the pole, to the left, than β Cas. If we look for stars with β 29° in the *Almagest*, we only find, in the constellations of Pegasus and Andromeda, λ Peg (Baily no. 323, on the chest) and υ And (Baily no. 352, on the left knee-bend). But these are insignificant fourth-magnitude stars and are not normally inscribed on astrolabes.

It could be that the origin for this confusion was a faulty manuscript of al-Battānī's *Zīj*. Maslama directly refers to al-Battānī in the title of his star table, and for al-Qaṭṭān M. Comes also assumes relationship to al-Battānī. In any way, in al-Battānī's fundamental stars (ed. Nallino, II, p. 182, no. 44) the name of β Cas (*al-kaff al-khaḍīb*) is wrongly given to α Oph, and just a line above, in no. 43, the star α And (correctly: the head of Andromeda) is wrongly called *Caput Orionis*. However that may be, in Andalusian tables of astrolabe stars the star name *al-kaff al-khaḍīb* is registered with aberrant coordinates pointing to no conspicuous star, or, in other tables, indicating α And.

To sum up, in Andalusian astrolabe star tables, Arabic and Hebrew, we find either a star named with the name of β Cas and having coordinates pointing to no conspicuous star, or a star with the name of β Cas and coordinates of α And, or a star with both name and coordinates of α And. On Andalusian and Maghrebi astrolabes β Cas is always located in its correct position, in Psc, with its correct name. On some astrolabes also α And is to be found, also in its correct place.

If we then turn to the early European, Latin, star tables and astrolabes – which are all dependant on Spanish-Arabic sources and models –, we find that the confusion is still greater. As far as astrolabes are concerned, they depend, in the Latin West, on certain types of star tables in a much more distinct way than in Arabic-Islamic astronomy. We shall not, therefore, consider these instruments here in greater detail, but rather restrict ourselves to the star tables.

The oldest form of the star table of Type III, in MS 7412 (cf. above), registers *alchadib* (sic) with μ in Trutina 16°. Trutina would be = Libra. This is obviously a miswriting for Truta (= Pisces). (Cf. in this list the star ι Cet, in the μ of which the copyist erroneously wrote Virgo and underneath the correct Truta = Pisces.) Later, the authors of the main version of Type III, uncertain about the meaning of the rare Trutina = Libra and Truta = Pisces, wrote, instead, Taurus (a well-known name beginning with a T) and located it at 22°. In the drawing of the rete of Khalaf ibn al-Muʿādh's Andalusian astrolabe in MS 7412, 19v, the pointer for β Cas was put too far to the left (i.e., with too small δ) and was filled in black and left without name; at the bottom of the page, against note *f*, the name is written as *alhadip* (corrected from *aldaldip*). In a drawing of a rete showing the 27 stars of Type III in MS Vat. Regin. 598 (11th c.), 120r[21], the pointer for β Cas is nearer to the centre, i.e. with a better δ, but too long, reaching into Aqr (instead of being very short, in Psc).

In this way, European astrolabists found in their texts a star with the name of β Cas (*alhadip* et varr.) and the odd μ of Tau 22°. It is significant that the author of the star table of Type VIII (sometime after 1246), who amalgamated in his list stars from Types III and VI, omitted this star altogether.

Therefore, only very few of the early Latin astrolabes have a pointer in the position of β Cas: cf. Stautz (note 15), p. 245 (here, the name ascribed to the star is *humerus equi*, which would be β Peg; in Stautz it is no. 28); and *ibid.*, p. 247 (pointer without name).

2. No. 22, aldiraan

In the old Arabic star lore each of the two pairs of stars, αβ Gem and αβ CMi, were regarded as a *dhirāʿ*, "fore-arm, or front paw", of Leo, one outstreched (*mabsūṭa*) and the other contracted (*maqbūḍa*)[22]. According to al-Ṣūfī α Gem is an astrolabe star with the name *muqaddam al-dhirāʿayn*, "the preceding one of the two *dhirāʿ*"[23].

In the *Almagest* α Gem has λ Gem 23°20, β 9°30, and accordingly in the star catalogue of al-Battānī's *Zīj* λ 94°30 [= Cnc 4°30], β 9°30. Correct ecliptical coordinates are also found in Kunitzsch (note 18), item IV, p. 198, no. 18, and in Types XII A, 18 [λ 50°40, misspelling in the MS for 7°40]; XIII, 15/16 (here, names and coordinates of α and β Gem are interchanged); XIV, 18; and XV, 9. For μ/δ the following values can be found: al-Battānī, fundamental star no. 38 [ed. Nallino, II, p. 182], μ 95°10 [= Cnc 5°10], δ + 32°49 (cf. also al-Ṣūfī, note 23, above); Type XIII, 15/16 (for 527 H = 1132-3) [name of α Gem in no. 16, coordinates under no. 15], μ 99°42 [= Cnc 9°42], δ + 32°57.

On photographs of Arab-Muslim astrolabes of the earlier centuries α Gem never seems to be represented. The reason is apparently purely

[21] Làmina X in Millás (note 5).

[22] Cf. Kunitzsch (note 10), nos. 82a-b, 83.

[23] Cf. Kunitzsch (note 11), Table 1, no. 25. In one text, al-Ṣūfī mentions for α Gem: μ Cnc 5°, δ + 32° (*ibid.*, Table 2, no. 10).

technical: the width of the band representing the ecliptic on the rete and the bar usually connecting the ring around the pole and the ecliptic did not leave the space for placing the appropriate pointer of the star. Its pointer should be (in μ) between those of α Ori and α CMi, and north of the ecliptic, i.e. inside the ecliptic ring. Only on more recent astrolabes with different, more elaborate, constructions of retes one might expect to find a pointer for α Gem[24].

Again in Al-Andalus (and subsequently in the Maghrib), there is a major confusion about *muqaddam al-dhirāʿayn*. But this time it can hardly be ascribed to some defect manuscript of al-Battānī with whom, both in the star catalogue and among the fundamental stars, the stars α Gem and α Hya are correctly registered[25].

The star is not contained in al-Qaṭṭān's table. Maslama (Type I A, 8; AD 978) mentions for *muqaddam al-dhirāʿayn* λ 130°20 [= Leo 10°20], β – 16°, coordinates which (according to Maslama's precession rate of Ptolemy + 12°40) do not correspond to any star in the *Almagest*. Further, he mentions for this star μ 125°49 [= Leo 5°49], δ + (*sic*) 2°25[26].

In all versions of the Latin star table of Type III (star no. 22, *aldiraan*) we find μ Leo 6°. In Type VIII, 18 (*aldiran*) – where stars from Types III and VI were put together, with μ and δ – we find the same μ, and δ – 6°[27]. These coordinates, which are in any way not correct, are rather "phantastic" and point to a region of the sky in which α Hya would be the brightest star.

[24] The description by Frank and Meyerhof of the stars on the rete of the Moghul astrolabe made by the sons of ʿĪsā ibn Allāhdād in Lahore (between 1605 and 1627) is unreliable. Perhaps on this instrument both α and β Gem are represented; cf. P. Kunitzsch, *Arabische Sternnamen in Europa*, Wiesbaden 1959, p. 62 with note 1. The article of Frank and Meyerhof was reprinted in F. Sezgin (ed.), *Arabische Instrumente in orientalistischen Studien*, IV (Frankfurt/Main 1991); but the reproduction of the face of the astrolabe, on p. 353 there, is too faint and does not allow to control the inscriptions.

[25] The name *muqaddam al-dhirāʿayn* is not used in al-Battānī's tables, but occurs only in the textual description of the Ptolemaic constellations, ed. Nallino, III, p. 188,22 = II, p. 125,12.

[26] Type II, 15 (*muchd' m̄ d'aha*) similarly has RA (or μ?) 219°13 [counted from Sgr 1°; therefore, it is = Leo 9°13], δ + 2°25.

[27] MS g and h have δ as north (+), instead of south (–). In XI, 10 (*aldiraan*) – where MSS e-m repeat the values from Type VIII – MS e places μ in Cnc (instead of Leo), and all MSS give δ as +, instead of – . VIII, 42 has another *aldiran* which, however, is derived from VI, 1 (*aldramin id est dextrum adiutorium cephei*) and designates α Cep; the true origin of the name *aldramin* (etc.) for this star is still unknown; cf. Kunitzsch (note 18), item XXI, p. 54f.; item XXII, p. 93.

In the *Almagest* α Hya has λ Leo 0°, β –20°30, and accordingly in the star catalogue of al-Battānī's *Zīj* λ 131°10 [= Leo 11°10], β –20°30. Further, as fundamental star no. 66 [ed. Nallino, II, p. 185], al-Battānī gives μ 125°58 [= Leo 5°58], δ –2°55. Ḥabash, in the Berlin MS of his *Zīj*, has λ Leo 11°44, β –20°30, μ Leo 6°15, δ – 2°26. *Al-Ṣūfī* registers α Hya as an astrolabe star with the old Arabic name *al-fard*, "the solitary one", or a name derived from the *Almagest*, *ʿunuq al-shujāʿ*, "the neck of the (snake of the sort called) shujāʿ"[28].

In Latin, Type VIII, 19 also registers *alfart*, α Hya[29], taken over from Type VI, 20 (AD 1246)[30], with μ Leo 13°, δ – 18°30[31], beside the dubious *aldiran* (VIII, 18)[32].

Thus in the West – in Muslim Spain and in Latin Europe – we have in astrolabe star lists a dubious star carrying the name of α Gem and coordinates roughly pointing to a region in which α Hya is the brightest star. In addition, in Latin star tables from 1246 on, beside this "ghost star", α Hya itself is registered with its own, appropriate, coordinates.

When we, then, inspect astrolabes to see how this material was treated on the instruments, we find that Latin astrolabes (from the oldest known specimens of around 1300 on and as long as they were produced according to the medieval tradition) normally follow, in the location of stars on the rete, the well-known "types" of Latin astrolabe star tables. This means that they entered the "ghost star" *aldiraan* slightly above the metal ring along the equator, in the sign of Leo, some place in μ between α CMi and α Leo. Astrolabes following the tradition of Type VIII would inscribe instead, or additionally, the true star α Hya in its proper place. This underlines how much astronomy, and the making and use of instruments, in the Latin Middle Ages was book science executed in the study rather than under the open sky.

For Arabic – and especially Andalusian and Maghrebi – astrolabes the matter is more complicated. On many (older) instruments the star *muqaddam al-dhirāʿayn* (or, sometimes, simply *al-dhirāʿān*) is inscribed in its dubious place, as a "ghost star"[33]. On several of them the correct

[28] Cf. Kunitzsch (note 11), Table 1, no. 42.

[29] The Arabic name is explained in Latin as *equs* [i.e., confusing Ar. *al-fard*, "the solitary one", with *al-faras*, "the horse", i.e. the name of the Greek constellation of Pegasus] *vel cingulus* [*sic* for *singulus*, a translation of *al-fard*].

[30] Here, the coordinates are given as λ Leo 15°, β – 22°, μ Leo 10°, δ – 5°.

[31] Better variants for δ in other manuscripts: – 5° and – 5°30.

[32] Types IX, 9, Xa², 8 and XI, 11 take the star over from VIII, 19, with varying deviant coordinates.

[33] See e.g. Stautz (note 15), p. 192 (no. 11; anonymous, AD 1054); p. 193 (no. 9;

α Hya is inscribed, either in addition to *muqaddam al-dhirāʿayn* or alone[34]. In the course of time, other astronomers obviously realized the confusion about *muqaddam al-dhirāʿayn*; they entered the star pointer in its place in Leo, above the ring along the equator, but identified it as a star in Cnc and gave it the name (*al-*) *zubānā*, "the claw (of Cancer)"[35]. Of the two claws of Cancer, the position would apply to the southern one, α Cnc. And indeed, on the astrolabe of Muṣṭafā Ayyūbī, 1114 H = 1702-3, the star pointer in this position is directly labelled *zubānā janūbī*, "the southern claw", thus plainly indicating α Cnc[36]. Sometimes things become more difficult. Thus, e.g., al-Baṭṭūṭī (on an astrolabe dated 1136 H = 1723-4) has in the area of our old "ghost star" a pointer labelled *dhaqan al-shujāʿ*, "the jaw, or chin, of the (snake called) shujāʿ", which designates ζ Hya; and in the place of α Hya is a pointer conventionally labelled *ʿunuq al-shujāʿ*, "the neck of the snake"[37]. On other late Islamic astrolabes one may even find in both places, the place of the "ghost star" and the true place of α Hya, labels carrying names which both refer to the same star, α Hya.

3. No. 27, egreget

The last two stars in the table of Type III are no. 26, *arrucaba*, μ Leo 19°, and no. 27, *egreget*, μ Cnc 26°.

Of these, *arrucaba* can be easily identified as Ar. *al-rukba*, "the

Muḥammad ibn Saʿīd al-Ṣabbān, AD 1073); p. 195 (no. 8; Aḥmad ibn Muḥammad al-Naqqāsh, AD 1079); p. 219 (no. 11; Abū Bakr ibn Yūsuf, AD 1208).

[34] See Stautz (note 15), p. 186 (no. 14; the sons of Ibrāhīm al-Iṣfahānī, AD 984); p. 191 (no. 11; Muḥammad ibn al-Ṣaffār, AD 1029); p. 194 (no. 9; Muḥammad ibn Saʿīd al-Ṣabbān, AD 1081); p. 196 (no. 9; Ibrāhīm ibn al-Sahlī, AD 1085); p. 197 (no. 11; Muḥammad ibn al-Sahlī, AD 1090); p. 198 (no. 10; Badr ibn ʿAbdallāh, AD 1130); p. 201 (no. 16; Muḥammad ... al-Iṣfahānī, AD 1221); p. 220 (no. 12; Abū Bakr ibn Yūsuf, AD 1213-4); p. 228 (no. 12; anonymous, undated). Further also Abū Bakr ibn Yūsuf, AD 1216-7; see R. d'Hollander (note 42), p. 76.

[35] See Stautz (note 15), p. 225 (no. 10; anonymous, undated); p. 230 (no. 12; Ibn Bāṣo, Granada, AD 1304-5); p. 232 (no. 12; anonymous, AD 1477); p. 233 (no. 12; Muḥammad ibn Faraj, Granada, AD 1476); al-Baṭṭūṭī, AD 1715-6 (catalogue Alain Brieux, Paris, 9-10 octobre 1980, Collection L. Linton, p. 106f., lot no. 171).

[36] Cf. L.A. Mayer, *Islamic Astrolabists and Their Works*, Geneva 1956, plate XIX (on the Ottoman maker, see p. 79).

[37] See catalogue A. Brieux (note 35), pp. 110-112, lot no. 173. The same order and names of the two stars are earlier found on several astrolabes by Muḥammad ibn Fattūḥ al-Khamāʾirī, Muslim Spain, between AD 1207 and 1240; cf. Stautz, pp. 221-223; to the same tradition belongs also the instrument *ibid.*, p. 224. Cf. further *ibid.*, pp. 228, 229.

knee" (*sc*. on the left fore-foot of Ursa Maior), θ UMa. In the *Almagest* it is the 11th star of UMa, Baily no. 19, λ Cnc 10°40, β 35°. The star is not mentioned by al-Ṣūfī as an astrolabe star and is not registered in the Arabic lists of astrolabe stars cited in this paper. It is also not mentioned among the 75 fundamental stars in al-Battānī's *Zīj*. In the Latin star table of Type VIII, 21 *alrucaba* (compiled from III, 26) is registered with μ Leo 20°, δ + 35° [in some manuscripts: 45°].

Since the early Andalusian tables of astrolabe stars cannot be of help here, we have to consult the instruments themselves, i.e. mostly astrolabes belonging to the Andalusian and Maghrebi tradition[38].

Here we are confronted with a delicate problem: these instruments show on the rete in the area concerned three stars: *yad al-dubb*, "the fore-foot of the (Greater) Bear", which would be ι UMa; *al-rukba* or *rukbat al-dubb*, "the knee (of the Bear)", θ UMa; and *al-rijl*, "the foot", which can be, according to position, either ι or μ UMa[39]. The area to be controlled is the small lower right quadrant north of the ecliptic, i.e. inside the ecliptic ring, with μ in Leo and, sometimes, also in Cnc. (The exact values of μ are difficult to be measured on photographs; also the ends of star pointers may be bent away from their original position in several cases.)

One star only: *rukba*, astrolabe of Muḥammad ibn Saʿīd al-Ṣabbān, AD 1081 (Stautz, p. 194); *rukbat al-dubb*, Muḥammad ... al-Iṣfahānī, AD 1221 (Stautz, p. 201). This is θ UMa.

Two stars, one of them *rukba*: Muḥammad ibn al-Ṣaffār, AD 1029: more to the left a star (no name inscribed on the pointer), μ in Cnc – and more to the right, with μ in Leo, *rukbat al-dubb*[40]; Abū Bakr ibn Yūsuf, AD 1208-9: to the left, μ near the end of Cnc, *al-rijl* – to the right, μ Leo ca. 17°, *al-rukba*[41]; Abū Bakr ibn Yūsuf, AD 1216-7: to

[38] On photographs and other illustrations in modern studies and catalogues the names inscribed on the star pointers of astrolabe retes are often not legible. Therefore the choice of examples cited below is relatively small.

[39] Stautz (note 15) is of no use here, because he identifies our two stars in all his examples – without regard to their names – as either ι or/and μ UMa alone; θ UMa is never considered by him.

[40] F. Woepcke, *Über ein ... zu Berlin befindliches arabisches Astrolabium*, Berlin 1858, Tafel II, Fig. 7, and p. 18. The second star is incorrectly called ψ UMa by Woepcke, instead of θ; the first, unnamed, star is ι UMa.

[41] F. Sarrus, *Description d'un Astrolabe, construit à Maroc en l'an 1208*, in: Mémoires de la Société du Muséum d'Histoire Naturelle de Strasbourg, t. 4, livraison 1, 1850 (as a separate extract, Strasbourg 1852); the rete is shown on Planche 4, Fig. 1; besides, I could use a very good private photograph. On p. 15, there is only a list of translated star names, without astronomical identification.

the left, μ in the beginning of Leo, *al-rijl* – to the right, μ in the middle of Leo, *al-rukba*[42]; Ibn Bāṣo, AD 1304: to the left, μ near the end of Cnc, *al-rijl* – to the right, μ in the former part of Leo, *al-rukba*[43]. In this group *rukba* can be identified as θ UMa. The star to the left of it, with smaller μ, then must be ι UMa, here called in all cases *al-rijl*.

Two stars, neither of them *rukba*: Anonymous, undated (Stautz, p. 225): more to the left, μ near the end of Cnc, *yad al-dubb* – more to the right, μ about in the middle of Leo, *rijl al-dubb*; Anonymous, AD 1477 (Stautz, p. 232): to the left, μ in Cnc, *yad al-dubb* – to the right, μ in the former part of Leo, *rijl al-dubb*; al-Baṭṭūṭī, AD 1723: to the left, μ in the former part of Leo, *yad al-dubb* – to the right, μ in the middle of Leo, *rijl al-dubb*[44]; Anonymous, undated (Maghreb, 18th c.?): to the left, μ in the beginning of Leo, *yad al-dubb* – to the right, μ about Leo 20°, *rijl al-dubb*[45]. In this group the star more to the left, i.e. with smaller μ, called *yad al-dubb*, can be identified as ι UMa. The star more to the right, with μ in Leo (about the location of *rukba* = θ UMa in the other group), called *rijl al-dubb*, can hardly be anything else but μ UMa.

Drawings of retes in Latin manuscripts: MS Vat. Regin. 598, 120r (cf. note 21, above): all the star pointers in this drawing are unnamed; here, the one more to the left of the two stars has μ near the end of Cnc, the one more to the right is about in the middle of Leo, but with great δ, i.e. much shorter and lying nearer to the pole than on the instruments described above; MS Paris 7412, 19v: the pointer more to the left, with μ slightly greater than the middle of Cnc, carries the name *Rigel*, the one more to the right, with μ near Leo 20°, is labelled *Arracuba*; this situation resembles Abū Bakr's astrolabe of 1208 (see above)[46].

In the light of the evidence described above it seems that, also in the star table of Type III, we have to identify no. 26, *arrucaba*, as θ UMa, and the star with the smaller μ (Cnc 26°), no. 27, *egreget*, as ι UMa[47].

[42] R. d'Hollander, *L'astrolabe – Les astrolabes du Musée Paul Dupuy*, Toulouse 1993, p. 76 (Planche VIII); name list on p. 59.

[43] S. Gibbs and G. Saliba, *Planispheric Astrolabes from the National Museum of American History*, Washington, D.C., 1984, p. 138, Fig. 99. Another astrolabe of Ibn Bāṣo, also of AD 1304 and with the same distribution of stars and names, is in Catalogue Brieux (note 35), pp. 88-89, lot no. 162.

[44] Catalogue Brieux (note 35), p. 112, lot no. 173.

[45] *Ibid.*, p. 113, lot no. 174.

[46] For the inscriptions on the drawing in MS 7412, cf. Kunitzsch (note 24), pp. 91f.

[47] This has now to be corrected in Kunitzsch (note 3), p. 28, and for all derivations

There remains the explanation of the curious name *egreget*. This word – contrary to the other 26 names in the list – does not show an open, direct relationship to any of the well-known Arabic star names here involved. For this star we found on the Arabic astrolabes described above two different names, *yad al-dubb* and *al-rijl*. Also in the drawing in MS 7412 it is called, in Latin, *Rigel*. We may therefore interpret *egreget* as a Latin corruption of a transliteration of Ar. *al-rijl*, here used for ι UMa[48].

from III, 27, such as VIII, 17, etc.

[48] In Kunitzsch (note 24), pp. 74-76 (no. 28), the same derivation had been proposed. But because, forty years ago, *al-rijl* was only known to me as a name for μ UMa, I there proposed that in the star table (and in the Latin drawings) the two names *arrucaba* and *egreget* be interchanged, i.e. the star with the smaller μ, more to the left, should be *arrucaba* = θ UMa, and the one with greater μ, more to the right, *egreget* = μ UMa. Now, knowing that also ι UMa, the star with the smaller μ, is called on Arabic astrolabes *al-rijl*, the proposed change of position is no longer required.

XX

The Chapter on the Stars in an Early European Treatise on the Use of the Astrolabe (ca. AD 1000)

The astrolabe was the most wide-spread astronomical instrument in the Middle Ages, both in the Islamic world and in Europe. The Europeans received the first knowledge of the instrument in the Christian area of North-Eastern Spain, around Barcelona. In the last two decades of the tenth century, about, they became acquainted with astrolabes and some texts related to the instrument from al-Andalus where at that time the leading astronomer was Maslama al-Majrīṭī (d. 398/1007-8). Writings on the astrolabe have only survived from several of his pupils, not from his own pen. But we have, in Arabic and in a (somehow confused) Latin translation, a table of 21 astrolabe stars set up for AD 978 by Maslama.[1] Subsequently Latin scholars in that area composed a number of treatises on

[1] Edited by P. Kunitzsch, *Typen von Sternverzeichnissen in astronomischen Handschriften des zehnten bis vierzehnten Jahrhunderts*, Wiesbaden, 1966, pp. 15-18, as "Type I", both the Arabic and the Latin versions. It should be added, however, that Maslama knew Ptolemy's *Planisphaerium*. He revised its Arabic text (which was then translated into Latin by Hermann of Carinthia, 1143) and added to it a number of notes and an extra-chapter as well as some chapters on the astrolabe. Maslama's notes were recently edited, in Arabic and in several Latin translations, by P. Kunitzsch and R. Lorch, *Maslama's Notes on Ptolemy's Planisphaerium and Related Texts*, Munich, 1994 (Sitzungsberichte, Bayerische Akademie der Wissenschaften, Phil.-hist. Klasse, 1994, 2). Here, in the Introduction, full information is given on editions and studies of the *Planisphaerium* and Maslama's contributions.

the description of the instrument, its construction and its uses. These writings may be called the "Old Corpus" on the astrolabe. The texts were edited by Millàs[2] and Bubnov[3]. One of these texts, *De mensura astrolabii*, inc. *Philosophi quorum sagaci* (called *h'* by Millàs), was accompanied by a table of 27 stars to be placed on the rete of the astrolabe. I have edited the table as "Type III" in 1966.[4] The oldest appearance of this table seems to be in MS Paris, BNF lat. 7412, fol. 5v. Here the table is still organized in the Arabic way: it has to be read from right to left (though a direct model of the table cannot be spotted in an Arabic source). It contains 26 stars (one star, η UMa, is omitted) with two coordinates: *mediatio coeli*[5] (here called *altitudo*) and another value (called *latitudo*) which may "best correspond to the maximum altitude of the stars at a geographical latitude of 39° (Valencia?)"[6]. The stars are mentioned with their Arabic names, in Latin transliteration; to several of the Arabic names tentative, often suitable, Latin translations are added.[7] In a re-organized form and with 27 stars, then, the table appears in the text *h'* and in endless repetitions in

[2] J.M. Millàs Vallicrosa, *Assaig d'història de les idees físiques i matemàtiques a la Catalunya medieval*, Barcelona, 1931.

[3] N. Bubnov (ed.), *Gerberti postea Silvestri II papae opera mathematica*, Berlin, 1899.

[4] Kunitzsch, *Typen*, pp. 23-30.

[5] I.e., the degree of the ecliptic culminating together with the star.

[6] A proposal of E. Dekker; see P. Kunitzsch and E. Dekker, "The stars on the rete of the so-called «Carolingian astrolabe»", in *From Baghdad to Barcelona. Studies in the Islamic Exact Sciences in honour of Prof. Juan Vernet*, Barcelona, 1996, II, p. 656 n. 8. The table was edited, in an unsatisfactory form, by W. Bergmann, in *Francia* 8 (1980), 84; a partial edition was given by M. Destombes in *Archives internationales d'histoire des sciences* 15 (1962), 27 (Table II, columns 2-4).

[7] See, e.g., *brachium* for *addirahan*, Arabic *muqaddam al-dhirāʿayn* (= Maslama's table, star no. 8; in the Latin version, star no. 16, also correctly *antecedens brachia*). This is a "ghost star", already in Maslama's table and then wandering through all versions of the star table of "Type III"; its Arabic name indicates α Gem, but its coordinates point to a region of the sky where α Hya would be the brightest star. Also on many Andalusī, Maghribī and Latin astrolabes the star, with this name, was subsequently inscribed in the approximate region of α Hya, according to the coordinates given in the star table.

manuscripts down to the 15th century. In this (main) form the table registers as *latitudo* the mediations, and as *altitudo* a special value never mentioned in Arabic texts and in later Latin astrolabe treatises; the value seems to be read off from an (Arabic) astrolabe in the hands of the first author of that type of the table. The *mediatio* (here: *latitudo*) could be read off on the graduated ecliptic ring of the rete of an astrolabe by putting a ruler through the centre of the instrument and through the star. To find the second coordinate, here called *altitudo*, the ruler was put at right angles through the first line at the position of the star until the graduated outer rim of the astrolabe. Here, the number of degrees between the meeting point of the first line on the rim and each of the two meeting points of the second line were counted, and this (i.e., in one of the two directions) is the value called *altitudo*.[8] In this form the table was taken over by Ascelinus of Augsburg (early 11th c.) in his treatise on the construction of the astrolabe,[9] and afterwards - obviously from Ascelinus - by Hermannus Contractus (1013-1054), also in a treatise on the construction of the instrument.[10]

Somewhere in time between the "Old Corpus" and Ascelinus may be located a text on the description and the uses of the astrolabe, commonly known as *De utilitatibus astrolabii*, Inc. *Quicumque astronomicae discere peritiam disciplinae*, edited by Bubnov[11] and called by him "J". Its author is unknown. Some scholars ascribed it - with little probability - to Gerbert of Aurillac (ca. 940/50-1003; as pope, Silvester II, 999-1003); Bubnov

[8] In some later reproduction of this star table, this coordinate was aptly called *longitudo ex utraque parte* ("Type XI", sources *a-d*, in Kunitzsch, *Typen*, pp. 67-71). The numerical values of this coordinate are different from those of the *latitudo* in MS 7412.

[9] Edited from only one, incomplete, manuscript by W. Bergmann, *Innovationen im Quadrivium des 10. und 11. Jahrhunderts*, Stuttgart, 1985, pp. 223-225; new edition, from five manuscripts (with additional notes from a sixth manuscript on a separate sheet added to the paper), by C. Burnett, in *Annals of Science* 55 (1998), 343 ff., see also Figs. 13-15.

[10] Edited, from MS Munich, Clm 14836 (written still during Hermannus' life-time), by J. Drecker, in *Isis* 16 (1931), 200-219.

[11] Bubnov, *Gerberti opera mathematica*, pp. 109-147.

edited it as a dubious work of Gerbert; others ascribed it, more cautiously, to a pupil of Gerbert. Chapter 17 of the treatise is entitled *De vocabulis Latinis et Arabicis stellarum et formationibus earundem*. In it the author tries to identify the stars of the table of "Type III", which were registered there only with their Arabic names, among the traditional classical constellations that were transmitted in Latin texts since late Antiquity.

The transmission of the text *J* is in itself utterly complicated. Bubnov edited it as a text of altogether 21 chapters. Some years ago, W. Bergmann[12] tried to demonstrate that there exist, in the numerous manuscripts, two versions of the treatise: the original one, in 19 chapters, and a version revised by Hermannus Contractus, in 21 chapters. However, afterwards it was observed[13] that this straightforward classification obviously cannot be maintained any more since, e.g., one manuscript (Munich, Clm 560) claimed by Bergmann for the basic 19-chapter version, does contain the 21 chapters and some other elements noted by Bergmann for the 21-chapter version and was copied in the early 11th century, i.e. before the time of Hermannus' activity. (Lately I inspected MS Clm 560 *in loco* and found these observations fully confirmed.) I shall not enter here into more details about this delicate problem, especially because I do not have copies from all the relevant manuscripts at hands. Moreover, I shall discuss a few selected items from the star descriptions in ch. 17 of the treatise which are in themselves doubtful.

The description begins in the north, with Benenaz = η UMa (to which some manuscripts add its second Arabic name, Alkaid, sometimes written as Alcaio), and proceeds towards the south, with some irregularities.

12 Bergmann, *Innovationen*.

13 Private communications of Prof. A. Borst, Konstanz, who has collected and collated all these manuscripts in connection with preparations for a new critical edition of Hermannus' scientific writings. Important in this respect is also the "Konstanz Fragment", four pages from a manuscript written at Reichenau around AD 1008 and containing the latter part of our chapter 17; see the edition by A. Borst, *Astrolab und Klosterreform an der Jahrtausendwende*, Heidelberg, 1989 (Sitzungsberichte der Heidelberger Akademie der Wissenschaften, Phil.-hist. Klasse, 1989, 1), esp. pp. 112-127. This very early fragment contains already elements ascribed by Bergmann to the revised version of *J*.

Several of the Arabic star names - which are written in all the sources in a great variety of misspellings - are confused. For Alrif (α Cyg) an alternative form Archeitus, Arrectus etc. is mentioned[14], the derivation or background for which is not obvious.

In Auriga two names are mentioned, *Menreb Alroech id est humerus* (of this, *Menreb id est humerus* belongs to Orion, not to Auriga; for Alroech a better spelling in some manuscripts is Alhaioch which is the correct Latinized Arabic name of α Aur), and *Rigel id est pes* (this is the correct name for β Ori and belongs there; it is as such correctly mentioned below, *in Geminis*, in the sign of Gemini).

The confusion about Telum, Aquila and Alhadib has been settled by Bergmann.[15] Apparently the intention was to assign Alhadib[16] to Telum (*sic* instead of Cassiopeia), and Alceir (α Aql) to Aquila (where some manuscripts, instead, refer to Cygnus a second time).

In the sign of Gemini two stars of Orion are mentioned: *Alhaioch Algeuze id est humerus* (α Ori; here wrongly the word Alhaioch - name for α Aur - is written, whereas the element Menreb belonging to α Ori was wrongly added above to the name of α Aur) and *Rigel id est pes* (β Ori; Rigel was wrongly mentioned above under Auriga also, and we shall find it a third time later, below).

In the sign of Leo two stars are listed: *Aldiraan id est frons*, and Calbalazeda (α Leo). For Aldiraan, a "ghost star", cf. note 7, above. Neither the Arabic name nor the coordinates (in the table) contain an

[14] See Bubnov, *Gerberti opera mathematica*, p. 137, 3. This star's description is omitted in Bergmann's parallel edition of the chapter in the two versions: Bergmann, *Innovationen*, p. 221.

[15] Bergmann, *Innovationen*, p. 77f.

[16] This star appears already in Maslama's star table (Kunitzsch, *Typen*, pp. 15-18, star no. 21, with the name of β Cas, but with confused coordinates - longitude and latitude alone, mediation and declination not given - rather pointing to the area of Pegasus). Also in the star table of "Type III" the coordinates of Alhadib are odd; in MS 7412, 5v the mediation is given as *Trutina* [leg. *Truta* = Pisces!] XVI°, which appears in the other versions as Taurus [*sic*] XXII°. Astronomers really trying to locate this star in the sky were of course perplexed.

allusion to *frons*, i.e. the Lion's forehead. It is obvious that here some reader, not content to be unable to find the identity of Aldiraan, tried to identify this star with some star in Leo; he chose *frons*, a name occurring in other texts belonging to the same period as the "Old Corpus" of astrolabe treatises, as the name of the 10th Lunar Mansion, *al-jabha*, "the (Lion's) Forehead", consisting of $\zeta\gamma\eta\alpha$ Leo. Since α Leo appears among the astrolabe stars separately, this reader probably aimed at ζ Leo. This interpolation appears in two forms: often *id est frons* is simply added to the unintelligible name Aldiraan; but in some manuscripts, instead of Aldiraan the true Arabic name of *frons* is found: Aliebaha (MS Munich, Clm 14763) or Algebaha (MS Ripoll 225[17]).[18]

After completing the stars in the zodiac the author arrives at the stars of the southern hemisphere. Here he seems to be confused, unable to identify the names with certain stars (though in the area of northern France, where these identifications were probably made, all of the astrolabe stars mentioned in the table of 27 stars could well be observed). He mentions ι and ζ Ceti, *quae aut raro aut numquam in nostris climatibus cernuntur*. Hereafter he continues: *Est et Rigel et Alhabor*. The second of these two stars is Sirius, α CMa, the brightest of the fixed stars, which must have been nicely visible in the author's place. The first star is enigmatic.

In manuscripts of the 21-chapter version it is called Rigel, i.e. with the name of β Ori mentioned already above, once in its correct place and once erroneously under Auriga. In manuscripts of the 19-chapter version -but also in MS Munich, Clm 560, and in the Konstanz Fragment (note 13) - we find, instead, the name Addeleni. Bergmann assumes that the revisor of J - in his opinion, Hermannus - has replaced the unintelligible Addeleni by Rigel (though this makes not much sense, because Rigel was already mentioned earlier, in its correct place, as β Ori). In its earliest form, it seems, the name was spelled Addelem (Paris, BNF lat. 7412, 8r; Munich,

17 *Apud* Millàs, *Assaig*, p. 155 note 1.

18 Cf. the Lunar Mansion name *Alcebata - frons*, from the *Mathematica Alhandrei summi astrologi*, MS Paris, BNF 17868 (10th c.), in Millàs, *Assaig*, p. 251.

Clm 560;[19] Leiden, Scal. 38). This then degenerated into other forms, such as Addeleni, Adelem and Addelen. Until the moment of writing these lines I am not able to explain this name (just as Archeitus/Arrectus, above). It does not echo any of the Arabic names of the 27 stars contained in the list of astrolabe stars accompanying the "Old Corpus" of astrolabe texts. One would not easily be ready to assume behind this word the name of any other star, outside the 27 stars of the tradition.

The text continues with three other names of stars on which the author "has nothing to say" (*Quid autem dicam de his dum a nobis minime videntur?*): Algomeiza Aldirnam (the first word is the name of α CMi; the second is wrongly repeated from above, under Leo); Ganamalgurab (γ Crv) and *Alcasal*[20] *vel Alhimech* (α Vir),[21] both of which are located far too deep in the south, *in Centauro*.

Four stars out of the 27 are totally omitted in the description: α Oph (Alhauui), β Per (Algol), θ UMa (Arrucaba) and ι UMa (Egreget).

In the present state of exploration of the history and transmission of the text J, I think, it would be premature to give definite explanations of how, when and by whom the various corruptions and deficiencies were brought into the text of this chapter. At least, most of the Arabic star names here mentioned can be traced in the tradition of the table of 27 astrolabe stars

[19] Bubnov (*Gerberti opera mathematica*, p. 138, App. under *az*) has wrongly registered from MS Mon H (= Clm 560) the spelling Addeiemz. This spelling caused me, in my book *Arabische Sternnamen in Europa*, Wiesbaden, 1959, p. 70f., no. 19, to explain Addeiemz (et varr.) as a corruption of the Arabic *al-jawzā'*, the name of Orion and also Gemini (in Latin, *alieuze* → *alienze* → *addeiemz*). After inspecting MS Clm 560 *in loco*, this explanation must be withdrawn. The true reading of the entire phrase in Clm 560 is: *Est et AdDELEMZALHABOR*. The copyist wrote Addelem; then he mistook the abbreviation "et" between the two Arabic names as a Z and wrote the two separate names together in one conglomeration. So we know that Clm 560 has the spelling Addelem and that this name cannot be derived from Arabic *al-jawzā'*, Orion.

[20] Clm 560 has Algazal, not registered by Bubnov, *Gerberti opera mathematica*, p. 138, App. under *bg*.

[21] The two words together render the full Arabic name of α Vir, *al-simāk al-aʿzal*, "the unarmed Simāk"; but the element Alcasal/Algazal is only rarely found in Latin in the "Old Corpus". The star table in BNF 7412, 5v, has: Alhazel.

("Type III"), except for Archeitus/Arrectus and Addelem/Addeleni for which no plausible explanation is ready at hand.

On the occasion of this paper I should like to mention another unexplained star name belonging to the same tradition. MS Paris, BNF 7412, contains, after a compilation of portions of texts from the "Old Corpus", on foll. 19v-23v drawings of the rete, the seven plates (constructed for the latitudes of the seven climates) and the back of an Andalusī astrolabe. Its inscriptions, in Arabic, in the Andalusī cufi ductus, were also copied by the Latin draftsman. On the back, he even copied the maker's name: Khalaf ibn al-Muᶜādh.[22] While the inscriptions of all the other parts of the instrument are copied in Arabic characters, the names of the 27 stars on the rete on fol. 19v are given in Latin transliteration. To each name in the rete a serial letter or mark is added, and at the bottom of the page for each of these letters or marks alternative spellings of the respective names are given. All the names here mentioned belong to the tradition of the 27 stars of "Type III".[23] Against Addiraan, in the rete, we read, written upside down: *vel liragenz* (or *-geni*). If the reading *-genz* is correct, one would assume here another corruption of Arabic *al-jawzā'*, Orion or Gemini. With the reading *-geni*, no explanation comes up. The first part of the word, *lira-*, remains also unexplained. For a better understanding other spellings of the name in other manuscripts must be awaited.

[22] Cf. the description by P. Kunitzsch, "Traces of a tenth-century Spanish-Arabic astrolabe", in *Zeitschrift für Geschichte der Arabisch-Islamischen Wissenschaften* 12 (1998), 113-120.

[23] These names were edited by Kunitzsch, *Sternnamen*, p. 90f. After having available a better photograph from the manuscript, I here can correct a few readings: at *a*, column 2, read: Alwagakba; at *b*, read Alkaio; at *l*, in the Arabic, read *munīr al-fakka*.

XXI

A Note on Ascelinus' Table of Astrolabe Stars

Summary

The treatise on the construction of the astrolabe of Ascelinus of Augsburg, edited by C. Burnett in *Annals of Science*, 55 (1998), 343 ff., contains, in two of the six known manuscripts, a table of 27 stars. This star table belongs to 'type III' in the tables edited by P. Kunitzsch in 1966. Here it is shown that now, after several decades of more research and edition of texts, a more detailed classification of the star table of 'type III' can be given. Four subgroups can now be distinguished. Specifically for Ascelinus' star table, the characteristic change in the names of three stars is demonstrated. Further, the additional designation of two stars as *caput* and *cauda serpentis* respectively in Ascelinus' version of the table is explained.

Contents

1. The star table of 'type III' reconsidered

In *Annals of Science*, 55 (1998), 343ff., C. Burnett has edited the treatise on the construction of the astrolabe of Ascelinus of Augsburg (early eleventh century AD). In two manuscripts the treatise is accompanied by a table of 27 stars to be marked on the rete of the instrument (*Annals of Science*, 55 (1998), 350, and Figures 13–15). This star table is, as rightly pointed out by Burnett (p. 344), another representative of the star table associated with the 'Old Corpus' of astrolabe treatises composed in northeastern Spain in the late tenth century. The table was edited by P. Kunitzsch in 1966 as 'type III'.[1] Studies and editions made in the following decades now allow a more detailed classification of the various subgroups of this star table of 'type III'.

Its first representative appears to be the table of 26 stars (one star, Benenaz = η UMa, is here omitted) in MS Paris, BNF lat. 7412 (eleventh century), 5v. This table still echoes an Arabic model; it is written in two columns, each comprising 13 stars, to be read, in the Arabic manner, from right to left. The title of the first column at right, above the names of the stars (Arabic, in Latin transliteration), has still preserved the Arabic: *alkewekib .i. stellę* (i.e. Arabic, *al-kawākib*, 'the stars').[2] In this earliest form the table registers under the title *altitudo* the mediations of the stars,[3] and

[1] P. Kunitzsch, *Typen von Sternverzeichnissen in astronomischen Handschriften des zehnten bis vierzehnten Jahrhunderts* (Wiesbaden, 1966), 23–30 ('Type III').

[2] The table was edited, in an unsatisfactory form, by W. Bergmann, in *Francia* 8 (1980), 84; a partial edition was given by M. Destombes, in *Archives Internationales d'Histoire des Sciences*, 15 (1962), 27, Table II, columns 2–4.

[3] *mediatio* indicates the degree of the ecliptic culminating together with a star.

under the title *latitudo* values that 'would best correspond to the maximum altitude of the stars at a geographical latitude of 39° (Valencia?)'.[4] The sequence of the stars in this form of the table differs from the unified sequence in the other three subgroups.

The main representative of 'type III' is the table of 27 stars as edited by Millás, forming part of the treatise *De mensura astrolabii* (h′), Inc. *Philosophi quorum sagaci studio*.[5] Here the mediations are registered under the title *latitudo*. Under the title *altitudo* is to be found a coordinate which only exists here, in 'type III', without parallels in the Arabic or in the later Latin astrolabe literature. In later star tables where these values were repeated ('type XI' in Kunitzsch (note 1), 67–71, sources a–d) this coordinate is appropriately called *longitudo ex utraque parte*. It seems to show values that were read off by the first author from an Arabic astrolabe in his hands; the *mediatio* (here, *latitudo*) can be read off on the graduation of the ecliptic ring of the rete of the astrolabe by putting a ruler through the centre of the instrument and the star; to obtain the second coordinate the author put the ruler at right angles through the first line at the position of the star and then counted, on the graduated outer rim, the number of degrees from the point where the first line met the rim, in both directions, left and right, to the two points where the second line met the rim. The number of degrees between the first and each of the second two meeting points gave the value called by him *altitudo*.[6]

The third subgroup of 'type III' is formed by the star table added to Ascelinus' treatise. It repeats the coordinates of the second subgroup, unchanged (except for scribal errors in the transmission of the Roman numerals). Characteristically, the names of the three stars, ζ Cet, ι Cet and δ Cap, are here written in a special way (see below).

The fourth subgroup, then, is again the same star table as used by Hermannus Contractus (1013–1054) in his treatise on the construction of the astrolabe.[7] It shows the same characteristics as Ascelinus' table and may be regarded as being taken over from him (Hermannus may have written his treatise around 1045 AD[8]).

The second to the fourth subgroups show the star table in the same form. The sequence of the stars, their (Arabic) names and their coordinates are basically the same (apart from scribal errors and variants).

2. Some comments on Ascelinus' star table

The star table of 'type III' as found in Ascelinus' treatise (and afterwards in the treatise of Hermannus Contractus) shows a characteristic modification of three of the 27 Arabic star names. The names of ζ Ceti, ι Ceti and δ Capricorni were each composed, in the Arabic, of two nouns, the first being in the nominative case and the second being added to the first in the genitive case: *baṭn qayṭus* ('the belly of κῆτος, Cetus'), *dhanab qayṭus* ('the tail of κῆτος, Cetus') and *dhanab al-jady* ('the tail of the young goat = Capricorn') respectively. In MS 7412, 5v, they appear as *patangaitoz*,

[4] Cf. P. Kunitzsch and E. Dekker, 'The stars on the rete of the so-called "Carolingian Astrolabe"', *From Baghdad to Barcelona. Studies in the Islamic Exact Sciences in Honour of Professor Juan Vernet* (Barcelona, 1996), II, 656, note 8.

[5] J. M. Millás Vallicrosa, *Assaig d'Història de les Idees físiques i Matemàtiques a la Catalunya Medieval* (Barcelona, 1931), 301 f., and Làmina VII (MS Ripoll, 225, 9v–10r).

[6] Cf. Kunitzsch and Dekker (note 4), 656 f. (this value is here called 'marginal longitude').

[7] Edited, from MS Munich, Clm 14836 (eleventh century, still written during Hermannus' lifetime), by J. Drecker, in *Isis*, 16 (1931), 200–219; the star table is on p. 209. (Drecker's rendering of the manuscript's readings is not always reliable.)

[8] Cf. A. Borst, *Astrolab und Klosterreform an der Jahrtausendwende* (Heidelberg, 1989); *Sitzungsberichte der Heidelberger Akademie der Wissenschaften, Phil.-hist. Klasse*, 1 (1989), 79.

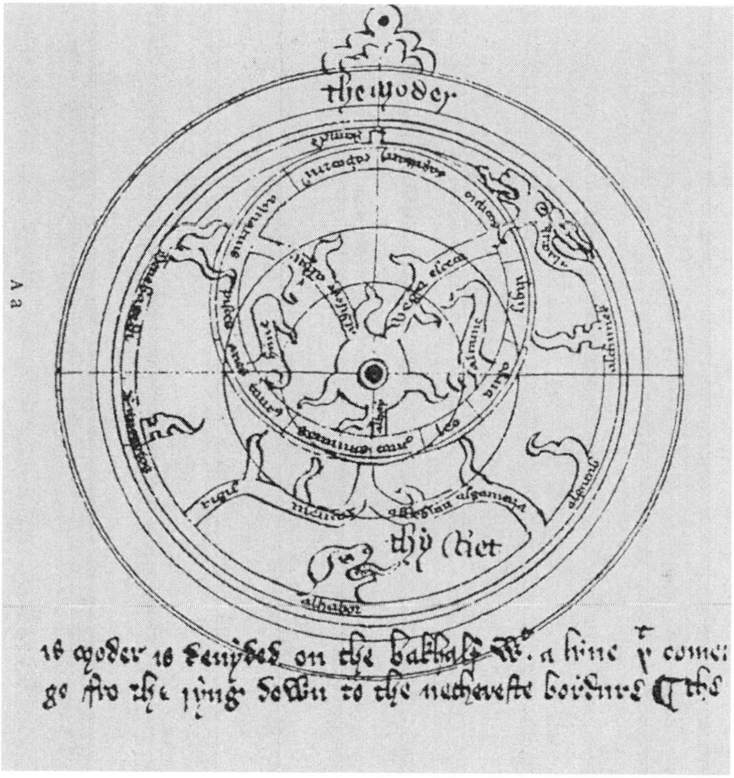

Figure 1. Drawing of a rete in the Chaucer MS, Cambridge University Library, Dd.3.53, here reproduced from Tomba (note 10), Figure 3.

denebgaitoz and *denebaliedi* respectively. Similarly, in the second subgroup (edited by Millás), they are *Pantangaitot*, *Denebgait* and *Denebalix* (with many variants in other manuscripts). Now, in Ascelinus and similarly in Hermannus we read *Gaitozpatan*, *Gaitozderep* and *Licdideneb* (again with various spellings). That means that someone (not necessarily Ascelinus himself, but perhaps a scribe of one of the sources lying behind) cut the three names into two pieces and recomposed them, now erroneously placing the genitive case in front of the nominative, of course in ignorance of the original Arabic forms hidden behind the transliterated names written in Latin. Thus, *Pantangaitot* (etc.) became *Gaitozpatan*, *Denebgait*[*oz*] (etc.) became *Gaitozderep*, and *Denebaliedi* (etc.) became *Liedi-* or *Licdideneb* (etc.). All these Arabic names were copied in endless variants by the Latin copyists, but the specific regrouping of the elements of these three names is characteristic of the version of the star table found in Ascelinus and in Hermannus.

In MS Avranches, Burnett (p. 350, notes 221 and 226; cf. also Figures 13 and 14) registers that a hand A^1 added to the name of α Sco, *Calbalagrab*: *caput serpentis*,[9]

[9] Misread as *caput Scorpionis* by W. Bergmann, *Innovationen im Quadrivium des 10. und 11. Jahrhunderts* (Stuttgart, 1985), 225, note i. Bergmann edited Ascelinus' treatise from MS Avranches alone (pp. 223–225) which he thought to be a unique copy of the text. Burnett was now able to use five

Figure 2. Front of a Latin astrolabe, fourteenth century, private collection of T. Tomba, Milan. Reproduced from Tomba (note 10), Figure 1.

and to the name of δ Cap, *Licdideneb*: *cauda serpentis*. These words were also added in MS London, here by the same scribe (cf. Figure 15, *ibid.*). At the first glance, one would understand these additions as additional alternative names for the two stars, although it would seem highly improbable that 'the heart of Scorpius' is also called 'the head of the serpent' and that 'the tail of Capricorn' is also called 'the tail of the serpent'. Both designations, nevertheless, are correct in their way, but they are no star names. Moreover, they indicate the location of the two stars on the rete of an astrolabe. On many (especially Latin) astrolabes the rete is bordered, along the tropic of Capricorn, by a metal band beginning at the upper left, near the position of the star δ Cap, continuing downwards and around a large part of the rete and ending in the upper right, near the position of the star α Sco. On many astrolabes this band was artificially worked out in the shape of a serpent, with its head near the position of α

manuscripts for his recent edition and to give the collation of a sixth manuscript in an Addendum on a separate sheet added to his paper.

Sco and the tail near the position of δ Cap. A nice specimen is the drawing of a rete in the Chaucer manuscript, Cambridge University Library, Dd.3.53 (see our Figure 1).[10] There even exists an astrolabe with a rete of nearly the same shape (see our Figure 2).[11]

Here it can be easily seen that α Sco is at *caput serpentis*, and δ Cap at *cauda serpentis*. More examples can be cited for this shape of the band along the tropic of Capricorn in the form of a serpent as follows: drawing of a rete in MS Vat. Regin. 598 (eleventh century), 120r (here, both ends of the band are formed as serpent's heads);[12] the drawing of the rete in MS BN 7412, 19v, shows the head of the serpent in an unclear manner, like a sketch rather and not fully worked out; a Latin astrolabe in the Museo di Palazzo Madama, Torino;[13] a Latin astrolabe in the Pinacoteca Ambrosiana, Milan (here, the serpent's head is on the left near δ Cap, and the tail on the right near α Sco);[14] a Latin astrolabe in the Germanisches Nationalmuseum, Nuremberg, inv. no. WI6;[15] a Latin astrolabe in the National Museum of American History, Washington.[16] There may exist many more drawings in manuscripts and surviving astrolabes with this type of rete. It seems that all the examples are Latin. So far, I have not been able to find the same decoration on an Arabic or other Muslim astrolabe. In any case, the meaning of Ascelinus' *caput* and *cauda serpentis* besides α Sco and δ Cap is now sufficiently clear and the reason behind these designations is well documented.

[10] See T. Tomba, 'Un astrolabio inglese medioevale della "tradizione Chaucer"', *Istituto Lombardo –Accademia di Scienze e Lettere, Rendiconti*, B, 126 (1992), 77, Figure 3.

[11] *Ibid.*, 126 (1992), 73, Figure 1 (private collection of T. Tomba, Milan).

[12] Millás (note 5), Làmina X.

[13] See T. Tomba, 'Tre astrolabi latini, del XIV secolo, conservati in Italia', *Estratto da: Rassegna di Studi e di Notizie*, 20 (22) (1996), Figure 1.

[14] See T. Tomba, 'Gli astrolabi della collezione Settala nella Pinacoteca Ambrosiana', *Atti della Fondazione Giorgio Ronchi*, 33 (2) (March–April 1978), Figure 7.

[15] See *Focus Behaim Globus* (Nuremberg, 1992), II, 578, Figure 1.74.1.

[16] See S. Gibbs and G. Saliba, *Planispheric Astrolabes from the National Museum of American History* (Washington, 1984), 150, Figure 99.

XXII

On Six Kinds of Astrolabe: a Hitherto Unknown Latin Treatise

The Latin manuscript in Florence, Biblioteca Nazionale, conv. soppr. J.2.10 (=San Marco 200), was described by A. Björnbo in 1912.[1] It is a collection of astrological, astronomical and other texts, written in different hands of the 13th to the 15th centuries. Some of the texts are incomplete and there are several blank folios and pages. The manuscript needs a re-examination *in loco*. From a microfilm kept in the Institute for the History of Science, University of Munich, we recently controlled the manuscript in search of a further copy of Ptolemy's *Planisphaerium* in Hermann of Carinthia's Latin translation from the Arabic. It is now clear that this text does not exist in the manuscript.

On the other hand, we could identify the text on fol. 65r–72r cited by Björnbo, *loc. cit.*, p. 196 (item 2), with the "incipit" *A phisicis spera gestibilis ad exemplum composita.*[2] It is part of Adelard of Bath's treatise on the astrolabe (*De opere astrolapsus*). Of this text Bruce George Dickey (Toronto) is currently preparing an edition. In the Florence manuscript the beginning is lost; the text opens on the top of fol. 65r (after two blank folios), in the middle of a sentence corresponding to the last line on fol. 88r in MS Salzburg a.V.2 (where the true beginning is on fol. 82r, inc. *Quod regalis generis nobilitas*...), and ends on fol. 71rb, line 7 from bottom, with the explicit as in MS Salzburg, fol. 101r, ... *quandiu te proficere non penitebit.*[3] The remaining bit of text until the bottom of fol. 72r does not form part of Adelard's treatise. It is a collection of astronomical definitions beginning *Residuum latitudinis lune* ..., and ending ... *tot horis distas* [*sic*] *uersus orientem et cetera.*

The text to which the present paper is mainly dedicated is in the Florence MS on fol. 175v–176r. It is entitled *Incipit liber de diuersis speciebus astralabiorum et earum* [*et*] *probatione utrum scilicet uera sit...*, and it ends at the bottom of the second page, *Explicit liber de diuersitate specierum astrolabiorum et de eorum utrum vera sit necne probatione*; the short treatise thus appears to be complete. Its hand can be attributed roughly to the 13th century. The text is divided into six *distinctiones* of varying length (the title for the first *distinctio* is absent). The word *distinctio* echoes the Arabic *faṣl*, similarly used to designate sections in books or treatises.

The title – on various, or six, kinds of astrolabe – is somewhat misleading. In the Arabic tradition the phrase "various kinds of astrolabe" usually refers to astrolabes with retes differently shaped from the retes of the normal (northern or southern) astrolabe. Here, however, the Latin author intends astrolabes of varying completeness, with almucantars drawn for each degree or for each second, third, fifth, sixth or tenth degree of altitude. In Arabic theory all of these would belong to one kind of astrolabe, *viz.* the normal northern (or southern) astrolabe.

The first *distinctio* describes these six kinds of astrolabe, indicating the number of almucantars and the graduation of the signs of the zodiac in the rete for each kind. The second *distinctio*, the longest, gives a description of the parts of the astrolabe and the circles and lines drawn on it, giving for most of them their names in Arabic. The great number of Arabic words is striking; among them there are many which do not occur in any other Latin astrolabe treatise known to us. The third *distinctio* relates three methods for verifying the exactness of the astrolabe. The fourth to sixth *distinctiones* refer to the use of the astrolabe, listing a number of uses without, however, describing any of the corresponding procedures.

The manuscript does not mention an author for the treatise. Because of certain elements (acquaintance with Arabic in general; the frequent transliteration of the Arabic article *al-* as *el*; the use of *scannum* for the sights on the alidade; the use of such verbs as *perpendere*, etc. – the latter two similarly appear in Adelard's astrolabe treatise) we are inclined to consider Adelard of Bath as a possible author. The treatise shows the author's close acquaintance with the Arabic literature on the astrolabe.[4] On the other hand, the treatise is somewhat different in content and arrangement from the usage followed in Arabic treatises on the use of the astrolabe: in Arabic, the first chapter is usually dedicated to the description of the parts of the instrument and their names; then follow the uses, normally beginning with a chapter on taking the sun's altitude. Astrological matters are dealt with in the last chapters. The drawing and division of the almucantars is normally included in the introductory chapter, on the instrument's parts, circles and lines. A chapter (or chapters) on the verification of the instrument's exactness does not occur in all of the Arabic treatises.

Our Latin text, on the other hand, begins with a detailed description of the various systems of dividing the almucantars and the zodiac. Then follows what in Arabic is usually the introductory chapter, on the instrument's parts, followed by a chapter on the instrument's verification. The uses – occupying in Arabic treatises from about forty to more than a hundred or even several hundred chapters – are here reduced to three *distinctiones* which do not describe any operations; they just list some of the problems that can be solved by means of the astrolabe, as if merely listing some chapter headings.

It therefore seems that our text is not the Latin translation of some Arabic treatise. Moreover it seems that it is an abbreviated extract from an Arabic source or Arabic sources, rearranged and written down by a Latin author as something like a guideline for further work of his own on the subject. If Adelard was indeed the author, our text could be regarded as a preliminary study, based on Arabic material, preceding his treatise on the astrolabe which, as is now generally assumed, was written in his late career, near A.D. 1150.

NOTES

1. A. Björnbo 1912, pp. 194–201.
2. Cf. also L. Thorndike & P. Kibre 1962, col. 154.

3. The Florence MS is not mentioned by C. Burnett in his catalogue of manuscripts of Adelard's works; see Burnett 1987, pp. 168f. (work no. 10).
4. In Adelard's astrolabe treatise there seems to be an allusion to Maslama al-Majrīṭī in connection with the excentric zodiac on the back of the astrolabe (MS Salzburg, fol. 90r: ... *quod in uno solo, videlicet doctoris Almirethi, astrolabio observatum esse repperi*). Cf. the introduction to Dickey's forthcoming edition.

Here follows the edition of the treatise from the Florence MS, fol. 175v–176r. Explanations of the Arabic names are given in the notes to the Latin text; as far as possible, reference is made to Kunitzsch 1983. For the Arabic terminology of the astrolabe in general, see Kunitzsch 1983; Morley 1856; Hartner 1939 and 1960.

<175v> Incipit liber de diuersis speciebus astralabiorum et earum[a] probatione utrum scilicet uera sit[b].

<A>strolabiorum igitur sex sunt species. Est enim quedam perfecta que dicitur *cullie*[1] id est universalis habens *almukantarat*[2] 90 que singulorum graduum interuallis ab inuicem distant, secundum hanc autem unumquodque signorum in 30 partes diuiditur. Est et aliud eorum quod dicitur *nuzfi*[3] id est dimidiale habens progressionarios 45 quibus bini ac bini gradus intersunt, huius autem omne signum in 15 partes secatum. Est[c] et eorum quod dicitur *thulchi*[4] id est ternale habens *elmukantaraz* 30 inter quos terni gradus numerantur, huius autem omne signum in 10 partes dissipatur. Est iterum eorum *khumci*[5] id est quinale habens 18 *elmukantarat* inter quos numero quini et quini gradus sunt, cuius omne signum in 6 partes diuiditur. Eorum quoque est *szuci*[6] id est senale habens 15 *elmukantarat* inter quos numero seni et seni gradus sunt, huius autem omne signum in 5 partes diuiditur. Eorum denique est *ashri*[7] id est decenalis habens 9 progressionarios inter quos deni et deni gradus numerantur, huius autem omne signum in 3 partes diuiditur.

Distinctio 2[a], de cunctis astrolabii membris memorandis operibusque in eis operandis singulorumque nominibus[d]. Primum ergo eorum est *elhelka*[8] quod dicitur iterum quasi subspendiculum cum quo subleuatur astrolabium[e] pro altitudine examinanda. Deinde *elorwa*[9] suspendicula mutuo amplexu coniuncta. Deinde *elvm*[10] id est mater[f] supra quam *elhogera*[11] id est solium[g] in 340 [*sic*] partes diuisa diciturque iterum quasi girus orbite estque circumdans laminas atque *elankabud*[12] atque subtus illis clauum qui tenet laminas ne circumferantur. Deinde lamine in quibus descripti sunt *almukantarat* prenominati qui locus super terram dicitur. Linee quoque horarum in eis descripte sunt scilicet linee ille breues lineate a circulo cancri ad circulum capricorni, sunt autem omnes 11 linee atque inter eas 11 hore continentur, locus autem hic[h] sub tellure dicitur. In eisdem quoque latitudines regionum sunt iterum designate apud circulum capitis cancri qui est circulus paruus. Circulus autem arietis est circulus medius, at uero capricorni circulus magnus qui est in extremitate laminarum. Deinde *elankabud* scilicet rethe mobile distinctum supra se signa stellasque fixas habens. Deinde *nitakelburus*[13] id est vinculum uel cinctorium[i] signorum ...[k] scilicet linea que exiens a capite arietis transit supra perpendicularem poli spere applicat apud capud libre. Est autem linea hec terminalis in zodiaco diuidens inter signa septentrionalia et australia. Australia enim sunt cuncta exeuntia a linea capitis arietis et libre recedentiaque ab illa ad australem partem scilicet uersus polum antarticum. Septentrionalia uero sunt omnia ingredientia in eam exeuntiaque a linea capitis arietis et libre ad partem septentrionalem scilicet uersus polum articum. Locum autem eum polum in quem de medio astrolabii foramen scilicet rote in quo est axis. Hii autem duo poli *kameni*[14] inueniuntur in spera preter astrolabium in quo non inuenitur nisi polus articus tamen nisi fuerit australe fueritque *khat heliztaxe*[15] in ipso ad partem australem. Polus ergo antarticus reperitur ibi solus. Deinde *elmantakate elburus*[16] id est zona signifera scilicet orbis in quo sunt signa 12. Deinde *elmuri elegzea*[17] id est numerator partium diciturque iterum iuxta capud capricorni est augmentum quod est in rethi supra uerticem capricorni factus ad deliberandum cum eo partes que sunt supra *helhogeram* id est solium[l]. Deinde *elmudir*[18] id est volutor siue girator et est scruplus qui est supra faciem rethis factus ibi propter circumferendum rethe cum eo. Deinde *elaidhade* id est <176r> *allidada*[19] habens duas aures que iterum perhibentur quasi duo scanna perforata per que deprehenditur radius solis in die aspiciturque ab eis ad astra per noctem estque premens dorsum astrolabii ex quo loco altitudo perpenditur. Deinde *elkod*[20] id est axis scilicet perpendicularis qui tenet repagulum et matrem et laminas et *elankabud.* Deinde *elferaz*[21] id est caballus scilicet qui intrat in perpendicularem axem. Deinde *khat elizere*[22] id est[m] videlicet linea recta porrecta ab oriente ad occidentem supra quam est ortus capitis arietis et libre estque secans laminam per medium et transiens super centrum. Deinde *khat nuzfe nahar elgenubie*[23] id est linea meridiana australis diciturque quasi linea recessuum itemque quasi linea medii celi estque linea que exit de subtus suspendiculo et procedens secat lineam *eliziree* et centrum lamine atque horas per medium applicatque ad clauum laminarum. Dicitur etiam ipsa extremitas que cadit supra clauum iuxta cardinem terre estque iterum linea meridiana septentrionalis. Deinde *khat elofok*[24] id est orizon orientalis atque orizon occidentalis scilicet primus *almuchatarat* super quem oritur oroscopus ex oriente occiditque oppositum ex occidente. Deinde *scenit elrool*[25] id est[n] videlicet circulus paruus qui inter lineam mediam progresionariorum inter quam scriptum est 90.

Distinctio 3[a], de probando astrolabio utrum sanum sit an infirmum. Respice ergo ad sectionem partium suarum atque partitionem progressionariorum suorum que si concordauerint erit incolume, si autem discordauerint erit inbecille. Similiter autem quemlibet[o] graduum suorum pone supra orizonta orientale et tunc respice ad oppositum eius qui si occiderit fueritque gradus 4[i] sicut gradus 10[i] erit studiosum, alioquin erit prauum. Preualet autem hiis quod probes illud per computationem *ezich*[26] ubicumque climatum fueris. Facito ergo cum isto deliberationem solis et scito horoscopum et quota instet horarum diei. Cum itaque per astralabium ista perpenderis eadem iterum excute per computationem. Si igitur consenserit id quod exierit per computationem ei quod per astrolabium erit astrolabium modis omnibus inreprehensibile.

Distinctio 4[a], de capienda altitudine solis in die atque perpendendo horoscopo ex altitudine, sed uel scilicet[p] de excudendis horis earumque minutis ex horoscopo atque de conuertendis horis temporalibus cognitis pro obliquis que sunt hore astrolabii ad horas[q] *elmustewias*[27] examinales videlicet horas *ezich* sed et arcubus atque conuertendis horis equalibus ad inequales iterum atque de faciendo circulo firmamenti ex quo ortum est dimidium orbis solis ad horam deliberationis gradus oroscopi sed et quantitate circuli in quam ex horis examinalibus earumque minutis.

Distinctio 5[a], in capienda altitudine stellarum fixarum atque vagarum noctu atque eliciendo horoscopo ex altitudine sed et conuertendis horis rectis ad curuas et iterum curuis ad rectas atque faciendo circulo firmamenti ab occasu medii corporis solis ad horam deliberationis gradus horoscopi sed et quantitate circuli[r] ex horis *elmustewiis* earumque minutis.

Distinctio 6[a], in disponendis domiciliis 12 atque cognoscendo gradu medii celi sed de faciendo ortu signorum secundum equinoctialem circulum atque exequando ortu signorum supra omnes regiones et climata atque conuertendis gradibus.

Explicit liber de diuersitate specierum astrolabiorum et de eorum utrum vera sint necne probatione.

NOTES TO THE LATIN TEXT

[a] earum: et *add.* MS [b] perhaps there follow one or two words illegible in the binding [c] Est: 3[a] species *add. in marg.* MS [d] nominibus: de membris astralabii et eorum operibus et nominibus eorundem *add. in marg.* MS [e] astrolabium: abstrolabium MS [f] id est mater: *supra* MS [g] id est solium: *supra* MS [h] hic: his MS [i] uel cinctorium: *supra* MS, cinctorium *corr. ex* cunctorium [k] here some text seems to be missing because what follows is not the description of the zodiac [l] id est solium: *supra* MS [m] blank in the line for one or two words [n] blank in the line [o] quemlibet: quilibet MS [p] uel scilicet: *supra* MS [q] horas: horam MS [r] circuli: horoscopi *add. et del.* MS

1. Ar. *kullī*, "total, full".
2. Ar. *al-muqanṭarāt* (plur.), "the almucantars, parallels to the horizon"; cf. *Kunitzsch* 1983, no. 31 (also in AB = Adelard's astrolabe treatise).
3. Ar. *niṣfī*, adjective related to *niṣf*, "half".
4. Ar. *thulthī*, adjective related to *thulth*, "a third".
5. Ar. *khumsī*, adjective related to *khums*, "a fifth".
6. Ar. *sudsī*, adjective related to *suds*, "a sixth".
7. Ar. *ᶜushrī*, adjective related to *ᶜushr*, "a tenth".
8. Ar. *al-ḥalqa*, "the ring"; cf. *Kunitzsch* 1983, no. 11.
9. Ar. *al-ᶜurwa*, "the handle"; cf. *Kunitzsch* 1983, no. 56.
10. Ar. *al-umm*, "the mother, mater".

11. Ar. *al-ḥujra,* "the rim"; cf. *Kunitzsch* 1983, no. 12.
12. Ar. *al-ᶜankabūt,* "the spider"; cf. *Kunitzsch* 1983, no. 1 (also in AB).
13. Ar. *niṭāq al-burūj,* "the belt of the signs, zodiac"; cf. *Kunitzsch* 1983, no. 37.
14. Perhaps Ar. [*al-*] *kawni,* "of the world" (genit.), i.e. the two poles of the world, a common designation in Arabic for the north and south poles (as distinguished from the two poles of the ecliptic).
15. Ar. *khaṭṭ al-istiwāʾ,* "the line of equality", a designation for the east-west line on the plates of the astrolabe frequently found in Arabic astrolabe treatises.
16. Ar. *minṭaqat al-burūj,* "the belt of the signs, zodiac"; cf. *Kunitzsch* 1983, no. 29.
17. Ar. *murī al-ajzāʾ,* "the indicator of the degrees"; cf. *Kunitzsch* 1983, no. 32a (also in AB).
18. Ar. *al-mudīr,* "the one which lets rotate", i.e. a knob on the rete for turning it around.
19. Both words are from Ar. *al-ᶜiḍāda,* "the rule, on the back of the astrolabe", *elaidhade* being the author's own transliteration and *allidada* being a vulgarized Latin form; cf. *Kunitzsch* 1983, no. 19 (also in AB).
20. Ar. *al-quṭb,* "the pole; axis"; cf. *Kunitzsch* 1983, , no. 40 (also in AB).
21. Ar. *al-faras,* "the horse"; cf. *Kunitzsch* 1983, no. 9 (also in AB).
22. The same as in note 15, *khaṭṭ al-istiwāʾ.*
23. Ar. *khaṭṭ niṣf al-nahār al-janūbī,* "the southern meridian line".
24. Ar. *khaṭṭ al-ufuq,* "the horizon line"; cf. *Kunitzsch* 1983, no. 55 (also in AB).
25. Ar. *samt al-ruʾūs,* "the path of the heads, i.e. the zenith"; cf. *Kunitzsch* 1983, no. 43a (in AB without a transliterated Arabic form).
26. Ar. *al-zīj,* "(a work containing) astronomical tables".
27. Ar. *al-mustawiya,* "equal" (used for the equal, or equatorial, hours); cf. *Kunitzsch* 1983, no. 35.

BIBLIOGRAPHY

Björnbo, A.

1912: "Die mathematischen S. Marcohandschriften in Florenz", *Bibliotheca Mathematica,* 3. F., 12, pp. 97–132, 193–224.

Burnett, C. (ed.)

1987: *Adelard of Bath. An English Scientist and Arabist of the Early Twelfth Century,* London: Warburg Institute Surveys and Texts, XIV.

Hartner, W.

1939: "The principle and use of the astrolabe", in A. U. Pope (ed.) 1939, pp. 2530ff., repr. in W. Hartner 1968, pp. 287–311, and in W. Hartner 1978.

1960: "Asṭurlāb", *Encyclopaedia of Islam,* new ed., I, Leiden, pp. 722ff., repr. in Hartner 1968, pp. 312ff.

1968: *Oriens-Occidens,* I, Hildesheim.

1978: *Astrolabica,* no. 1, Paris.

Kunitzsch, P.

1983: *Glossar der arabischen Fachausdrücke in der mittelalterlichen europäischen Astrolabliteratur,* Göttingen.

Morley, W. H.
1856: *Description of a planispheric astrolabe,* London, repr. in F. Sezgin 1990, pp. 249ff.
Pope, A. U. (ed.)
1939: *A Survey of Persian Art,* III, Oxford.
Sezgin, F. (ed.)
1990: *Arabische Instrumente in orientalistischen Studien,* I, Frankfurt am Main.
Thorndike, L. & P. Kibre
1962: *A Catalogue of Incipits of Mediaeval Scientific Writings in Latin,* London (2nd ed.).

XXIII

ZUR PROBLEMATIK DER ASTROLABSTERNE: EINE WEITERE UNBRAUCHBARE STERNTAFEL

SUMMARY. – The Latin manuscript at Fermo (Italy), Biblioteca Comunale, no. 85 (late 13th cent.), contains *inter alia* the treatise on the construction of the astrolabe by Hermann the Lame (d. 1054). The star table regularly added to the treatise is that of 'Type III' (according to Kunitzsch [1]). Here however, instead, a star table of a hitherto unnoticed type is added which may have been compiled around or after 1300, since 24 of its 52 stars are marked as being "verified by the torquetum". Its coordinates, called *longitudo* and *latitudo*, are explained in a note as representing *mediatio coeli* and declination, respectively. Analysis reveals that the table was compiled from several other known star tables: the major part of the stars are taken from 'Type VIII' (which itself was compiled from Types III and VI), but thirteen stars were collected from Types XII and XIII. The unknown author of the present table retained the ecliptical coordinates from Types XII and XIII established for the epochs AD 1066/1067 and, presumably, 1181/1182 (but in reality 1132/1133) and listed them, unchanged and unnoticed, in the table although its coordinates were explicitly declared to be *mediatio* and declination. Five stars are listed twice without noticing their identity. The edition of the star table comes at a time when new intensified work on the astrolabe and its stars is in progress. The table with all its defects and inconsistencies may thus serve as another evidence to warn everybody to take the data of the stars on medieval – especially Western – astrolabes straightforward as they are and to use them as a basis for the identification of the stars and the dating of the instruments.

Das Astrolab ist gegenwärtig wieder stark ins Blickfeld der Forschung gerückt. An der Universität Frankfurt am Main hat Prof. D. King mit einem team ein umfangreiches Projekt zur Erfassung und wissenschaftlichen Beschreibung aller erhaltenen orientalischen und europäischen Astrolabien bis um 1550 begonnen [1]. R.K.E. Torode wandte sich einer systematischen Auswertung der Daten von Astrolabsternen zu [2]. Prof. A. Borst, Konstanz, bereitet eine kritische Edition der naturwissenschaftlichen Schriften Hermanns des Lahmen (1013-1054), darunter sein Astrolabtraktat, vor [3]. Der Zeitpunkt erscheint da-

[1] Cf. King (1991).

[2] Cf. Torode [1] und [2]; hierzu die Kritik von Dekker (1992).

[3] Cf. einen Hinweis in dieser Zeitschrift, *AIHS*, *40* (1990), 163.

her passend, hier eine neu gefundene Astrolabsterntafel vorzuführen und dabei gleichzeitig erneut auf eine Reihe von Problemen um die Astrolabsterne hinzuweisen.

Auf der *rete* des Astrolabs wird bekanntlich eine Anzahl von Fixsternen markiert, die zur nächtlichen Einstellung des Instruments benötigt werden. Naturgemäß handelt es sich dabei um die wichtigsten, hellsten Fundamentalsterne; auch kleinere Sterne können hinzukommen, um eine gleichmäßige Bestückung rund um den gesamten Himmel zu gewährleisten. Die Astrolabien der frühen Zeit im Orient enthalten ca. 15-20 Sterne. Allmählich nimmt ihre Zahl zu; um 964 registriert aṣ-Ṣūfī 44 Astrolabsterne [4], im 16. Jahrhundert und später können es bis über 70 werden.

Viele islamische Astronomen haben Tafeln von Astrolabsternen aufgestellt. Für die Verwendung auf dem Astrolab bedurfte man der Koordinaten *mediatio coeli* (μ) und Deklination (δ). Die großen Sternkataloge (im *Almagest* sowie die daraus abgeleiteten) enthielten aber nur ekliptikale Koordinaten, Länge (λ) und Breite (β). Die für das Astrolab benötigten Werte μ und δ mußten also durch Umrechnung aus λ und β gewonnen werden. Bei der Herstellung von Astrolabien wird man hauptsächlich zwei oder drei Vorgehensweisen annehmen dürfen: der Astrolabmacher bediente sich für die Anbringung der Sternzeiger auf der *rete* einer vorliegenden Sterntafel; oder er imitierte kurzerhand – mehr oder weniger genau – andere Instrumente; in seltenen Fällen mag ein Astrolabmacher, der dazu fähig war, die Positionen der Sterne selbst berechnet haben.

Für den ostarabischen Raum ist bisher noch nicht erforscht worden, wie weit einzelne Astrolabien von bestimmten Sterntafeln abhängen. Im westarabischen Raum ist eine gewisse Stabilität im Gebrauch der Sterne zu beobachten; in einem Fall (Abū Bakr aus Marrākuš, Astrolab von 605 H = 1208/1209, in Strasbourg) wurden sogar exakt jene 27 Sterne verwendet, die die frühe lateinische Liste vom Typ III (s.u.) verzeichnet, die indirekt auf Maslama al-Maǧrīṭī (978) zurückgeht [5].

In Europa wurde das Astrolab als Errungenschaft von den Arabern übernommen. Aber die Gesamtlage ist hier anders. Nach dem 'Karolingischen Astrolab' (Ende 10. Jh., jetzt im Institut du Monde Arabe, Paris) [6] klafft eine lange Lücke; es gibt dann erst wieder (erhaltene) Instrumente von ca. 1300 und später. Die schriftliche Beschäftigung mit dem Astrolab war dichter: ein Anschub aus dem spanisch-arabischen Raum brachte gegen Ende des 10. Jahrhunderts eine erste Welle in Gang, die bis ins 11. Jahrhundert (Hermann der Lahme u.a.) fortwirkte; Übersetzungen aus dem Arabischen in Spanien im 12. Jahr-

[4] Cf. Kunitzsch [4].

[5] Cf. Kunitzsch [2], 90f.; Befort-Debeauvais.

[6] Cf. Destombes; dazu jetzt noch: *El Legado Científico Andalusí*, 61f. (M. Viladrich), 192f. (J. Samsó).

hundert setzten eine zweite große Welle in Gang. Die Übersetzungen regten eine größere Zahl westlicher Astrolabschriften an, die sich bis in die Renaissance hinziehen (z.B. Johann Stoeffler, 1512 bzw. 1513). Wenn man die mittelalterlichen westlichen Astrolabien untersucht, stellt man fest, daß auf ihnen die Sterne in viel stärkerem Maße als im Orient von bestimmten Sterntafeln abhängen, die uns ebenfalls vorliegen. Das geht sowohl aus der Nomenklatur als auch aus den Sternpositionen hervor. Auch hier ist selbstverständlich damit zu rechnen, daß Astrolabmacher häufig mehr oder weniger gut andere Astrolabien nachmachten und nicht auf geschriebene Tafeln zurückgriffen.

Besonders bei den mittelalterlichen europäischen Astrolabien konnte die Abhängigkeit von bestimmten Sterntafeln zu großen Fehlern bei der Positionierung der Sterne auf der *rete* führen. Nicht nur, daß Schreibfehler bei den Zahlenwerten in dem gerade von einem Astrolabmacher benutzten Textexemplar auf ein Instrument übertragen werden konnten und daß gewisse Texte und Sterntafeln noch Jahrhunderte nach ihrer Entstehung unverändert weiter abgeschrieben wurden und somit deren Werte nicht mehr für die Epoche später angefertigter Instrumente paßten, auch die ursprüngliche Tafel selbst konnte bereits von Anfang an Fehler enthalten, die auf die danach gebauten Instrumente übergingen und dic in den späteren Textabschriften sowie auf imitierten Astrolabien ständig weitergetragen wurden.

So kann es zu regelrechten 'Geistersternen' kommen, die einmal durch irgendein Versehen falsch registriert waren und die dann jahrhundertelang durch die Texte und über die Instrumente geistern. Ein markantes Beispiel ist der Stern, den – soweit bisher bekannt – zuerst Maslama 978 in seiner Astrolabsterntafel registriert mit dem Namen von α Gem, *muqaddam aḏ-ḏirāʿain*, aber mit Koordinaten, die auf eine Himmelsgegend weisen, in der α Hya der hellste Stern ist; in die westliche Tradition ist er als *aldiraan*, *aldiran* usw. eingegangen (s. auch unten, Tabelle, nr. 23).[7] Dieser Stern findet sich anschließend – unter dem genannten Namen und etwa an der bezeichneten Stelle – auf zahllosen spanisch-arabischen, nordafrikanischen und mittelalterlich-europäischen Instrumenten. Es scheint, in Bezug auf die Fixsterne (bei denen ja, außer der Präzession, im vorkopernikanischen Weltbild keine Veränderung zu erwarten war) blieben die Astronomie und die Instrumentenpraxis im Mittelalter – mit ganz wenigen Ausnahmen – bloße Papier- und Buchwissenschaft, die nie durch einen Blick an den Sternhimmel mit der Realität abgeglichen wurde.

Zu den einflußreichsten Sterntafeln in der mittelalterlichen europäischen Astrolabtradition gehören die Tafeln der Typen III, VI und VIII.[8] Typ III

[7] Cf. Kunitzsch [1], Anm. zu Typ I A, 8; III, 22.

[8] Es handelt sich um die Typen von Sterntafeln, die bei Kunitzsch [1] ediert und besprochen sind.

(mit 27 Sternen) entstand im Umkreis der ältesten westlichen Astrolabtexte im späten 10. Jahrhundert in Nordostspanien und steht in gewisser Beziehung zu den Astrolabarbeiten von Maslama und seiner Astrolabsterntafel von 978. Als Koordinaten enthält Typ III μ (hier *latitudo* genannt!) sowie einen Wert, der auf einem Vorlageastrolab ausgemessen worden sein muß [9]. Der Wert von μ in III entspricht meist demjenigen in Maslamas Tafel (diese ist Typ I A bei Kunitzsch [1]), auf volle Grad nach oben aufgerundet.

Typ VI enthält 40 Sterne mit λ und β, die von Johann von London 1246 in Paris mit der Armillarsphäre beobachtet wurden (die Koordinaten weichen also stets ein wenig von denen im *Almagest* ab).

Typ VIII wurde von einem unbekannten Verfasser in der zweiten Hälfte des 13. Jahrhunderts (vor 1276) gebildet, indem er aus III und VI insgesamt 49 Sterne (darunter mehrere unerkannte Duplikate) zusammenstellte und einheitlich auf μ und δ einrichtete. Bei den Sternen, die aus III stammen, ist μ viermal = III und zeigt ansonsten Werte, die um + 1° bis 6° oder – 1° bis 11° von denen in III abweichen.

Für alle drei Tafeltypen sind Astrolabien bekannt geworden, die deren Sterne (und – besonders im Fall von VIII – deren Fehler und Monstrositäten) repräsentieren [10]. VIII hat dabei die weiteste Verbreitung gefunden, da diese Tafel dem meistverbreiteten mittelalterlichen europäischen Astrolabtraktat ("Ps.-Messahalla") [11] beigegeben wurde.

Angesichts dieser Ausgangslage erscheint es methodisch unzulässig, die Sterne besonders auf mittelalterlichen europäischen Astrolabien von den Instrumenten abzunehmen und hiernach astronomisch zu identifizieren. Will man die astronomische Genauigkeit dieser Sternpositionen ermitteln, müßte man vielmehr zuerst die Werte der zugrunde liegenden Sterntafeln analysieren. Bei den Astrolabien selbst ließe sich dann lediglich feststellen, wieweit sie mit den schriftlichen Vorgaben übereinstimmen bzw. davon abweichen (sofern die Identität einer zugrunde liegenden Sterntafel bekannt ist). In und seit der Renaissance verbessern sich die Sternpositionen auf europäischen Instrumenten wesentlich, weil von da an die Astronomen auf bessere Textquellen, zum Teil di-

9 Vom Zentrum des Astrolabs wird durch einen Stern eine Gerade nach außen, bis zum Außenrand, gezogen. Derjenige Grad, bei dem die Gerade die Ekliptik schneidet, ist μ. Nun wird durch den Stern rechtwinklig zu der ersten eine zweite Gerade gezogen, ebenfalls bis zum Außenrand. Die Anzahl von Grad, die sich nach jeder Seite zu ergibt, wenn man von der Schnittstelle der ersten Geraden auf dem Außenrand (der ja in 360° geteilt ist) jeweils nach links und rechts weiterzählt bis zu den beiden Schnittstellen, an denen die zweite Gerade den Außenrand schneidet, ergibt die zweite Koordinate von Typ III (dort *altitudo* genannt; in Typ XI heißt sie sehr treffend *gradus longitudinis ex utraque parte*). Cf. auch North, III, 159-161; Bergmann, 45-53.

10 Cf. Kunitzsch [1], 25, 40, 52.

11 Zu diesem s. Kunitzsch [3], Text X.

rekt auf die griechischen Originale zurückgreifen oder neue eigene Messungen durchführen.

Bei der geschilderten nicht sehr optimistischen Lage hinsichtlich der Astrolabsterne und der entsprechenden Sterntafeln ist es interessant, hier eine neu bekannt gewordene Sterntafel zu veröffentlichen, die ebenfalls wieder grundlegende Fehler enthält und die, falls danach Astrolabien hergestellt worden wären, die merkwürdigsten Sterne auf deren *rete* projiziert hätte.

Die lateinische Handschrift in Fermo (Italien), Biblioteca Comunale, nr. 85 (spätes 13. Jh.), enthält u.a. auf fol. 76vb-79va den Traktat von Hermann dem Lahmen über die Konstruktion des Astrolabs. Hermann hat hierfür die Sterntafel vom Typ III aus der vor ihm liegenden Literatur übernommen. In unserer Handschrift (fol. 78va) ist statt dessen jedoch eine andere Sterntafel eines eigenen, bisher nicht bekannten Typs eingesetzt. (Ob die Sterntafel von derselben Hand geschrieben ist wie der Text selbst, läßt sich nach der mir vorliegenden Xerox-Kopie [12] nicht entscheiden.)

Die Tafel umfaßt 52 Sterne. Gemäß einer unten beigefügten Notiz soll die *longitudo* genannte Koordinate μ darstellen, *latitudo* soll δ sein. 24 von den 52 Sternen sind mit Punkten markiert; diese sollen nach jener Notiz "mit dem Torquetum verifiziert" sein. Da das Torquetum erst 1284 oder kurz vorher aufkam [13], könnte die Sterntafel auf um oder nach 1300 anzusetzen sein. Der Verfasser wird nicht genannt, bleibt also unbekannt.

Die weitere Untersuchung ergab: die Tafel ist keine Neuschöpfung des Verfassers, sondern schlicht eine Kompilation aus mehreren auch uns bekannten älteren Sterntafeln.

Die Mehrzahl der 52 Sterne ist aus Typ VIII übernommen (der seinerseits aus III und VI zusammengemischt war); dabei sind, wie die Edition unten zeigen wird, größtenteils die Koordinatenwerte (μ und δ) von VIII unverändert beibehalten worden. Wo Abweichungen vorliegen, dürfte es sich meist um reine Überlieferungsfehler (Schreib-, Lese- oder Flüchtigkeitsfehler) handeln; Anzeichen für eine systematische Abänderung sind nicht erkennbar. 12 Sterne sind ferner aus Typ XII, einer ist aus Typ XIII übernommen, alle ebenfalls mit unveränderten Zahlen der Koordinatenwerte – nur: in XII und XIII bezeichneten diese Zahlen λ und β, hier dagegen werden sie ohne Unterschied mit den anderen Sternen zusammen als μ und δ registriert! Der Wert von λ in XII (Ptolemäus + 14°7) war für 459 H = 1066/1067 berechnet [14], derjenige von

[12] Ich verdanke Herrn Koll. A. Borst, Konstanz, diese Kopien wie auch die Zustimmung, die Sterntafel separat zu edieren.

[13] Cf. Poulle, 33.

[14] Die Sterntafel in einem vielleicht von Azarquiel (az-Zarqāllu, gest. 1100) stammenden arabischen Astrolabtraktat ist in allen Einzelheiten identisch mit dem lateinischen Typ XII in Kunitzsch [1], sie ist berechnet für 459 H = 1066/1067; cf. Kunitzsch [3], Text IV, 196ff.

λ in XIII paßt auf 527 H = 1132/1133.[15] In unserer Tafel werden diese Zahlen indes als μ für eine Zeit um oder nach 1300 angeboten. Es läßt sich leicht ausmalen, was für eine absurde *rete* entstanden wäre, wenn ein Astrolabmacher die Sterne nach den Zahlen dieser Tafel, genommen – wie ausdrücklich angegeben – als μ und δ, angebracht hätte.

Bei der Übernahme ist dem Verfasser ferner entgangen, daß er mehrere Sterne aus den verschiedenen Quellen doppelt aufgenommen hat. Nr. 1 und 2 waren schon in der Quelle, Typ VIII, ein unerkanntes Duplikat, ebenso nr. 27 und 29. Neben nr. 3 aus XIII hat er unbemerkt denselben Stern ebenfalls aus XII übernommen (nr. 5). Neben nr. 43 (aus VIII) hat er denselben Stern auch aus XII aufgenommen (nr. 49), desgleichen neben nr. 47 (aus VIII) noch nr. 48 (aus XII). Wahrscheinlich ist auch der unidentifizierbare Stern nr. 7 lediglich ein Duplikat zu nr. 11. Das Nichterkennen dieser Duplikate mag hauptsächlich durch die äußerliche Verschiedenheit der zumeist arabischen Sternnamen bedingt sein, die ein gewöhnlicher Lateiner um 1300 natürlich nicht erkennen und auseinanderhalten konnte; hinzukommen die Unterschiede bei den Koordinaten. Auch dieser Sachverhalt beweist, daß hier reine Schreibtischarbeit geleistet wurde, die mit aktueller Himmelsbeobachtung und echter Astronomie nichts zu tun hatte.

Die gleiche Inkompetenz wird sichtbar bei den mit dem Torquetum 'verifizierten' Sternen. So findet sich unter den so markierten Sternen z.B. nr. 22, der inmmitten der mit μ und δ verzeichneten Sterne in der Tabelle die Ziffern von λ und β hat, und der nicht identifizierte Stern nr. 7, der wohl lediglich ein Duplikat zu nr. 11 ist. Die übrigen 'verifizierten' Sterne stammen aus VIII und sind teilweise mit sehr fehlerhaften Koordinaten aufgezeichnet, trotzdem wurden sie danach 'verifiziert'. Man fragt sich, was für eine 'Verifizierung' das gewesen sein soll.

Als Hilfsmittel für echte astronomische Arbeiten ist die Tafel mit den genannten Defekten also weitgehend unbrauchbar.

Unsere merkwürdige kompilierte Tafel mit ihren Unzulänglichkeiten steht übrigens nicht allein da.

Auch ihre Quelle, die Sterntafel von Typ VIII, war aus anderen, älteren Tafeln kompiliert. Immerhin war es dem unbekannten Verfasser von VIII gelungen, die Koordinaten in seiner Tafel einheitlich auf μ und δ umzustellen; von seinen Quellen hatten III μ und dazu *altitudo* (cf. o. Anm. 9) und VI λ und β. Andererseits war es auch ihm unterlaufen, die Identität einer Reihe von Ster-

[15] Typ XIII hat λ = Ptolemäus + 15°7, müßte also für eine Epoche 66 Jahre später bestimmt sein, also 1132/1133. In diese Zeit fällt das Hiǧra-Jahr 527 Die in den Handschriften von XIII genannte Jahreszahl 577 H könnte also ein alter Schreibfehler für 527 sein. Damit wäre das von einem alten Benutzer (in Ms. l) aus 577 errechnete christliche Datum 1181 hinfällig.

nen in beiden Quellen nicht zu erkennen, so daß er mehrere Duplikate in seine eigene Tafel einfügte (eine solche Dublette hat dann auch unsere Tafel unerkannt als nr. 1 und 2 übernommen). Nach den Angaben in VIII sind seit ca. 1300 in Europa sehr viele Astrolabien gebaut worden. Die Sternpositionen auf deren *rete*s sind entsprechend ungenau, abgesehen von den Duplikaten und sonstigen starken 'Ausrutschern', die auf den Instrumenten als Geistersterne erscheinen.

Noch schlimmer steht es um die Sterntafel, die Rodolfus Brugensis, ein Schüler von Hermann von Kärnten, in seinem 1144 geschriebenen Astrolabtraktat bietet [16]. Er nennt die Koordinaten in seiner Tafel gewohnheitsmäßig *longitudo* und *latitudo*. Die Zahlenwerte bei den einzelnen Sternen repräsentieren jedoch in Wirklichkeit ganz verschiedene Daten. Auch er hat seine Tafel kompiliert. Bei den aus III entnommenen Sternen hat er deren dortige Werte von μ und *altitudo* (cf. Anm. 9) beibehalten. Andere Sterne sind aus XII übernommen mit ihren dortigen Werten von λ und β (wobei zu berücksichtigen ist, daß λ in XII für die Epoche 1066/1067 berechnet war!). Die ganze Mischung wird zusätzlich entstellt durch Schreibfehler (in der Handschrift sind die Ziffern in römischen Zahlen geschrieben) sowie durch Vertauschungen von Werten innerhalb der Zeilen.

Bei diesen kompilierten Sterntafeln, besonders in unserer Tafel und bei Rodolfus Brugensis, erhebt sich die Frage, warum die Autoren überhaupt solche Kompilationen vorgenommen haben und warum sie die verschiedenen Ausgangswerte in ihren Tafeln unverändert mit Zahlenwerten stehen ließen, die nicht ihrem System und ihrer Zeit entsprachen. Es wäre dann ja doch leichter gewesen, eine andere Tafel – und davon gab es genug – einheitlich mit zusammenpassenden Koordinaten komplett zu übernehmen und auf die jeweilige Epoche umzurechnen.

Es sieht so aus, als sei die fachliche Kompetenz der Autoren schlicht ungenügend gewesen. Auch die Duplikate hätten bei wirklicher Himmelsbeobachtung und sorgfältiger Bearbeitung erkannt werden müssen. Orientalische Sterntafeln sind demgegenüber meist zuverlässiger, es gibt keine Dubletten, und die Koordinaten sind genau definiert und auf die jeweilige Epoche umgerechnet.

Die dargestellten Beispiele zeigen, wie brüchig und unzuverlässig die Grundlagen waren, auf die sich mittelalterliche europäische Astrolabmacher stützen mußten. Es ergibt sich daraus die methodische Notwendigkeit, bei der Analyse mittelalterlicher westlicher Astrolabien äußerst behutsam vorzugehen und die Sternpositionen auf den Instrumenten nur als Annäherungswerte zu verstehen, die aus verschiedenartig fehlerhaften Quellen oder (bei nachgebauten Astrolabien) Vorlagen hervorgegangen sind.

Es folgt nun die Edition der Sterntafel aus Ms. Fermo 85.

[16] Ediert in Kunitzsch [3], Text X, 58ff.

XXIII

204

(Nr.)	(Tierkreis-zeichen)	(Sternnamen)	longitudo		(Himmels-hälfte)	latitudo		(Moderne Bezeichnung)
			gr.	min.		gr.	min.	
1.	aries alhamel	pantaikator ∴	13	0	–	*14*	*0*	ζ Cet*
2.		batenkaiton	20	0	–	14	10	ζ Cet*
3.		mire	*2*	*25*	+	*26*	7	α And*
4.		mirach ∴	*7*	*0*	+	33	*30*	β And
5.		capud mulieris	*1*	*37*	+	*36*	*0*	α And*
6.		enif	21	47	+	*32*	*30*	α Ari
7.	taurus artur	algeuse ∴	10	20	+	47	0	———
8.		aldebaran ∴	*29*	*0*	+	*14*	20	α Tau
9.		menkar	*6*	*0*	+	*1*	*0*	α Cet
10.		capud algol	*13*	47	+	*23*	*0*	β Per
11.		algenif	*10*	*0*	+	*49*	*0*	α Per
12.	gemini algeuse	alhaioth ∴	*6*	*0*	+	43	40	α Aur
13.		rigil ∴	10	40	–	*10*	*0*	β Ori
14.		humerus orionis	*15*	7	+	*8*	*0*	α Ori
15.		cauilla ...	*4*	27	–	*17*	*50*	γ Ori
16.		lucidum	*28*	*47*	–	*10*	*50*	γ Gem
17.	cancer alzeacal	alhabor ∴	*3*	*0*	–	*15*	30	α CMa
18.		algomeza ∴	*13*	*17*	+	7	10	α CMi
19.		egreger ∴	*24*	*0*	+	48	40	μ UMa
20.		ratageuse	*9*	47	+	*33*	15	α Gem
21.		merkep	*21*	*0*	–	22	*30*	ρ Pup
22.	leo allaceilt	alfert ∴	*14*	7	–	*20*	*50*	α Hya
23.		aldyran	*6*	*0*	–	*6*	*0*	———
24.		vrsa ∴	15	0	+	*35*	*0*	θ UMa
25.		cor leonis ∴	*20*	*0*	+	*15*	50	α Leo
26.		collum leonis	*16*	*17*	+	*8*	*50*	γ Leo
27.	virgo	corvvus ∴	*1*	37	–	*11*	50	——— *
28.		deneblezet ∴	*15*	*0*	–	*19*	40	β Leo
29.		algorap ∴	*22*	7	–	*13*	40	γ Crv*
30.	libra	alramech ∴	*27*	*0*	+	22	30	α Boo
31.		bennenas ∴	*19*	*0*	+	*53*	*0*	η UMa
32.		alchimech ∴	12	0	–	7	50	α Vir
33.	scorpius	azaben yet	*26*	*0*	+	2	0	δ Oph
34.		alfeca	18	8	+	*29*	30	α CrB
35.		cor scorpionis ∴	29	0	–	*23*	30	α Sco
36.	sagittarius	rascaben	15	0	+	50	50	γ Dra
37.		occultus	*25*	*58*	+	*0*	*42*	ν Sgr
38.		alhayre	14	25	+	*15*	*0*	α Oph
39.	capricornus	delfin ∴	*29*	*0*	+	*12*	*30*	ε Del
40.		genu sagittarii	*1*	*7*	–	*18*	30	α Sgr
41.		altayr ∴	*16*	*0*	+	6	41	α Aql
42.		vega ∴	*3*	*0*	+	37	50	α Lyr
43.	aquarius	libedinep	*6*	*0*	–	24	50	δ Cap*
44.		os piscis	*21*	50	–	*23*	*0*	α PsA
45.		aldyra	*10*	*0*	+	*59*	*0*	α Cep
46.		cauda galline	*23*	*17*	+	*60*	*0*	α Cyg
47.		misida equi	*13*	*0*	+	7	*0*	ε Peg*
48.		tangea	*19*	*20*	+	22	*50*	ε Peg*
49.		cauda equi	*14*	*0*	–	40	0	δ Cap*
50.	pisces	humerus equi alferas ∴	*6*	*0*	+	23	33	β Peg
51.		sceach ∴	0	10	–	*19*	*0*	δ Aqr
52.		summitas caduce	*19*	47	+	25	*20*	β Cet

latitudo stellarum hic positarum est distantia ab equatore et longitudo est gradus cum quo celum media<n>t. et stelle punctate sunt verificate per torquetum.

Kommentar

Für die Koordinaten werden folgende Abkürzungen verwendet: λ (ekliptikale) Länge, β (ekliptikale) Breite, μ *mediatio coeli*, δ Deklination. Bei β und δ bedeutet + 'Nord', – 'Süd'. < bedeutet 'übernommen aus'.

Wie in der Notiz am Ende der Tabelle angegeben, soll *latitudo* hier δ bezeichnen, *longitudo* soll μ sein. Bei den aus 'Typ XII' und 'Typ XIII' übernommenen Sternen (nr. 3, 5, 10, 15, 16, 22, 26, 37, 40, 44, 46, 48, 52) stellen die Zahlenwerte jedoch β und λ dar. Alle Zahlenwerte, die mit denen in der jeweiligen Quelle identisch sind, erscheinen oben in der Tabelle *kursiv*. Die mit ∴ gekennzeichneten Sterne wurden laut Notiz am Ende der Tabelle "mit dem Torquetum verifiziert". Ein Stern * hinter der modernen Bezeichnung markiert Duplikate.

Die Anordnung der Kolumnen in der Handschrift ist ungewöhnlich: hinter den Sternnamen erscheinen zuerst Grad und Minuten von *latitudo* (= δ), dann Grad und Minuten von *longitudo* (= μ), und hiernach als letztes die Angabe der Himmelshälfte (+ oder –). In der Edition ist die übliche Reihenfolge (zuerst μ, danach δ) hergestellt.

In der Edition wird bei Namen und Wörtern einheitlich Kleinschreibung durchgeführt; in der Handschrift erscheinen *G M S* immer, *A* teilweise als Majuskeln, alle anderen Buchstaben nur als Minuskeln.

Eine Erklärung der (meist arabischen) Sternnamen wird hier nicht gegeben; sie ist zu finden bei den Tabellentypen und Sternen in Kunitzsch [1], auf die unten ständig verwiesen wird.

1. < VIII, 3 (δ gleich) [< III, 16]. Tierkreiszeichen: *aries*, und dazu auch dessen arabischer Name *alhamel* = arab. *al-ḥamal*.
2. < VIII, 2 [< VI, 3]. μ 20°0 erscheint in VIII, 3 und ist dort so aus III, 16 übernommen. Unsere nrr. 1 und 2 waren schon in Typ VIII unerkannte Duplikate.
3. < XIII, 13 (μ = λ dort; δ: dort β = 26°0).
4. < VIII, 1 (μ = VIII, 1 b-il-n; δ dort = 32°30) [< VI, 4].
5. < XII, 14 ($\mu\delta$ = $\lambda\beta$ dort). Der Stern wurde bei der Übernahme offenbar nicht als Duplikat von nr. 3 erkannt.
6. < VIII, 4 (δ = VIII, 4 cfghin) [< VI, 6].
7. Zweifelhafter Stern. Der arabische Name *algeuse* gehört zu Orion oder Gemini, die Koordinaten ähneln denen von nr. 11. Vielleicht ein unerkanntes Duplikat zu nr. 11, mit korruptem Namen (vgl. α Per in XII, 20a und XII A, 15). Daß gerade dieser dubiose Stern auch "mit dem Torquetum verifiziert" wurde, ist bezeichnend! Tierkreiszeichen: *taurus*, dazu *artur* = arab. *aṯ-ṯaur*.
8. < VIII, 9 (μ gleich; δ: im Ms. über die 4 von 14 klein eine 5 geschrieben; in VIII, 9: 14°30) [< III, 19].
9. < VIII, 6 ($\mu\delta$ gleich) [< VI, 7; hier in Ms. g $\mu\delta$ ebenfalls gleich].
10. < XII, 15 ($\mu\delta$ = $\lambda\beta$ dort).
11. < VIII, 7 ($\mu\delta$ gleich) [< VI, 8; hier in Ms. g μ ebenfalls gleich].

12. < VIII, 10 (μ gleich) [< III, 25]. Tierkreiszeichen: *gemini*, dazu *algeuse* (im Ms. über *g* ein kleines *a*) = arab. *al-ǧauzāʾ*.
13. < VIII, 11 (δ gleich) [< III, 15].
14. < VIII, 12 (δ gleich; μ dort 15°0) [< III, 20].
15. Name: im Ms. hinter *cauilla* ein unkenntlicher Buchstabe mit Abkürzungszeichen; vgl. XII, 26: *cauilla sagittarii* = β Sgr und 33: *cauilla dextera* = β Tau. Koordinaten < XII, 16 (*humerus orionis sinister*), $\mu\delta$ = $\lambda\beta$ dort, also als γ Ori zu identifizieren.
16. < XII, 34 ($\mu\delta$ = $\lambda\beta$ dort).
17. < VIII, 13 (μ gleich; δ dort = −15°0) [< III, 14]. Tierkreiszeichen: *cancer*, dazu *alzeacal* = arab. *as-saraṭān*.
18. Hier scheint bei den Koordinaten eine Mischung vorzuliegen: μ = λ in XII, 6; δ in VIII, 15 = 7°0 (so auch VI, 16 g). Vgl. ferner μ in VIII, 15 und VI, 16 g: 13°0. Vom Namen her erscheint keine eindeutige Zuordnung möglich, da dieser überall in zu vielen Varianten auftritt.
19. Eine Identifizierung als ι UMa ist weniger wahrscheinlich (cf. Kunitzsch [2], 74ff., nr. 28). < VIII, 17 (μ gleich) [< III, 27].
20. < VIII, 14 (dort μ 9°0, δ 33°0) [< VI, 15].
21. < VIII, 16 ($\mu\delta$ gleich) [< VI, 17].
22. < XII, 21 ($\mu\delta$ = $\lambda\beta$ dort – aber so “mit dem Torquetum verifiziert”!). Tierkreiszeichen: *leo*, dazu *allaceilt* = arab. *al-asad*.
23. < VIII, 18 ($\mu\delta$ gleich) [< III, 22; dort ebenfalls μ = 6°0]. Ein unidentifizierbarer “Geisterstern”, der seit Maslama al-Maǧrīṭī’s arabischer Sterntafel von 978 in vielen westlichen Texten sowie auf vielen westarabischen (maghrebinischen und spanisch-arabischen) und europäischen Astrolabien sein Unwesen treibt. Der Name bezeichnet eindeutig α Gem, die Koordinaten weisen grob auf eine Himmelsgegend, in der α Hya der hellste Stern ist. Cf. die Anmerkungen zu Typ I A, 8; III, 22.
24. < VIII, 21 (δ gleich) [< III, 26]. Hier übernahm der Autor aus VIII statt des arabischen Namens *alrucaba* nur dessen lateinische Interpretation.
25. < VIII, 20 (μ gleich; δ dort 15°0) [< VI, 21].
26. < XII, 20 ($\mu\delta$ = $\lambda\beta$ dort).
27. < VIII, 22 (dort Duplikat zu nr. 25; μ 1°0, δ – 11°0) [< VI, 24].
28. < VIII, 24 (μ gleich; δ dort + 19°30) [< VI, 22; in Ms. g μ ebenfalls gleich, δ + 20°0].
29. < VIII, 25 (dort μ 22°0, δ – 13°30) [< III, 13].
30. < VIII, 28 (μ gleich) [< III, 1].
31. < VIII, 27 ($\mu\delta$ gleich; dort in der Edition nur 9°0; mehrere Hss. haben aber 19°0 – im Apparat nicht erwähnt) [< III, 10].
32. < VIII, 26 (δ dort – 7°0) [< III, 12].
33. Das Wort *azaben* steht irrtümlich hier (gehört zu nr. 36). < VIII, 31 (μ

gleich; δ dort $-$ 3°0) [< VI, 28; dort in Ms. g $\delta = -$ 2°0]. δ muß + sein.

34. < VIII, 29 (δ dort 29°0) [< III, 2].
35. < VIII, 32 (δ dort $-$ 23°0) [< III, 11].
36. < VIII, 34 [< VI, 31].
37. Name verlesen aus *oculus sagittarii*. < XII, 40 ($\mu\delta = \lambda\beta$ dort).
38. < VIII, 33 (δ gleich) [< III, 3].
39. < VIII, 37 ($\mu\delta$ gleich – cf. Mss. bcikmn) [< III, 5].
40. < XII, 25 ($\mu = \lambda$ dort; δ: dort $\beta = -$ 18°0).
41. < VIII, 36 (μ gleich) [< III, 4].
42. < VIII, 35 (μ gleich) [< III, 9].
43. < VIII, 40 (μ gleich) [< III, 18]. Cf. nr. 49.
44. < XII, 13 (μ: dort λ = 21°7; $\delta = \beta$ dort).
45. < VIII, 42 ($\mu\delta$ gleich) [< VI, 1]. Der in VI noch verschiedene Name ist in VIII und so hier an denjenigen von nr. 23 angeglichen.
46. < XII, 27 ($\mu\delta = \lambda\beta$ dort).
47. Name: *misida*, verderbt aus *musida*; statt der arabischen Form wurde nur die lateinische Interpretation aus VIII übernommen (cf. nr. 24). < VIII, 43 ($\mu\delta$ gleich) [< VI, 37].
48. < XII, 36 ($\mu\delta = \lambda\beta$ dort). Bei der Übernahme wurde offenbar nicht erkannt, daß 48 ein Duplikat zu nr. 47 ist.
49. Name: aus VIII nur die lateinische Interpretation, nicht der arabische Name übernommen, dabei das zweite Wort (*capricorni*) zu *equi* verlesen. < VIII, 44 (μ gleich; δ: starker Schreibfehler statt $-$ 19°30 oder $-$ 19°0) [< VI, 35]. Bei der Übernahme wurde offenbar nicht erkannt, daß 49 ein Duplikat zu nr. 43 ist.
50. < VIII, 46 (μ gleich) [< III, 6].
51. < VIII, 45 (δ gleich) [< VI, 38].
52. < XII, 37 ($\mu = \lambda$ dort; δ: dort $\beta = -$ 20°20). Zu *caduce* (statt *caude*) cf. XII, 37, Ms. g. δ (+) müßte $-$ sein.

LITERATUR

BERGMANN = W. Bergmann, *Innovationen im Quadrivium des 10. und 11. Jahrhunderts*, Stuttgart, 1985.

BEFORT-DEBEAUVAIS = P.-A. Befort, F. Debeauvais, "Beschreibung des in Strasbourg befindlichen Astrolabs von Abū Bakr al-Marrākušī (dat. 605 H = 1208/1209)" [in Vorbereitung] (Auszüge wurden auf der Jahrestagung der Société Internationale de l'Astrolabe, Frankfurt am Main, Mai 1992, mitgeteilt).

DEKKER = E. Dekker, "Of Astrolabes and Dates and Dead Ends", in *Annals of Science*, 49 (1992), 175-184.

DESTOMBES = M. Destombes, "Un astrolabe carolingien et l'origine de nos chiffres arabes", in *AIHS*, *15* (1962), 3-45.

El Legado Científico Andalusí (Ausstellungskatalog), Madrid, Museo Arqueológico Nacional, 1992.

KING = D. King, "Medieval Astronomical Instruments: A Catalogue in Preparation", in *Bulletin of the Scientific Instrument Society*, no. 31 (1991), 3-7.

KUNITZSCH [1] = P. Kunitzsch, *Typen von Sternverzeichnissen in astronomischen Handschriften des zehnten bis vierzehnten Jahrhunderts*, Wiesbaden, 1966.

KUNITZSCH [2] = P. Kunitzsch, *Arabische Sternnamen in Europa*, Wiesbaden, 1959.

KUNITZSCH [3] = P. Kunitzsch, *The Arabs and the Stars*, Northampton, 1989.

KUNITZSCH [4] = P. Kunitzsch, "Al-Ṣūfī and the Astrolabe Stars", in *Zeitschrift für Geschichte der Arabisch-Islamischen Wissenschaften*, 6 (1990), 151-166.

NORTH = J.D. North (ed.), *Richard of Wallingford, An Edition of His Writings* ..., I-III, Oxford, 1976.

POULLE = E. Poulle, *Les instruments astronomiques du Moyen Age*, Paris, 1983 (= Astrolabica, 3) [zuerst erschienen 1967].

TORODE [1] = R.K.E. Torode, "A Mathematical System for Identifying the Stars of an Astrolabe and Finding Its Age", in *Études 1987-1989*, Paris, 1989 (= Astrolabica, 5), 53-76.

TORODE [2] = R.K.E. Torode, "A Study of Astrolabes", in *Journal of the British Astronomical Association*, *102* (1992), 25-30.

XXIV

CORONELLI'S GREAT CELESTIAL GLOBE MADE FOR LOUIS XIV: THE NOMENCLATURE

Of the Franciscan cleric and famous Venetian cartographer and globe-maker Vincenzo Coronelli (1650-1718) it is known that he added to certain types of his celestial globes Arabic constellation and star names not only in Latin transliteration well-known since the Middle Ages, but also Arabic names in Arabic script. To these globes carrying inscriptions in Arabic script belong the huge globe of diameter 385 cm made at the request of cardinal César d'Estrées as a present for king Louis XIV, a unique piece, and the globes of diameter 110 cm of which more than sixty exemplars are known to exist until now in European collections. The Louis XIV globe has not been accessible for detailed research since long time[1]. But its recent publication in form of a CD-Rom by the Bibliothèque nationale de France[2] now allowed us to study its contents and to present, in the following, a complete edition of the celestial nomenclature found on it.

The Louis globe was made in Paris in the years 1681-83 in the presence and under the supervision of Coronelli who sojourned there for these years. The globe is of the "convex" type, showing the sky as seen from the outside. The paintings of the constellation figures are due to Jean-Baptiste Corneille (1649-1695). It is evident that Coronelli employed a native Arab collaborator who supplied him with the Arabic names of the celestial objects as well as with a calligraphic model from which the names were copied onto the surface of the globe. The late 17th century was a period rich in Orientalist interests and activities in Paris. It was the time of activity of such famous orientalists as B.

[1] A few names were mentioned in Kunitzsch 1997, p. 17, but the readings – based on some selected photographs – can now be improved from the CD.

[2] *Coronelli, Les globes de Louis XIV.* Collection Bibliothèque nationale de France, *Sources.* Coordination scientifique: Monique Pelletier, conservateur général, directeur du département des cartes et plans à la Bibliothèque nationale de France. Paris 1999. – I wish to express my sincere gratitude to Prof. M. Folkerts, director of the Institute for the History of Science, University of Munich, who was kind enough to make available to me all the relevant labels printed on paper so that I could study them in detail at home.

d'Herbelot (1625-1695) and A. Galland (1646-1717), and one known Arab present there was the Syrian Salomo Negri (ca. 1665-1729). There may have been others, too. The Arabic collaborator for the Louis globe must have had some knowledge of Arabic astronomical literature, because it is evident that he used for the rendering of the names of the 48 classical constellations the names given in the "Book on the 48 Constellations" of the Persian-Arabic astronomer ʿAbd al-Raḥmān al-Ṣūfī (903-986); for the non-Ptolemaic, new, constellations this collaborator created Arabic names himself, translating, as it seems, from the French. From the almost elegant styling of the Arabic letters on the globe it cannot be excluded that it was the Arabic collaborator himself who put them down on the globe (though, in some cases, wrong vowels for the case endings of some nouns occur). The Arabic script on the 110 cm globes, to the contrary, shows all signs indicating that there the Arabic names were copied by European draftsmen from an Arabic model sheet. The Greek names on the Louis globe often deviate from the classical names in Ptolemy's *Almagest*; here the Greek collaborator seems to have more freely translated from the French, also with the 48 Ptolemaic constellations.

Coronelli's celestial globes carrying inscriptions in Arabic script may be classified as follows:

1) The Louis XIV globe of 1681-83. It must be added that this globe shows an unfinished state. The stars on it are – with a few exceptions – not marked by designations or Bayer letters or catalogue numbers and do not show arrows indicating precession (the epoch for which the globe was made is the birth year of Louis XIV, 1638)[3]. There are Arabic names for nearly all of Coronelli's 73 "constellations" (including the two Magellanic clouds). In addition to the constellation names there are added, at random, selected other Arabic star names: I found the names of 3 (out of 28) Arabic lunar mansions and two other Arabic star names.

2) The 110 cm globes. These were intended as a more practicable, reduced, form of the Louis globe. According to the recent study of M. Milanesi based on important first hand sources not considered by earlier authors, their classification must now be established as follows:

a) Convex globe, finished in Venice, 1689, single exemplar in the Biblioteca Marciana in Venice. It was produced from the proofs of J.-B. Nolin's engravings (the copper plates themselves had been

[3] Cf. Milanesi, p. 148f. On Salomo Negri (above), cf. Fück, pp. 96 f.

kept by Nolin in Paris[4]). The inscriptions are added by hand. The Arabic nomenclature is different from that of the Louis globe. (Due to circumstances *in loco*, I could only control some constellations, on a visit in 1993.) This globe was called "Type I" in my previous studies on Coronelli[5].

b) Concave globe, Venice, 1691-92[6]. The Arabic (and other) nomenclature is different from the Louis globe and partly different from the Marciana globe. This globe was called "Type III" in my previous studies (cf. below). To this day I have examined *in loco* or on photographs 14 exemplars of this type.

c) Convex globe, Venice, 1699[7]. The Arabic (and other) nomenclature is identical with that of the concave globe of 1691-92, save for printing errors. This globe was called "Type II" in my previous studies. To this day I have examined *in loco* or on photographs 11 exemplars of this type. Before now I had assumed that the inscriptions of "Type III" (as of 1693, Paris) were corrected and improved against those of "Type II" (as of 1692, Venice). In the light of M. Milanesi's recent findings my former order of Types II and III must now be reversed: "Type II" has to be dated Venice, 1699; it was produced after my former "Type III" and shows clear deterioration in the inscriptions. There is, thus, a line of declining quality in the inscriptions from the Marciana globe through the concave globe of 1691-92 and to the convex globe of 1699.

The 110 cm globes carrying Arabic inscriptions, therefore, will be here cited from now on as "the Marciana globe" (abbreviated as M), "the concave globe" (abbreviated as cv) and "the convex globe" (abbreviated as cx).

The main purpose of the present paper is to edit the nomenclature found on the Louis XIV globe. It is given below in the following order: first the 48 classical, Ptolemaic, constellations (21 northern, 12 zodiacal and 15 southern constellations); then 14 new constellations formed after the travels of discovery in the southern hemisphere; then 8 other constellations invented shortly before Coronelli's time (several

[4] From these plates Nolin produced a separate edition of the 110 cm globe in Paris around 1693. It does not contain the inscriptions in Arabic script and is, therefore, not considered in the present study. Cf. Milanesi, p. 148.

[5] Kunitzsch 1992; 1994; 1995; 1997.

[6] See Milanesi, pp. 149-151.

[7] See Milanesi, pp. 151-153.

of which no longer in use in modern astronomy)[8]. Coronelli himself numbered his constellations to 73 (including the two Magellanic clouds)[9]. Of these, Eridanus is drawn on the Louis globe as a constellation figure, but left without any names; and the River Tigris (found on the 110 cm globes) is completely omitted. After these are listed some purely Arabic star names found on the globe (in very small script[10]) in Arabic alone, without any Western names or designations added; and, finally, the Arabic designations of a number of circles represented on the globe. Further, the Arabic names of the Louis globe are compared to the names used on the 110 cm globes. Of these, I could only compare some names from the Marciana globe (M). Of the concave (cv) and convex (cx) globes I could compare a number of photographs and the complete set of names in the *Libro dei globi* which were reprinted from cx[11]. Since the nomenclature on cv and cx is identical (save for some misprints), the nomenclature of these two is fully covered. Here it becomes evident that Coronelli had the Arabic nomenclature formed three times anew, for the Louis globe (which is largely based on al-Ṣūfī's book of 964[12]), for the Marciana globe (which appears to be newly formed, as far as my examples go), and for the cv and cx

[8] A survey of the introduction of all "new" constellations is given by Dekker, pp. 97-99. For the Arabic nomenclature of Ptolemy's constellations, see Kunitzsch 1974 and Ptolemäus (1986). Indigenous Arabic stellar names are displayed in Kunitzsch 1961.

[9] V. Coronelli, *Epitome cosmographica*, Venice 1693, pp. 40-44.

[10] The script is so small that the names can only be found accidentally and with great difficulty on the CD, even when using the greatest enlargement. There may exist more names of this sort on the globe which escaped my eyes.

[11] The Bayerische Staatsbibliothek, Munich, owns the reprint (Amsterdam 1969) of the *Libro dei globi*, Venice 1693 (1701). In addition, there is a separate set of gores of the convex ("Ottoboni") globe in a bound volume entitled *Struttura d'un globo celeste* and dated Venice, 1700. This set lacks the map of the northern circumpolar stars. On the other hand, the *Libro* (and the reprint) lack two maps, one showing the gores with the constellations Tri, Ari and Psc, and the other with CMa and Vol (but in the separate collection they are there).

[12] Sezgin p. 214 lists eight manuscripts of al-Ṣūfī's book in the Bibliothèque Nationale, Paris. It cannot be excluded that one or the other of these existed in Paris at the time of Coronelli's work and was used by his Arabic collaborator. – The use of al-Ṣūfī's book for the Arabic nomenclature on Coronelli's globe made for Louis XIV may be counted as the seventh point of contact between the Europeans and al-Ṣūfī's astronomical work, after Colom's celestial globe of about 1635 which formed a sixth point of contact (cf. Kunitzsch 1997, p. 14 note 20, and Kunitzsch 1989, item XI, p. 64ff.).

globes which are identical between themselves, but partly different from the Marciana globe and largely different from the Louis globe. In the edition, apart from the abbreviations M, cv and cx explained above, the following other two symbols will be used: S for the Arabic text of al-Ṣūfī's book as edited in Hyderabad, 1954; and Schj for Schjellerup's French translation of al-Ṣūfī's book, 1874 (repr. 1986).

Now follows the edition of the complete nomenclature etc. found on Coronelli's celestial globe made for Louis XIV.

The 48 Ptolemaic constellations

The 21 northern constellations

1. Ursa Minor (UMi)

LA PETITE OURSE – VRSA MINOR – ΑΡΚΤΟΣ ΜΙΚΡΑ – *al-dubb al-aṣghar* ("The Lesser Bear").

2. Ursa Maior (UMa)

LA GRANDE OURSE – URSA MAJOR – ΑΡΚΤΟΣ ΜΕΓΑΛΗ – *al-dubb al-akbar* ("The Greater Bear"). In both Bears, cv and cx omitted the article *al-* in front of the word *dubb*.

3. Draco (Dra)

LE DRAGON – DRACO – ΔΡΑΚΩΝ – *al-tinnīn* ("The [Great] Serpent"; here vocalized as *al-tanīn*. Sic also cv, cx).

4. Cepheus (Cep)

CEPHÉE – CEPHEUS – ΚΗΦΕΥΣ – *qīfāwus ay al-multahib* ("Cepheus, i.e. The Burning One"; here vocalized as *al-multahab*. S 45 / Schj 60). cx *qīfūs* (a new transliteration).

5. Bootes (Boo)

LE BOUVIER – BOOTES – ΒΟΩΤΗΣ – *al-ʿawwāʾ wa-yusammā al-ṣayyāḥ* ("The Howler, also called The Screamer"; here vocalized as *wa-yusmā al-ṣayāḥ*. S 50 / Schj 64). cv, cx have *būwut aw al-rumḥ*, "Bootes or The Lance", the latter as in Grotius and Blaeu (this, in turn, is an error for *al-rāmiḥ*, "The One Armed with a Lance", the traditional Arabic name for the star α Boo).

6. Corona Borealis (CrB)

COURONNE BOREALE – (no Latin and Greek names) – *al-iklīl al-shamālī* ("The Northern Crown"; here vocalized as *al-aklīl al-shumālī*). cx *tāj al-gharbī*, "The Northern Crown", an independent new translation[13].

7. Hercules (Her)

HERCULES – ΗΡΑΚΛΗΣ (but Pt 'Ο ἐν γόνασιν) – *al-jāthī wa-yusammā al-rābiḍ* ("The Kneeling One, also called ..."; here vocalized as *al-jātī wa-yusmā*; *al-rābiḍ* results from a misspelling or misreading in the source for *al-rāqiṣ*, "The Dancer". S 59 / Schj 70). cv, cx *hirkūl*, transliterated from French Hercule.

8. Lyra (Lyr)

LA LYRE – LIRA – ΛΥΡΑ – *al-sulyāq wa-yusammā al-lūr* ("the Sambyke [i.e. a Byzantine harp, correctly to be spelled in Arabic as *al-salbāq*, but variously misspelled in all Arabic astronomical sources], also called Lyra"; spelling again *wa-yusmā*; *al-lūr* stands for the more complete *al-lūrā*, Lyra. S 67 / Schj 75). cv, cx *al-nasr al-wāqiʿ*, "The Falling Eagle", the traditional Arabic name for the star α Lyr.

9. Cygnus (Cyg)

LE CYGNE – CYGNUS – ΚΥΚΝΟΣ (but Pt "Ορνις) – *al-ṭāʾir wa-yusammā al-dajāja* ("The Bird, also called The Hen"; here spelled as *wa-yusmā al-dijāja*. S 70 / Schj 78). cx *wazza*, "Goose" (an independent new translation).

10. Cassiopeia (Cas)

CASSIOPÉE – CASSIEPEA – ΚΑΣΣΙΕΠΕΙΑ – *dhāt al-kursīy* ("The Lady with the Chair". S 76 / Schj 82). cv, cx *kassībiya* (a new transliteration).

11. Perseus (Per)

PERSEE – PERSEUS – ΠΕΡΣΕΥΣ – *birshiyāwus wa-huwa ḥāmil raʾs al-ghūl* ("Perseus, i.e. The Carrier of the Demon's Head". S 81 / Schj 86). M alone ΠΕΡΣΕΥΣ, سيوس ... (apparently a new transliteration, not fully readable on the globe). For cv, cx see no. 11a.

[13] Grammatically incorrect form (the noun lacks the article); cf. also cv, cx here in nos. 1, 2, 20, 38, 39 and 47. On the rare term *gharbī* in the sense of "northern", cf. Kunitzsch 1994, p. 94 note 14. A similarly rare case is also *qiblī* for "southern", in no. 60.

11a. The Head of Medusa

TESTE DE MEDUSE – (no Latin and Greek names) – *ra'ʾs al-ghūl* ("The Demon's Head", i.e. the Arabic rendering of Ptolemy's γοργόνιον). cv, cx have no name for Perseus as such, but only the designation of Medusa's Head (also without Latin and Greek names), Arabic *raʾs al-ḥaʿl* (sic, obviously a miscopying from Grotius' and Blaeu's *raʾs al-juʿal*, "The Dor-Beetle's Head"; this, in turn, was mentioned by Grotius as a conjectured Arabic original for the Latinized *Caput Algol*, the name of the star β Per, which, however, in reality derives from the well-documented *raʾs al-ghūl*).

12. Auriga (Aur)

LE CHARTIER – AURIGA – ΗΝΙΟΧΟΣ – *mumsik al-aʿinna* ("The Rein-Holder". S 89 / Schj 91). cv, cx *māsik al-ʿinān*, "The Rein-Holder", apparently a new translation.

13. Ophiuchus (Oph)

SERPENTAIRE – SERPENTARIUS – ΟΦΙΟΥΧΟΣ – *al-ḥawwāʾ* ("The Snake-Carrier". S 95 / Schj 95). cv, cx *ufiyūkhūs* (in cx, the dot over *kh* is omitted), a new transliteration.

14. Serpens (Ser)

LE SERPENT – SERPENS – ΕΓΧΕΛΥΣ (Pt Ὄφις) – *al-ḥayya* ("The Snake"). cv, cx *al-ḥanash*, "The Snake", a new translation.

15. Sagitta (Sge)

LA FLECHE – SAGITTA – ΒΕΛΟΣ (Pt Ὀϊστός) – *al-sahm* ("The Arrow". S 108 / Schj 104). cx: no Arabic name.

16. Aquila (Aql)

L'AIGLE – AQUILA – ΑΕΤΟΣ – *al-ʿuqāb* ("The Eagle". S 110 / Schj 105). cx *al-nasr*, "The Vulture, or Eagle" (obviously a new translation).

17. Delphin (Del)

LE DAUPHIN – DELPHINUS – ΔΕΛΦΙΝ (Pt Δελφίς)– *al-dalfīn*. On cv, cx *dilfīn*.

18. Equuleus (Equ)

LE PETIT CHEVAL – EQUVLEUS – ΗΜΙΤΟΜΟΣ ΙΠΠΟΣ (Pt Ἵππου προτομή) – *al-jaḥfala* ("The Horse's Lip", i.e. properly the name of the star ε

46

Peg). cx *qiṭʿat al-faras al-aṣghar*, "The Section of the Smaller Horse" (cf. S 118 / Schj 111: *qiṭʿat al-faras*); here at the end vocalized as *al-aṣghara* (sic also in M, where the first words of the name were not readable).

19. Pegasus (Peg)

PEGASE – PEGASUS – ΠΗΓΑΣΟΣ (Pt Ἵππος) – *al-faras al-aʿẓam* ("The Greater Horse". S 120 / Schj 112). cx has no names for this constellation (but the picture is there); M *al-faras al-mujannaḥa*, "The Winged Horse" (*j* written without the dot).

20. Andromeda (And)

ANDROMEDE – ANDROMEDA – ΑΝΔΡΟΜΕΔΗ – *al-marʾa al-musalsala* ("The Chained Woman". S 125 / Schj 116). In cv, cx spelled as *amraʾatu 'l-musalsalatu*, اِمْرَأَتُ اَلْمُسَلْسَلَةُ.

21. Triangulum (Tri)

LE TRIANGLE BOR. – TRIANGULUM BOR. – ΤΡΙΓΩΝΟΝ – *al-muthallath* ("The Triangle").

The 12 constellations of the zodiac

22. Aries (Ari)

LE BELIER – ARIES – ΚΡΙΟΣ – *ṣūrat al-ḥamal* ("The Constellation of the Ram". S 139 / Schj 125). cx only *al-ḥamal*, "The Ram".

23. Taurus (Tau)

LE TAUREAU – TAURUS – ΤΑΥΡΟΣ – *ṣūrat al-thawr* ("The Constellation of the Bull"). cv, cx only *al-thawr*, "The Bull".

24. Gemini (Gem)

LES GEMEAUX – GEMINI – ΔΙΔΥΜΟΙ – *al-tawʾamayn* ("The Twins", sic, in the genitive case). cv, cx *al-jawza*, sic pro *al-jawzāʾ*, the old Arabic name for the constellation; Greek name omitted in cx.

25. Cancer (Cnc)

L'ECREVISSE – CANCER – ΚΑΡΚΙΝΟΣ – *al-saraṭān* ("The Crab"). cv, cx: Greek and Arabic names omitted.

26. Leo (Leo)

LE LION – LEO – ΛΕΩΝ – *al-asad* ("The Lion").

27. Virgo (Vir)

LA VIERGE – VIRGO – ΠΑΡΘΕΝΟΣ – *al-ʿadhrā wa-hiya al-sunbula* ("The Virgin, i.e. the Ear of Corn". S 187 / Schj 158. The normal spelling is العذراء, but here and in cx it is ألْعَذْرَي).

28. Libra (Lib)

LA BALANCE – LIBRA – ΣΤΑΘΜΟΣ (Pt Χηλαί and Ζυγός) – *al-mīzān* ("The Balance").

29. Scorpius (Sco)

LE SCORPION – SCORPIUS – ΣΚΟΡΠΙΟΣ – *al-ʿaqrab* ("The Scorpion"). Both in cv and cx the Greek name is written with RP in the middle, instead of ΡΠ (in M the spelling is correct). In cx alone, the Arabic name of Lupus was added to Scorpius, and the Arabic name of Scorpius to Lupus.

30. Sagittarius (Sgr)

LE SAGITTAIRE – SAGITTARIUS – ΤΟΞΕΥΤΗΣ (Pt Τοξότης) – *al-rāmiḥ wa-yusammā al-qaws* (the first word is an error for *al-rāmī*, "The Archer": "The Archer, also called the Bow"; vocalization again *wa-yusmā*. S 214 / Schj 175). cv, cx only *al-qaws.*

31. Capricornus (Cap)

LE CAPRICORNE – CAPRICORNUS – ΑΙΓΟΚΕΡΩΣ – *al-jady* ("The Young Goat"; here vocalized as *al-jidī*; in cv, cx vocalized as *al-jidā*).

32. Aquarius (Aqr)

LE VERSEAU – AQUARIUS – ΥΔΡΟΧΟΟΣ – *sākib al-māʾ wa-huwa al-dalw* ("The Water Pourer, i.e. the Bucket". S 231 / Schj 185). cx only *al-dalw.*

33. Pisces (Psc)

LES POISSONS – PISCES – ΙΧΘΥΕ (sic, in the dual case; Pt plural, Ἰχθύες) – *al-samakatayn wa-huwa al-ḥūt* ("The Two Fish, i.e. the Fish"; the first word is given in the genitive case, the second name is here misspelled as *al-ḥūth.* S 245 / Schj 194). cx only *al-ḥūt*, with a decorative star over the last letter (which is thus truly a *t*, not to be read as a *th*).

The 15 southern constellations

34. Cetus (Cet)

LA BALEINE – CETUS – ΚΗΤΟΣ – *qīṭus* ("Cetus"; here vocalized as قيطس. S 257 / Schj 199). cv, cx *qīṭūs.*

35. Orion (Ori)

ORION – ΩΡΙΩΝ – *al-jabbār wa-huwa al-jawzā'* ("The Giant, i.e. *al-jawzā'*". S 264 / Schj 204). cv, cx only *al-jabbār*.

36. Eridanus (Eri)

The figure of the constellation is drawn on the globe, but no names are added to it. M and cv, Arabic *nahr ayrīdān* ("The River Eridanus", transliterated from French Éridan). The Arabic name is omitted in cx.

37. Lepus (Lep)

LE LIEVRE – LEPUS – ΛΑΓΩΟΣ – *al-arnab* ("The Hare"). The Greek name is miswritten with Δ (instead of Λ) in M, and with Τ (instead of Γ) in cv, cx. In the Arabic name, the dot under the last letter is omitted in M and cx, but exists in cv.

38. Canis Maior (CMa)

LE GRAND CHIEN – CANIS MAJOR – ΚΥΩΝ ΜΕΙΖΩΝ (Pt only Κύων) – *al-kalb al-akbar* ("The Greater Dog", here vocalized as *al-kalbu 'l-akbari* (sic). S 285 / Schj 217). cv, cx add further Greek ΑΣΤΡΟΚΥΩΝ; the Arabic name is here spelled *kalbu 'l-akbaru*, without article in front of the first word.

39. Canis Minor (CMi)

LE PETIT CHIEN – CANIS MINOR – ΠΡΟΚΥΩΝ – *al-kalb al-mutaqaddim* ("The Preceding Dog"; in this case, the name is not derived from al-Ṣūfī, where the constellation is called *al-kalb al-aṣghar*, "The Lesser Dog", S 293 / Schj 223. The name *al-kalb al-mutaqaddim* appears only in al-Bīrūnī, *al-Qānūn al-mas'ūdī*, ed. Hyderabad, III (1956), p. 1106, but we would hardly believe that it was taken from here; rather it is a new translation, imitating the Greek Procyon). cv, cx write *kalbu 'l-aṣgharu*, similar to the name of CMa.

40. The Ship Argo

LE NAVIRE ARGO – ARGO NAVIS – ΑΡΓΩΝΑΥΣ – *safīnat arghūs* ("The

Ship Argo", combining the traditional name *al-safīna*, "The Ship", with a new transliteration of the Greek/Latin name Argo, perhaps even expressing the Greek genitive, 'Ἀργοῦς). cv, cx have the same Arabic name as the Louis globe, but omit the dot of *gh*, in *arghūs*.

41. Hydra (Hya)

L'HYDRE – HYDRUS – ΥΔΡΟΣ – *al-shujāʿ* ("The Serpent [of the kind called *shujāʿ*]"; here spelled *al-shajāʿ*. S 308 / Schj 232). M Arabic *afʿā*, "Snake", spelled as أَفعَاء; cv, cx *shaḥḥāḥ*, obviously from a misreading of *shujāʿ* in the source.

42. Crater (Crt)

LA COUPE – CRATER – ΚΡΑΤΗΡ – *al-bāṭiya* ("The Vessel, or Jar". S 318 / Schj 238). cv, cx Arabic *al-ka's*, "The [Wine-]Cup" (cf. the Arabic *Almagest*, translation of al-Ḥajjāj).

43. Corvus (Crv)

LE CORBEAU – CORVUS – ΚΟΡΑΞ – *al-ghurāb* ("The Raven").

44. Centaurus (Cen)

LE CENTAURE – CENTAURUS – ΚΕΝΑΥΡΟΣ [sic, with the omission of T] – *qinṭūrus* (vocalized as قِنْطُورُس). cv, cx Arabic (with wrong dots) *qīṭūrus*.

45. Lupus (Lup)

LE LOUP – LUPUS – ΛΥΚΟΣ (Pt Θηρίον) – *al-dhīb* ("The Wolf"); the traditional Arabic name, however, was *al-sabuʿ*, "The Rapacious Animal", cf. S 329 / Schj 245. *al-dhīb* is also found on M, cv, cx; on cx it was erroneously given to Scorpius, and the name of Scorpius to Lupus.

46. Ara (Ara)

L'AUTEL – ARA – ΒΩΜΟΣ (Pt Θυμιατήριον) – *al-majmara* (sic, "The Censer"). cv, cx Arabic *majmara aw madbaḥ* (sic vocalized, for *mijmara aw madhbaḥ*), "The Censer or Sacrificial Altar". S 339 / Schj 250 has *al-mijmara*.

47. Corona Australis (CrA)

COURONNE AUS – CORONA AUS – ΣΤΕΦΑΝΟΣ ΝΟΤΙΟΣ – *al-iklīl al-janūbī* ("The Southern Crown"), here vocalized as أَلأَكْلِيل آلْجَنُوبِي. In cv, cx the

Arabic is written أَكْلِيلُ ٱلْجُنُوبِي (sic, without dot under *j*; the two dots under *ī* of *aklīlu* are omitted in cx). In the Greek, both cv and cx have ΒΟΡΕΙΟΣ ("northern") instead of ΝΟΤΙΟΣ ("southern").

48. Piscis Austrinus (PsA)

POISSON MERIDI.[AL] – PISCES NOTIUS – ΙΧΘΥΣ ΝΟΤΙΟΣ – *al-ḥūt al-janūbī* ("The Southern Fish"), here vocalized as ٱلْحُوث ٱلْجُنُوبِي. On M, the Arabic name is *ḥūt niṣf al-nahārī* ("The [Meridian=] Southern Fish"), here written as حُوتُ نِصْفَ ٱلنَّهَايِ (sic). In cv, cx the second and third words are written in one line, and the first word under them. The second word has been deformed so heavily that it cannot be identified; in the third word the consonant *r* has been omitted, just as in M.

14 new constellations in the southern hemisphere

49. Apus (Aps)

L'OYSEAU INDIEN – AVIS INDICA – ΟΡΝΙΣ ΙΝΔΙΚΗ – *ṭā'ir al-janna* ("The Bird of Paradise", here vocalized *al-jinna*). cx Arabic, in a more literal translation, *ṭayr hindī*, "Indian Bird".

50. Chamaeleon (Cha)

LE CHAMELEON – CAMŒLEON – ΧΑΜΑΙΛΕΩΝ – *al-ḥirbā'* ("The Chamaeleon"). On cx *ḥirbā'*, without the article.

51. Columba (Col)

LA COLOMBE – COLUMBA – ΠΕΡΙΣΤΕΡΑ – *ḥamāmat Nūḥ* ("Noah's Dove"; *Nūḥ* is wrongly declined as *Nūḥa*, for the genitive, instead of *Nūḥin*). cv, cx Arabic only *ḥamāma*, "Dove".

52. Crux (Cru)

LA CROIX – CRUX – ΣΤΑΥΡΟΣ – *al-ṣalīb* ("The Cross"). On cx, without the article *al-*, *ṣalīb*.

53. Dorado (Dor)

LA DORADE – DORADO – ΔΩΡΑΣ – *al-durād* (a simple transliteration from the French). cx Arabic *dūrus*.

54. Grus (Gru)

LA GRUE – GRUS – ΓΕΡΑΝΟΣ – *al-kurkīy* ("The Crane"). On cx no

Greek and Arabic names, but M has Greek ΓΕΡΑΝΟΣ , Arabic *ghurnūq*, "Crane", miswritten as غرتوق.

55. Hydrus (Hyi)

L'HYDRE – HYDRUS – ΥΔΡΟΣ – *al-ḥanash* ("The Snake"). On cv, cx the Greek name is ΥΔΡΟΣ ΜΕΣΗΜΒΡΙΝΣ, and the Arabic name *afʿā*, "Snake", spelled أفعاء.

56. Indus (Ind)

L'INDIEN – INDUS – ΙΝΔΟΣ – *al-hindī* ("The Indian"). On cv, cx Arabic *hindī* هندي, "Indian", without the article.

57. Musca (Mus)

L'ABEILLE – APIS – ΜΕΛΙΣΣΑ – *al-naḥla* ("The Bee"). On cv, cx the Greek is miswritten with Α instead of Λ , and the Arabic is *naḥla*, without the article.

58. Pavo (Pav)

LE PAON – PAVO – ΤΑΩΣ – *al-ṭāwūs* ("The Peacock"). The same on cv, cx, but without the article *al-*.

59. Phoenix (Phe)

LE PHENIX – PHOENIX – ΦΟΙΝΙΞ – *al-nādira* ("The Rare, Extraordinary [Bird]"). M, cv, cx all have Arabic *al-ʿuqāb*, "The Eagle".

60. Triangulum Australe (TrA)

TRIANGLE AUSTRAL – TRIANGULUM AUS. – ΤΡΙΓΩΝΟΝ ΜΕΣΗΜΒΡΙ-ΝΟΝ – *al-muthallath al-janūbī* ("The Southern Triangle"); here spelled ألْمُثَلَّث الجنوبي (some points were omitted, for lack of space). On cv, cx the Arabic is *muthallath qiblī*, "Southern Triangle", with a rarer word for "southern".

61. Tucana (Tuc)

TOUCAN – TOUCAN – ΤΟΥΚΑΝΟΣ – *al-ṭūqān* ("The Toucan", here simply transliterated from the French name). On cv, cx without the article, *ṭūqān* (vocalized as *ṭūqānu*).

62. Volans (Vol)

LE POISSON VOLANT – PISCES VOLANS – ΙΧΘΥΣ ΠΕΤΟΜΕΝΟΣ – *al-*

52

samaka al-ṭayyāra ("The Flying Fish"; here spelled as أَلسمكة آلطّيّارة.
In cx, the *alif* (for *ā*) in the second word was omitted, so the Arabic name is here spelled as *al-samaka al-ṭayyara.*

8 other non-classic constellations

63. Camelopardalis (Cam)

LA GIRAFFE – GIRAFFA – ΚΑΜΗΛΟΠΑΡΔΑΛΙΣ – *zurāfa* ("Giraffe"). The same on cv, cx.

64. Coma Berenices (Com)

LA CHEVELURE DE BERENICE – COMA BERENICES – ΚΟΜΗ ΒΕΡΕΝΙΚΗΣ – *ḍafīrat birnīzah* ("Berenice's Tuft of Hair"; the first word is, colloquially, misspelled as *ẓafīra*). M, cv, cx Arabic *shūshat birnīz*, "The Tuft of Hair of Berenice".

65. Monoceros (Mon)

LA LICORNE – VNICORNU – ΜΟΝΟΚΕΡΩΣ – *waḥīd al-qarn* ("The One with One Horn"). The same Arabic on cv, cx.

66. Antinous (not in modern astronomy)

ANTINOUS – ΑΝΤΙΝΟΟΣ – *antīnūwus* (simply transliterated), spelled as أَنْتينووس. The same Arabic is on cv, cx.

67. The Jordan River (not in modern astronomy)

LE FLEUVE IOURDAIN – IORDANIS FLUVIUS – ΙΟΡΔΑΝΗΣ ΠΟΤΑΜΟΣ – *nahr al-urdunn* ("The River Jordan"). The Arabic name exists on cv, but is omitted on cx.

68. The Lily (not in modern astronomy)

FLEUR DE LYS – (no Latin, Greek and Arabic names). On cv, cx Greek ΚΡΙΝΟΝ, Arabic *zanbaqa*, "Lily".

69. The Rhomboid (not in modern astronomy)

LE ROMBOIDE – ΠΛΑΓΙΟΝ ΤΕΤΡΑΓΩΝΟΝ – *tarbīʿa muqarrana* ("Horned Quadrilateral"[14]; the second word is here spelled مُقَرَّنة, with a decorative

[14] The reading in Kunitzsch 1994, p. 97, in III, 6 (*muqarrafa*, with *f*), was caused by the bad styling of the Arabic letters in cx. It must be corrected to *muqarrana*, with *n*, as found on the Louis globe.

form of the *tā' marbūṭa* at the end). The same Arabic on cx.

70. The Sceptre (not in modern astronomy)

LE SCEPTRE – SCEPTRUM – ΣΚΗΠΤΡΟΝ – *qaḍīb al-mulk* ("The Staff of Kingship"). On cv, cx the Arabic is vocalized as *qaḍīb al-malik*, "The King's Staff".

71. The River Tigris (not in modern astronomy)

Neither the picture of the constellation nor its name(s) appear on the Louis XIV globe. But the constellation is drawn on cx, and its Greek name there is ΤΙΓΡΗΣ ΠΟΤΑΜΟΣ; but there is no Arabic name.

The two Magellanic Clouds

72. The Greater Magellanic Cloud

LE GRAND NUAGE – NUBECULA MAJOR – ΝΕΦΕΛΙΟΝ ΜΕΓΑΛΟΝ – *al-ghayma al-kubrā* ("The Greater Cloud"). On cx, the Arabic is *ghayma kabīra*, "Great Cloud".

73. The Lesser Magellanic Cloud

LE PETIT NUAGE – NUBECULA MINOR – ΝΕΦΕΑΙΟΝ [sic] ΜΙΚΡΟΝ – (*al-ghayma al-ṣughrā*, "The Smaller Cloud"; the Arabic script is almost erased near the turning point of the South Pole). On cx, the Arabic is *ghayma ṣaghīra*, "Small Cloud".

Some purely Arabic star names, without names added in other languages

1. *al-sharaṭayn* (sic, in the genitive case), appears twice, each time written beside an unnamed star in the head of Aries. It is the name of the 1st Arabic lunar mansion, consisting of βγ, or βα, Ari[15].
2. *al-buṭayn* ("The Little Belly"), appears also twice, each time written against an unnamed star in the body of Aries. It is the name of the 2nd Arabic lunar mansion, consisting of εδϱ Ari.
3. *al-naʿām al-wārid* ("The Ostriches Entering [the River to Quench Their Thurst]"), appears above the point of the arrow of Sagittarius.

[15] For the Arabic lunar mansions, cf. Kunitzsch 1989, item XX, and Kunitzsch 1961.

It is the name of the four stars γδεη Sgr forming part of the 20th Arabic lunar mansion.
4. *al-ʿayyūq* (not translated), the traditional Arabic name of Capella, α Aur. Here written near an unnamed star which is probably meant to be α Aur.
5. *al-ṣuradān* ("The Two Great Shrikes", here misspelled as أَلصُّرْدَانْ), Arabic star name, according to al-Ṣūfī for the two stars $\theta^1\iota$ Sgr. It is written under the belly of Sagittarius, near the hind foot.

Arabic names for some circles on the globe

The equator: *dāʾirat muʿaddil al-nahār* (wrongly vocalized *muʿaddal*).
The polar circle: *dāʾirat al-quṭb*.
The tropic of Cancer: *dāʾirat al-saraṭān*.
The tropic of Capricorn: *dāʾirat al-jady* (here vocalized *al-jidī)*.
The circle of equinoxes: *dāʾirat al-iʿtidālayn*.
(I did not find an Arabic name for the ecliptic.)

At last, a final remark. In the above quotations from the Louis globe the French, Latin and Greek names are truly rendered as they appear on the globe. The Arabic names on the globe are written in fine *naskhī* script, with full vocalization, also for the case endings of the nouns. (A speciality of the Arabic collaborator is his constant adding of a *hamza* to the article *al-*: أَلْ.) In the edition, the Arabic names are normally given in transliteration (that is, omitting the case endings). Only in cases of certain deviations are inflexion endings reported or words rendered in Arabic type. Quotations from M, cv and cx are generally made when their terminology is different from that of the Louis globe. On these three globes there appear frequent confusions, in the Greek, among the letters Α - Δ - Λ, Γ - Τ and Θ - Ο, only some of which are here reported. The Arabic names in M, cv and cx were copied by unexperienced Western hands from a calligraphic model sheet, also in *naskhī* script and fully vocalized. Also here, the quotations are normally in transliteration and only in cases of special interest in Arabic type.

Postscript: Mme. Hélène Richard, successor of Monique Pelletier in the directorate of the Département des Cartes et Plans, Bibliothèque nationale de France, Paris, informs me in a letter of April 28, 2000, that "Les noms portés sur les grands globes de Marly sont bien peints à la main". This would mean for the Arabic names that they were put on the globe by the Arabic collaborator, because the style of the letters looks so "genuine" that one would hardly believe that they were executed by a Western hand.

Bibliography

Dekker = E. Dekker, Der Himmelsglobus – eine Welt für sich, in: *Focus Behaim Globus.* Nuremberg 1992, vol. 1, pp. 89-100.

Fück = J. Fück, *Die arabischen Studien in Europa*, Leipzig 1955.

Kunitzsch 1961 = P. Kunitzsch, *Untersuchungen zur Sternnomenklatur der Araber*, Wiesbaden 1961.

Kunitzsch 1974 = P. Kunitzsch, *Der Almagest. Die Syntaxis Mathematica des Claudius Ptolemäus in arabisch-lateinischer Überlieferung*, Wiesbaden 1974.

Kunitzsch 1989 = P. Kunitzsch, *The Arabs and the Stars*, Northampton (Variorum Reprints), 1989.

Kunitzsch 1992 = P. Kunitzsch, Zu den dreisprachigen Inschriften einiger Himmelsgloben von V. Coronelli, in: *Der Globusfreund* 40/41 (1992), 67-71 (with engl. transl. by C. Embleton, *ibid.*, pp. 72-76; with plates 11-14).

Kunitzsch 1994 = P. Kunitzsch, The Arabic Nomenclature on Coronelli's 110 cm Celestial Globes, in: *Zeitschrift für Geschichte der arabisch-islamischen Wissenschaften* 9 (1994), 91-98.

Kunitzsch 1995 = P. Kunitzsch, European Celestial Globes with Arabic Insciptions, in: *Der Globusfreund* 43/44 (1995), 135-142 (German version, *ibid.*, pp. 143-150; with plates 15-16).

Kunitzsch 1997 = P. Kunitzsch, *Neuzeitliche europäische Himmelsgloben mit arabischen Inschriften*, München 1997 (Sitzungsberichte Bayer. Akademie d. Wissenschaften, Phil.-hist. Kl.,1997, 4).

Milanesi = M. Milanesi, Coronelli's Large Celestial Printed Globes: a Complicated History, in: *Der Globusfreund* 47/48 (1999/2000), 143-160 (with German transl. by R. Schmidt, *ibid.*, pp. 161-169; with plates 19-24).

Ptolemäus = *Claudius Ptolemäus, Der Sternkatalog des Almagest,* I: *Die arabischen Übersetzungen*, ed. P. Kunitzsch, Wiesbaden 1986.

Sezgin = F. Sezgin, *Geschichte des arabischen Schrifttums*, VI: *Astronomie, bis ca. 430 H.*, Leiden 1978.

al-Ṣūfī = *Kitāb ṣuwar al-kawākib al-thamāniya wa-l-arbaʿīn*, Hyderabad 1373/1954.

Schjellerup = H.C.F. Schjellerup, *Description des étoiles fixes*, St.-Pétersbourg 1874; repr. Frankfurt/Main 1986 (French transl. of al-Ṣūfī's book).

XXV

Rätselhafte Sternnamen

Sternnamen sind ein kultur- und wissenschaftsgeschichtlich höchst interessantes Phänomen. In ihnen spiegelt sich die Geschichte der Astronomie aus vier Jahrtausenden, und sie machen die Wanderwege sichtbar, auf denen astronomisches Wissen von einer Kultur zur anderen weitergereicht wurde.

Die Masse der in der heutigen internationalen Astronomie noch gelegentlich verzeichneten oder benutzten Sternnamen stammt aus dem Griechischen, dem Altlateinischen, dem Arabischen und dem mittelalterlichen Latein. Das entspricht in groben Zügen dem Verlauf der Tradierung des astronomischen Wissens von den Griechen über die arabisch-islamische Kultur an das mittelalterliche Europa, mit einer Neubelebung der Antike in der Renaissance[1]. Vereinzelte sumerisch-babylonische[2] oder chinesische[3] Namen, aber auch weitere arabische[4], wurden dem Komplex erst in jüngster Zeit hinzugefügt, entnommen aus der Forschungsliteratur über diese Kulturen.

Auch Winfried Petri hat gelegentlich über Sternnamen geschrieben[5], und so sei ihm zu seinem hohen Geburtstag dieser Beitrag, verbunden mit den besten Wünschen für die kommenden Jahre, gewidmet.

Wie entstellt und verdorben auch immer – sorgfältiges Quellenstudium und streng methodische Analyse machen Herkunft und Bedeutung der Sternnamen sichtbar, sofern sie in einer bestimmten Kulturtradition stehen[6].

1 Literatur über Sternnamen: R. H. Allen, *Star Names, Their Lore and Meaning*, New York (Dover Edition) 1963 (unveränderter Nachdruck der Erstauflage von 1899. In englischsprachigen Ländern noch heute viel benutzt, aber unzuverlässig bei allen Angaben mit arabischem Bezug und heute überholt mit den meisten Angaben über altägyptische und sumerisch-babylonische Dinge); P. Kunitzsch, *Arabische Sternnamen in Europa*, Wiesbaden 1959; P. Kunitzsch - T. Smart, *Short Guide to Modern Star Names and Their Derivation*, Wiesbaden 1986.

2 Z.B. Nunki (sumerisch), σ Sagittarii (nach neuerer Kenntnis jedoch eigentlich α Carinae oder ein anderer Stern im Bereich Vela–Puppis–Carina).

3 Z.B. Tsih (oder, in tschechischer Orthographie bei Klepešta, Cih), γ Cassiopeiae.

4 Z.B. Ankaa, α Phoenicis, oder Muhlifain, γ Centauri.

5 W. Petri, Eine arabische Sterngruppe unter den Füßen des Suhayl, *Die Sterne* 38 (1962), 74–77.

6 Z.B. Azelfafage (π^1 Cygni), aus arab. *as-sulaḥfāt*, "Schildkröte", einer paraphrasierenden Bezeichnung des ptolemäischen Sternbilds Lyra (mit falscher mittelalterlicher Übertragung auf das Sternbild Cygnus); oder

Daneben tauchen aber von Zeit zu Zeit immer mal wieder Namen auf, die sich keiner bekannten Tradition und keiner an den Überlieferungsprozessen beteiligten Sprache zuordnen lassen – im wahrsten Sinne des Wortes also rätselhafte Sternnamen. Über sie soll hier kurz referiert werden.

Aus dem Mittelalter sind dazu zwei Stellen zu nennen. Dem fälschlicherweise dem arabischen Autor Māšā'allāh (Messahalla) zugeschriebenen Astrolabtraktat mit dem Incipit *Scito quod astrolabium est nomen grecum ...*[7] werden in vielen Handschriften zwei Sterntafeln beigegeben, die eine davon (entstanden zwischen 1246 und 1276) umfaßt 49 Sterne mit den Koordinaten *mediatio coeli*[8] und Deklination[9]. Die Tafel enthält üblicherweise zwei Namenkolumnen, die eine (überschrieben *nomina stellarum*) gibt die zu jener Zeit verbreiteten vulgarisierten arabischen Namen an, die andere (überschrieben *ymagines stellarum*) gibt hierzu lateinische Übersetzungen bzw. Positionsbeschreibungen der Sterne. In einer späten Handschrift des 15. Jahrhunderts (St. Gallen, Stadtbibliothek, Vad. 412, fol. 93v–95r) ist diesem Verzeichnis ganz links, als erstes, eine dritte Namenkolumne hinzugefügt (Überschrift: *Noua nomina stellarum*), die nun lauter völlig phantastische, undeutbare Namen für diese Sterne angibt. Ich zitiere daraus hier nur einige Beispiele besonders bekannter Sterne:

für Aldebaran, α Tauri:	Emepsa
für Rigel, β Orionis:	Gamax
für Alhabor, Sirius, α Canis Maioris:	Ycomaqua
für Benenaz, η Ursae Maioris:	Midoaetsa
für Calbalacrab, α Scorpii:	Nobuscuo
für Wega, α Lyrae:	Nisothoea

etc.[10]

Alle diese "Namen" sind weder als griechisch noch als arabisch oder lateinisch zu identifizieren. Sie sind anscheinend künstlich erfunden

Beteigeuze (englisch meist Betelgeuse; α Orionis), aus arab. *yad al-ǧauzā'*, "Hand der *ǧauzā'*", der altarabischen Bezeichnung des Sterns.

7 Vgl. hierzu P. Kunitzsch, On the Authenticity of the Treatise on the Composition and Use of the Astrolabe Ascribed to Messahalla, *Archives Internationales d'Histoire des Sciences* 31 (1981), 42–62 (nachgedr. in P. Kunitzsch, *The Arabs and the Stars*, Northampton 1989, Nr. X).

8 Das ist derjenige Grad der Ekliptik, der mit dem Stern zusammen kulminiert, d. h. den Meridian passiert.

9 Diese Sterntafel ist ediert bei P. Kunitzsch, *Typen von Sternverzeichnissen in astronomischen Handschriften des zehnten bis vierzehnten Jahrhunderts*, Wiesbaden 1966, S. 51 ff., "Typ VIII".

10 Die komplette Liste bei Kunitzsch (wie Anm. 9), S. 53 Anm. 6.

und ähneln den Kunstnamen oder Dämonennamen, die häufig in manchen Zweigen der Magie oder Astrologie anzutreffen sind.

Der zweite mittelalterliche Fall betrifft ein Verzeichnis von 72 Sternen mit ekliptikalen Koordinaten (*longitudo* und *latitudo*) in einer englischen Handschrift des 15. Jahrhunderts (Cambridge, Trinity College, R. 15. 18, fol. 83r-v), kurz hinter dem Astrolabtraktat von Geoffrey Chaucer (der auf fol. 80v endet)[11]. Der Verfasser dieser Sterntafel hat sehr sorgfältig gearbeitet. Er hat offensichtlich den ptolemäischen Sternkatalog in der lateinischen Übersetzung Gerhards von Cremona aus dem Arabischen (ca. 1150–1180)[12] benutzt. Die Koordinaten entsprechen genau den ptolemäischen Werten, die Längen sind in den meisten Fällen um 18°49′ vermehrt. Auch bei den Sternnamen bzw. den Positionsbeschreibungen der Sterne scheint häufig der Wortlaut von Gerhards Almagestübersetzung hervor. Als Eigennamen erscheinen viele der bekannten, seit langem eingebürgerten arabischen Namen. Bei 23 Sternen setzt der Verfasser jedoch gänzlich unbekannte und unerklärliche Namen ein[13]. Neun dieser merkwürdigen Namen beginnen mit *Al-*, dem seit dem Mittelalter wohlbekannten arabischen Artikel, und geben sich so den Anschein arabischer Herkunft. In Wirklichkeit steckt darin jedoch kein echter arabischer Name; z. B. *Allozen id est sequens earum que est in dorso leonis* (δ Leonis); *Almanap id est coxa dextra centauri* (δ Centauri); *Alwarap id est super genu sinistrum centauri* (β Centauri); etc. Daneben erscheinen aber ebenso Formen, die keine Assoziation an irgendeine der aus der Tradition bekannten Sprachen hervorrufen, z. B. *Beken id est lucida sub transtro nauis* (γ Velorum); *Biuacaba super caput saltatoris* (α Herculis); *Zeball id est in dextro brachio Aquarij* (γ Aquarii); etc. Es scheint demnach so, als habe ein namen- und sprachverliebter Astronom hier Lücken füllen wollen, indem er für Sterne, die in seinen Vorlagen keine Eigennamen besaßen, selbst Namen erfand, die – zumindest teilweise – in ihrem Äußeren den vielfach korrumpierten echten Namen der Überlieferung recht ähnlich sehen.

Ein drittes Beispiel fällt in die moderne Zeit. Hier ragt eine Gruppe von folgenden 14 Namen heraus:

Achird	η Cassiopeiae
Arich	γ Virginis
Haris	γ Bootis

11 Ediert bei Kunitzsch (wie Anm. 9), S. 103 ff., "Typ XVI".

12 Herausgegeben von P. Kunitzsch: Claudius Ptolemäus, *Der Sternkatalog des Almagest. Die arabisch-mittelalterliche Tradition*, II: *Die lateinische Übersetzung Gerhards von Cremona*, Wiesbaden 1990.

13 Diese sind in der Edition bei Kunitzsch (Anm. 11) durch beigesetzten Stern * gekennzeichnet.

Hassaleh	ι Aurigae
Hatysa	ι Orionis
Heze	ζ Virginis
Kaffa	δ Ursae Maioris
Kraz	β Corvi
Ksora	δ Cassiopeiae
Kuma	ν Draconis
Reda	γ Aquilae
Sarin	δ Herculis
Segin	ϵ Cassiopeiae
Tyl	ϵ Draconis .

Soweit ich feststellen konnte, treten sie zum erstenmal bei A. Bečvář 1951 in Erscheinung[14], danach wieder 1959 bei demselben[15] und 1968 bei J. Klepešta, der sich eng an Bečvář anschließt[16]. Später findet man sie u. a. in den Namenlisten von P. Kalaja, 1977[17], und Hoffleit–Jaschek, 1982[18].

Einer von diesen 14 Namen, Haris, läßt sich tatsächlich als echt arabisch nachweisen, verkürzt aus *ḥāris aš-šamāl*, "Hüter des Nordens", als wortgetreue Wiedergabe einer griechischen Benennung 'Αρκτοφύλαξ für das Sternbild Bootes (nicht bei Ptolemäus im Almagest)[19]. Der Name wird auch bei Allen erwähnt[20], und so erscheint es nicht ausgeschlossen, daß Bečvář ihn von dort entlehnt und aus eigener Willkür auf γ Bootis angewendet haben könnte.

Die 13 übrigen Namen dagegen entziehen sich jeder Erklärung und lassen sich keiner der in den Traditionsprozessen eine Rolle spielenden Sprachen zuordnen. Wir können nicht erkennen, warum und woher Bečvář – falls er wirklich der erste Benutzer ist – diese seltsamen, unerklärlichen Namen eingeführt hat. Hinzu kommt, daß mehrere der betroffenen Sterne aus der Tradition bereits eingeführte

---

14 A. Bečvář, *Atlas Coeli, Skalnaté Pleso II*, Prag 1951; Namenverzeichnis S. 277–279.

15 A. Bečvář, *Atlas Coeli II* (Textband), Prag 1959; Namenverzeichnis S. 345–350.

16 *Taschenatlas der Sternbilder*, Text von J. Klepešta, illustriert von A. Rükl, 7. Auflage Hanau 1982 (1. Ausgabe Prag 1968); Namenverzeichnis S. 284–289.

17 P. Kalaja, Stjärnornas namn, *Astronomisk Tidsskrift* 10 (1977), 49–61.

18 D. Hoffleit – C. Jaschek, *The Bright Star Catalogue*, 4th ed., New Haven, Conn., 1982; Namenverzeichnis S. 462–468.

19 Vgl. P. Kunitzsch, *Der Almagest. Die Syntaxis Mathematica des Claudius Ptolemäus in arabisch-lateinischer Überlieferung*, Wiesbaden 1974, S. 77 f., 175 f.

20 Dover-Ausgabe (vgl. oben Anm. 1), S. 97 (für das Sternbild) und 101 (ebenfalls für den Stern α), beidemale mit dem häufigen arabischen Schreibfehler *ḥāris as-samā'* ("Hüter des Himmels", statt "... des Nordens").

Eigennamen besaßen, so daß kein Grund für eine Neubenennung vorgelegen hätte.

Vielleicht gelingt es in Zukunft noch einmal, mit Hilfe tschechischer Kollegen Näheres über Bečvářs Arbeitsmethoden und seine Quellen in Erfahrung zu bringen und auf diese Weise die Hintergründe um diese rätselhaften Sternnamen zu erhellen.

XXVI

FINDINGS IN SOME TEXTS OF EUCLID'S *ELEMENTS* (MEDIAEVAL TRANSMISSION, ARABO-LATIN)

Texts used

1. Arabic

* al-Ḥajjāj, I-VII.Def.7, ed. Copenhagen (Besthorn-Heiberg and Junge-Raeder-Thomson) [here called H]
* Isḥāq/Thābit VII-IX, ed. De Young [here called I/T]
* Some extracts from H and I/T in Klamroth (*Zeitschr.D.Morgenl. Ges.* 1881) [here called Kl]

2. Latin

* Version attributed to Gerard of Cremona, ed.Busard (Leiden 1983 - publication not yet distributed) [here called GCr]
* Version 'Adelard I', ed. Busard (Toronto 1983) [here called AI]
* Anaritius (= quotations from al-Ḥajjāj's version, with comm. by al-Nayrīzī, Lat. transl. by Gerard of Cremona), ed.Curtze (1899) [here called An]

3. Greek

* Euclides, *Elementa*, vol.ii, Post I.L.Heiberg ed. E.S.Stamatis, Leipzig 1970 [here called Gr]

(The sigla are also used for the respective authors or translators.)

Sections and items compared

1. GCr:

V.Def 12
V.7; 15

116

VI.Def 3; Prop.19; 33
VII (all)
VIII.1-22; 28
IX.11-12; 30-31
X.17-19 (additions by Thābit only, as cited by Kl 306; not existent in GCr)
XI.Def 9
XII.6

2. AI:
V.Def 12; Prop.7; 15
VI.18 (= GCr 19); 32 (= GCr 33)
VII.1
VIII.6; 11; 16 (= GCr 17); 18 (= GCr 18)
XI. Def 11 (= GCr 9)
XII.6

General observations on the texts and their relations

Arabic

H and I/T differ constantly. The differences may affect single terms in long periods otherwise identical, or may result in different wordings of longer periods. Also the use of letters designating quantities etc. may differ.

The two groups A and B observed by De Young in VII and VIII must be explained by supposing that group A gives the wording of al-Ḥajjāj, while group B gives the wording of Isḥāq/Thābit. There are two sets of testimonies for al-Ḥajjāj's wording beyond book VI, i.e. the text of Definitions 1-7 of book VII (in ed. H, MS Leiden), and the full text of Propositions VIII.20 + 21 cited in De Young p.196 annot.37 from MS S (= Escorial) explicitly under the name of al-Ḥajjāj. The proofs of these two propositions appear as additions in GCr (ed. Busard) 206, 59ff. (= GCr VIII.21: ... *probatur aliter*, i.e. anonymous, not citing the name of al-Ḥajjāj), and 207,50ff.

(= GCr VIII.22: ... *probatio aliter invenitur*, also anonymous), respectively. In De Young's edition, the number of MSS in group A decreases in the course of the work (6 MSS for book VII against 4 MSS group B; 3 MSS only for book VIII, until VIII.16 De Young = VIII.17 GCr ed.Busard, against 7 MSS group B; and all MSS, ten in number, only group B, from VIII.17 De Young = VIII.18 GCr Busard onwards). Group A had disappeared after VIII.17 De Young, and this allowed the addition, in MS Escorial, of the different wording of al-Ḥajjāj for VIII.20 + 21 De Young. Further, GCr, who is known to follow the I/T version, is generally on the side of group B in all the places controlled. From these facts it may be concluded that what appears as "group A" was the wording of al-Ḥajjāj, while "group B" transmitted the genuine I/T version. (Similar observations are made in the Arabic transmission of the *Almagest*, where some MSS of the I/T-version occasionally fall into the H-wording; cf. Kunitzsch, *Der Almagest* [1974], p.136ff., for *Almag.* VII,1 etc.). This result turns out opposite to De Young's conclusions (p.27f.) who mentions "possible relationship between Group B and the Ḥajjāj translation".

Books XI-XIII: Whereas numerous Arabic MSS are reported to contain the I/T-version completely, until XIII, it was apparently only MS Copenhagen that explicitly states (see Klamroth p. 304) that at the end of X the translation of Isḥāq ends, and that the subsequent text is the translation of al-Ḥajjāj. Nevertheless, there seems a general opinion today to be prevailing (cf. Klamroth l.c.) that the text of XI-XIII in MS Cop. were practically identical with the I/T-versions transmitted in other MSS. The full Arabic text of XI-XIII not being available in a printed edition, we may, for the time being, use Klamroth's citation of the wording of XII.6 (p.316f.), i.e. from the middle of this controversial section XI-XIII, from two MSS: Copenhagen (= K with Klamroth), and Oxford Thurston 11 (= O with Klamroth). Here it becomes sufficiently evident that the texts in these two MSS are not at all identical, but rather that they show the same signs of differences that are well known to exist between the H- and the I/T-versions. Klamroth has registered three different readings from O against K, which make it clear that, here, O (which is an undisputed I/T-MS) is different from K (which, then, must be an H-MS). While the two variants no.1 and 3 in Klamroth are of minor importance, variant no.2 is of an essential technical interest: K says "Prism ABGDEZ has been divided into three equal pyramids", while O says "From prism ABGDEZ three equal pyramids have been cut off" (this last formula was also read and translated by GCr, ed. Busard 377,51 - who however, transformed the passive verb into an active one: *divisimus ex serratili abgdez tres equales piramides*; and also by AI, ed. Busard p.339,196, who retained the passive verb construction: *Erunt ... divise de mansore a b g d e <z> tres pyramides equales*). For XII.6, therefore, it can be stated that MS O (= Thurston 11, an I/T-MS) transmits an I/T-wording which is also found in GCr (who is known to have followed, generally, I/T) and in AI (who, equally, normally follows the I/T-version). On the other hand, the wording in MS Copenhagen is different from the first one, and therefore in full congruence with the MS's own statement can be attributed to H. So, in XII.6 we doubtless have the "normal" situation of transmission also found in other parts of the 'Elements': Arabic version I/T in MS Thurston 11 and also represented in GCr and AI, and H-version in MS Copenhagen as stated by the MS itself and as corroborated by the evidences in the other texts cited. It would not seem unreasonable to extend these findings to the whole of XI-XIII thus making it very probable that Arabic MS Cop. indeed presents the text of the H-version, while other Arabic MSS, as Thurston 11, and both the Latin GCr and AI represent the text of the I/T-version. Of course, further detailed investigations have to be made into all parts of all the texts involved.

The general impression conveyed by the Arabic texts discussed here is not essentially different from the transmission of the text of the *Almagest* earlier studied by myself. There have survived two main versions, one ascribed to al-Ḥajjāj, and the other to Isḥāq/Thābit. These two can be more or less different, so that somebody not sufficiently familiar with the situation would often register the variations in, say, I/T as mere reading variants in a unique version, say H. For reasons still undiscovered and not sufficiently discussed the (more recent) I/T has often retained, or continued to use, the wording of the (earlier) H, or was content to introduce but minor changes in single words. It must, however, also be considered that scribes, or owners of MSS, in their turn have frequently added words, or portions of the text, from other versions to their MSS, so that the picture is somewhat obscured with regard to the true original wording in every single instance in every text.

Latin

The difficulty in discerning the different Arabic versions also affects the interpretation of the Latin translations. For the closer the two different Arabic versions are in a given place, the more difficult it will be to distinguish between the various Latin translations of them. An additional difficulty lies in the non-literal style of certain translators. Some, too, are inconsistent in their methods of translation.

The investigation of the controlled items cited above shows that GCr basically follows the Arabic I/T-version. The source MS for his translation must, however, have included several borrowings from the H-version, because we find, inserted in the I/T-text translated by GCr, isolated words, or formulas, or sometimes even passages of several words, which are recognizably in the H-wording.

The same applies to AI. Whereas in general this version appears close to the I/T-version, here also isolated words or longer formulas and full paragraphs clearly can be recognized as being in the H-wording.

Details cannot be repeated here. I have added a great many pencilled notes to Busard's editions of GCr and AI demonstrating in detail the relations between these and H or I/T.

As for GCr: Gerard has become known for his extreme literalness in his translations from the Arabic. (There is abundant evidence from his translation of the *Almagest*, in my book *Der Almagest*, 1974). The edited text of GCr leaves me with the impression that this obviously is not - or at least not in every single place - the original text of Gerard, and that it has suffered a strong "redaction" giving it a more standardized and uniform wording, and a better Latin style (not so severely "Arabicized" as the original Gerard used to be in following his Arabic model so extremely literally). This impression may be corroborated by the fact that all the surviving MSS of GCr used by Busard for his edition, and also four additional MSS found by M.Folkerts (cited by Busard p. XXII n.103), are dating from the 14th century or even later; no single MS of 13 c. or earlier (cf. the *Almagest*: one MS late 12th c., a good number 13th c., etc.). Maybe this is due to the fact that the original Gerard text was revised (in much the same way as the Arabic texts were revised by al-Ṭūsī and others), perhaps sometime in the 13th c., so that the surviving transmission - dating only from 14 c. on - had its starting point sometime in the 13th c. Gerard's style in translating a text of the 'Elements' type can be studied in his translation of 'Anaritius'. On the other hand, the wording of An cannot be immediately compared to GCr since An, whenever citing Euclid's own words, has them in the H-wording while GCr is translated from the I/T-wording. Nevertheless, the comparison of An shows sufficient similarities so as to agree that the underlying (Latin) text of our present GCr was a translation by Gerard.

As for AI: Here also we find many signs that AI might be a translation from the Arabic. But AI, likewise, has sections of a complete literalness against the Arabic, while other sections show a high degree of "literary Latin" transformation, and it would appear strange to imagine that the same author gave such utterly literal imitations of Arabic wording in some places, and rather elegant Latin transformations of the underlying contents in other places.

Some notes on various individual items

A. Regarding version GCr

(1) The term for the single "books" in GCr is always *pars*. This is different from Gerard's habit (cf. *Almagest*: *dictio*, for al-Ḥajjāj's *qawl*; Kunitzsch, *Der Almagest*, p.130f.). *Pars* is probably one of the elements of the later "redaction" of Gerard's original text. In An we find *pars*, *liber*, and no term at all (*super septimum*), where the Arabic (MS Leiden) has consistently *maqāla*. All the Arabic texts consulted for this analysis consistently use *maqāla* (used for longer sections of books in Arabic) which word, however, does not induce a translation *pars*. (Cf. also Klamroth, p.284).

(2) In some instances the edited text of GCr shows a wording coming closer to the Greek than to the Arabic: V.7 (Busard 123,47) *eque multiplicia* which corresponds literally to ἰσάκις πολλαπλάσια (G 12,1); the Arabic here has the usual *aḍʿāfan mutasāwiya* (accus.) (AI/Busard 151,164: *multiplicitas equalis*, literally from the Arabic, but converted into the sing.). The same Arabic term is given in GCr in other places, in literal correspondence to the Arabic, as *multiplicia equalia* (cf. V.4, ed. Busard 121,3+4 = G 7,17+18. AI/Busard 149, 104+105 again has: *multiplicitates equales*). *eque* (= ἰσάκις), if it was thus put deliberately after the Greek, naturally cannot have been introduced by Gerard whose proper manner was to follow the Arabic as closely as possible and whose translation, therefore, was *multiplicia equalia*.

(3) By comparison it has been found that Gerard's *si* corresponds to Arabic *in* ("if"), his *cum* to Ar. *idhā* ("when"), and his *quando* to Ar. *matā* ("when"). But obviously the later redaction has given preference to *si* over *cum*, so that in many of the propositions opened by *si*, the Arabic has *idhā*. Equally it has been observed that in most cases where our present GCr has *sed*, the Arabic merely has *wa* ("and"); only rarely does the Arabic have *wa-lākin* ("but", = *sed*); also obviously a result of the later redaction. Further: In GCr/Busard 193,27 and

194,6+7 (= VIII.5) we have *composita* and *aggregata*, two different Latin equivalents for the single Arabic *muʾallafa* ("composed"), and in ib. 198,29-31 (= VIII.12) we have *congregatur*, *adunatur*, and *colligitur*, three different Latin equivalents for the single Arabic *al-mujtamiʿ* ("what comes together, what assembles"); it appears quite un-Gerardian to vary so startlingly in the translation of a single Arabic term, for normally Gerard uses fixed Latin terms for one Arabic term; obviously we have here another sign of the later redaction, or of some mixture from different (Latin) material.

(4) Arabic words in GCr: I have noticed extremely few (in Gerard's translation of the *Almagest*, their number also was relatively small, cf. Kunitzsch, *Der Almagest*, 1974, p. 107ff.):

m u t a s c i t a, var. *mutuascita*, *mutua scita* (V.Def.12; ed. Busard 118,18): = Arab. *muttasiqa* ("being in harmony", with the Arabic feminine termination *-a* because referring to the plural noun *al-maqādīr*, "quantities"). This was also Gerard's reading in Anaritius ed. Curtze p.165,31: *mutasicha*. While this Arabic word is derived from the root *w-s-q*, the Arabic text of Anaritius (ed. H, p.20,-2) has instead *munassaqa* (from root *n-s-q*), "put in good order, well arranged". The difference between the two in the Arabic script is minimal (one dot over the second letter: = *n*, two dots: = *t*; to this may be added that very often in Arabic MSS those dots are totally omitted).

n a d h i r a t u (V.Def.12, ib. 118,19), var. *inordinatu*: this word is not contained in the edited text of al-Ḥajjāj (MS Leiden, l.c.); it seems to be a gloss, or an extract from a parallel Arabic translation, and is here added merely as a parallel term to the preceding *mutascita in proportione*. It is = Arabic *naẓīra* (again with the feminine termination *-a*, because it refers to the plural *al-maqādīr*), "corresponding ones" (scil. *in proportione*). The *-u* at the end is a remnant of the Arabic termination *-un* of the nominative case and was retained by Gerard here, as is known from him in some other texts. In context, the term *al-naẓīr lahu* was translated without difficulty, e.g. *(latus) quod ei refertur*, in VIII.17, Busard 202,46, etc.; cf. also AI/Busard 240,374 (here = VIII.16): *(latera) se respicientia*.

m e g u a r e m (XI.Deff.7; 9; 10; Busard 337,43; 56; and 338,17), in all three places *inter meguarem*, var. *maguarem*, *meguam*: the Arabic text of XI.Def.9 GCr is cited by Klamroth p.296 (without variants, the transmission seems to be uniform in both the H- and the I/T-versions, as very often in the definitions). While GCr's *meguarem* gives the immediate impression of being Latinized (into a Latin accusative, after *inter*), the spelling of the word has to be corrected to *meguarein*, this being the complete transliteration of Arabic *miḥwarayn* ("two poles" - cf. Klamroth p.296 n.1 -, in the Arabic casus obliquus after the Arabic preposition *bayna*, "between", in the phrase *bayna miḥwarayn*, "between two poles"). This is true Gerard - he can be observed also in his other translations to adopt Arabic terms in their inflected Arabic forms and to insert them without change into his Latin text.

B. Regarding version AI

(1) Observations on the translator's treatment of Arabic words. Cf. XII.6, Busard 339,187+189: *el mansor abgdhz*, transcribed from Arabic (Klamroth p.316) *manshūr abjdhz*, "prism abgdhz". In this Arabic phrase, the letters are appended to the preceding noun in the sense of a genitive, therefore the preceding noun is in the "status constructus", i.e. used without the definite Arabic article *al*. While in the Arabic this construction is strictly and correctly applied, in the Latin nevertheless the article *al* (here: *el*) was added to the preceding noun; this may be the work of the translator (whose knowledge of Arabic was not safe enough - it is known that most of the translators relied on the help of an Arab or Jewish assistant in their translations), or a copyist who wished to give the Arabic word a traditional shape, i.e. being accompanied by the well-known *el* or *al* before it. --- Note the

spelling *el* for the article, this is a spelling found elsewhere in works of Adelard (Tables of al-Khwārizmī) whereas Gerard and others applied *al* normally. --- On the same page 339,196: *Erunt ... divise de mansore abgdhz*, we observe that the translator has inflected the Arabic word according to Latin inflexion rules, which is also a known feature of Adelard's work (Tables of al-Khwārizmī). In the same proposition (XII.6), there can be mentioned *Omnem el mansorem* (Busard 339,185), where *el* is again added by the Latin translator or some copyist, and *mansor* inflected following the Latin rules, as an accusative. The corresponding Arabic wording (in Klamroth p.316) is *kullu manshūrin* (literally = *omne prisma*, in the nominative case; the word *manshūr* here not having the definite article *al* in the original Arabic).

(2) AI VIII.11: Here the contents of two propositions are contracted into one (cf. GCr VIII.11+12). The explanation of some Arabic words must be differentiated accordingly. Busard 235,263 (*duplicata*), App. MS.B: *muthetene biltekerir*, MS.L: *mutenebiltekerir*, is = Arabic *muthannāt bi'l-takrīr*, "twofold in repetition", cf. De Young p.457,6; whereas Busard 235,264 (*triplicata*), App. MS.B: *muthene biltekerir*, MS.L: *mutenebiltekerir*, in this instance must be = Arabic *muthallatha bi'l-takrīr*, "threefold in repetition", cf. De Young p. 461,3-4. In VIII.18 (Busard 242,421) there is again *triplicata (repetendo)*, App. MS.L: *mutenethabiltekerir* = Ar.*muthallatha bi'l-takrīr*, "threefold in repetition" (Busard's table p.394 has to be amended accordingly).

(3) VIII.6, Busard 231,151+154+164: *ultimum*. Here the translator has mistaken the unvowelled Arabic word *ākhar* ("*alium*" !) for *ākhir* ("*ultimum*"); cf. GCr/Busard 194,15+23+36, who has correctly *alium*.

C. On the introductory notice to the translation of al-Ḥajjāj

The Arabic text of al-Ḥajjāj's translation as transmitted in al-Nayrīzī's commentary (MS Leiden, ed. H) has at the beginning a notice on the details of how this translation was made (ed. H, p.4). The styling of this notice is close to a similar notice at the opening of al-Ḥajjāj's Arabic version of the *Almagest* (also in a Leiden MS, cf. Kunitzsch, *Der Almagest*, 1974, p.131f.). Since it appears hardly possible to assume that these two notices were added to the beginnings of the two translations by the translator himself, it would seem far more plausible that they were added by some redactor, or copyist, or bookseller, and henceforth became a constituent part of the text. Thus we find them transmitted in respective MSS in later centuries, 12th or 11th c. (*Almagest*, MS Leiden, not dated, but before 1218/9 A.D.), and 1144 (Euclid, MS Leiden). In the *Elements*, this notice was expanded to tell the additional details about the composition of al-Nayrīzī's commentary, which proves that the notice on the translations already existed when al-Nayrīzī wrote his commentary (al-Nayrīzī died ca. 922 A.D.). In the *Almagest*, the notice was translated into Latin by Gerard together with the text itself, as being a part thereof (see Kunitzsch, l.c.). On the other hand, Gerard's translation of Anaritius does not include the introductory matter; on the whole, the Latin text edited by Curtze differs remarkably from the Arabic text in MS Leiden, it is often shorter than the Arabic, and in many instances purely restricted to al-Nayrīzī's commentary alone where MS Leiden also relates al-Ḥajjāj's wording of Euclid's text. As for GCr, that is the version of the *Elements* ascribed to Gerard, the introduction is completely omitted here too. We may assume that Gerard, in translating the *Elements*, would have followed nearly the same lines as in his translation of the *Almagest*; thus he would certainly have translated the introduction of the *Elements*, too. Its omission in our present text of GCr, therefore, may perhaps be understood as another sign of the later redaction, which deliberately cut off all the paraphernalia not essentially belonging to Euclid's text itself, and perhaps equally deliberately was eager to strip the text of the discernible Arabic elements and to "Latinize" or "westernize" it as far as possible.

D. 'Iohanicius Babiloniensis' and 'iezidi'

GCr X.30 contains two additions. First: a second proof, said to be shorter, "from another work" (*secundum alium librum*). The text of this anonymous additional proof is found, almost verbatim, in Anaritius p.313,16-314,2 (corresponding to GCr/Busard 256,44-257,4). Since the second portion of the commentary on book X (ed. Curtze p.252-386), where this proof occurs, does not belong to the Arabic author Anaritius (who used the Arabic translation of al-Ḥajjāj), we have no safe statement as to the nature of the Arabic text used in that second portion. Since, however, GCr's main text (ed. Busard 255,45-256,36) is in the I/T-wording (I compared the Arabic MSS Escorial, Copenhagen, and Uppsala known to present book X in the I/T-version), it follows that the anonymous second proof in GCr (which is identical in Curtze's edition l.c.) is in another wording, most probably that of al-Ḥajjāj. (The three Arabic MSS mentioned above do not contain this second proof.) Cf. also the situation in VIII.21+22 described above, where the additional, shorter, anonymous proofs are safely in the Ḥajjāj-version. Second: after some introductory lines (Busard 257,6-10), there follows a second addition said to have been found in some Greek manuscripts before Prop.31, and ascribed to one 'Ioanicius Babiloniensis' (text: Busard 257, 22-60). Like the aforementioned additional proof, also this second addition is not contained in the three Arabic MSS cited above. There is, however, good reason to assume that 'Iohanicius Babiloniensis' is identical to Yūḥannā ibn Yūsuf ibn al-Ḥārith ibn al-Biṭrīq al-Qass, who, according to Ibn al-Nadīm's *Fihrist*, gave lectures on Euclid's book and on other works on geometry (perhaps 1st half of 10th c. Cf. Sezgin, *Gesch.d.arab.Schrifttums* V, 298); he is also said to have translated from the Greek, and there are several works of his own on related matters, either still surviving in manuscripts, or mentioned in the classical Arabic biographies. A transliteration of his name (Yūḥannā) into Latin as *Iohanicius* appears quite appropriate, while the appellation *Babiloniensis* cannot be derived from the forms of his name cited in the Arabic sources. It might be added that the author of Scholion II (GCr/ Busard 438,34-439,20), *Quidam qui vocabatur Iohannes*, could well be the same as 'Iohanicius Babiloniensis' of X.30. (Note that in the famous mediaeval Latin medical treatise *Isagoge Iohannitii* ... the name 'Iohannitius' stands for Ḥunayn [ibn Isḥāq]; the *Isagoge* was translated from Arabic into Latin at an early date, not by Gerard nor in his time. In Gerard's translation of Jābir ibn Aflaḥ's *Elementa astronomica* the name of Ḥunayn appears as *Hunani* [in the genitive case] which should perhaps be read *Hunain*, to correspond with the Arabic vocalization of the name, not with the [Latin] genitive termination *-i*.)

X.40 contains an addition which is said *esse cuiusdam qui vocabatur iezidi* (introductory remarks: GCr/Busard 264,25-31; text: 264,33-265,27). There is no author by the name of [al-] Yazīdī known among the numerous Arabic commentators of Euclid's *Elements*. There is, however, a strong suspicion that GCr's *iezidi* is merely the result of the Latin translator's misreading of the Arabic name al-Nayrīzī. This well-known commentator's name is given in the Leiden MS in almost all instances as al-Narīzī (omitting the *y* in the first syllable). The consonants of this spelling, *n-r-y-z-y*, can very easily be mistaken as *y-z-y-d-y* in the Arabic script. It is therefore almost likely that GCr's *iezidi* is nothing else than a bad spelling of al-Narīzī. The omission of the Arabic article *al* without any obvious reason is observed in Gerard also elsewhere (cf. my book *Der Almagest*, p.221f., no.23: the star name *alioze* = Ar. *al-jawza* or *al-jawzā*ʾ; p.190: the constellation name *aliouze* = Ar. *al-jawzā*ʾ, Gemini; but p.195: *iouze*, which is again the same Arabic name *al-jawzā*ʾ, here = Orion). A text of a closely related kind, partly in the same wording, accompanied by a diagram almost identical to that of GCr 264, is found in Anaritius/Curtze p.250,18-252,11 (there as an addition to X.76 = Heiberg X.78 = GCr X.76). Note that this comes at the end of the first section of the commentary on book X, which certainly belongs to al-Nayrīzī. Although the wording of the two texts in GCr and An is not identical throughout the section, it might be assumed that the Euclid MS from which Gerard translated contained this extract from al-Nayrīzī, with the wording slightly different from that in An 250-252.

E. Scholion I (GCr/Busard 437,14-438,32)

In a short introductory remark given by the Latin translator it is said that the following section was added after Prop.21 (of book XIII which itself, however, is not quoted by number) *in scriptis que transtulit Ysaac*, i.e. in the translation of Isḥāq. An inspection of MSS Escorial, Copenhagen, and Uppsala has shown that the text of what appears as 'Scholion I' in GCr, at the end of the entire work, is indeed found as an addition to the last proposition of book XIII in MS Escorial, fol.175v-176r. The other two MSS do not have an addition in this place. The introductory remark to the addition in MS Escorial reads: "I found in some copies of Euclid's Book at the end of book XIII [that] he says: Apart from the five figures that I have mentioned, there is no solid figure surrounded by surfaces of equal sides and angles that are equal to each other. Proof ...". GCr's wording is slightly different from the Arabic in MS Escorial, but that may be due to variations usual in the transmission of mediaeval MSS, both Arabic and Latin.

[The above notes were derived from an inspection of the original text material (cited at the beginning). Studies by modern scholars on the mediaeval transmission of Euclid's *Elements* were not consulted.]

XXVII

"The Peacock's Tail": On the names of some theorems of Euclid's *Elements*

In the course of the vast transmission of Euclid's *Elements* (Greek – Arabic – medieval Latin) some theorems were given individual names (Burnett 1985: 479: "nicknames"). In "Excursus II", Heath 1926: 415-418 discussed some of them without, however, arriving at the correct explanation in all instances. Since, through the almost complete collection of microfilms of involved manuscripts kept in the Institute of Prof. M. Folkerts, Munich, it was now possible to extend the investigation of these names to the entire Arabic-Latin stage of transmission, a new analysis could be tackled[1]. The results are here presented as a token of gratitude to H.L.L. Busard who has contributed so much to our knowledge of the medieval Western Euclidean tradition through his several valuable editions.

The names discussed in the following refer to five theorems: I,5; I,47; III,7; III,8 and IV,10.

While for one name ("the theorem of the Bride", I,47) Greek origin has been proved, the other four plus a second name for I,47 seem to be of Arabic origin. In the West, their introduction seems to be connected with Adelard of Bath. We find them in manuscripts of "Adelard I" (not mentioned in the apparatus of Busard's edition, 1983). Afterwards they occur mainly in "Adelard II" and the derived version of Campanus. Isolated quotations are also found in other versions (perhaps added in the manuscripts by later hands). In late medieval and Renaissance times speculations arose on the meanings of these names which, naturally, often failed to identify the true background. The names – mostly a choice of only one or some of them – are found in the following versions of the *Elements*[2]:

Arabic: Isḥāq–Thābit (F 6.2), 4 MSS.; an anonymous abbreviation of Book I in MS Paris BN ar. 2500[3]; the *Taḥrīr* of Naṣīr al-Dīn al-Ṭūsī[4]; the printed version ascribed to al-Ṭūsī, Rome 1594[4,5].

---

[1]I gratefully acknowledge the kind help of Prof. M. Folkerts and Dr. R. Lorch, Munich, in tracing some of the sources here used.

[2]Reference is made here to the exhaustive bibliographical list in Folkerts 1989. "F x" refers to the various sections in this list; the manuscripts are cited in abbreviated form. Below, in the discussion of the names, the manuscripts are cited with very short abbreviations.

[3]This MS has to be removed from F 6.2.

[4]Not listed in F; cf. Sezgin 1974: 113. The MSS inspected are: Istanbul, Saray, Ahmet III, 3453 (673 H = A.D. 1274-75); Munich, or. 848; Oxford, BL Marsh 621 (671 H = A.D. 1272-73).

[5]The version of Ibn Sīnā (980-1037) does not cite any of the names. See Ibn Sina, *Al-Shifa. Mathématiques, Géométrie (Usûl Al-Handasah)*. Ed.: I. Madkour. Texte établi par

Adelard I (F 7): Bruges 529 (s. XIII) [I,5]; Oxford, Trinity Coll. 47 (s. XII) [I,5; III,8; IV,10]; Paris BN lat. 16201 (s. XII) [I,5.47].

Gerard of Cremona (F 9): Madrid, Bibl. Univ. 102 [117-z-6] (s. XIV) [IV,10][6].

Adelard II (F 11): 14 MSS (see below, footnote 7).

Adelard III (F 12.1): Oxford, BL Digby 174 (s. XII) [III,7.8; IV,10]; Venice, BN Marciana f.a. 332 [1647] (s. XIII^2) [I,5; III,7.8; IV,10].

Anonymous commentary (F 12.3): Vat. Reg. lat. 1268 (s. XIV in.) [I,5].

Campanus (F 12.4): Cambridge, Univ. Lib. Add. 6866 (s. XIII/XIV) [I,5.47; III,7.8]; Erfurt, Amplon. fol. 37 (s. XIII/XIV) [I,5.(47?); III,7.8]; Florence, BN Magliabecchi XI 112 (A.D. 1259) [III,8]; Madrid, BN 8989 (s. XIII) [I,47]; Munich, Clm 28209 (s. XIII/XIV) [I,47; IV,10]; New York, Columbia University Plimpton 156 (ca. A.D. 1300) [I,47; III,7.8]; Vienna, Österr. Nationalbibl. 2465 (s. XIV in.) [I,5]. This is a choice of the oldest among the 130 MSS listed in F 12.4.

Albertus (Magnus?) (F 17): Vienna, Dominikanerkloster 80/45 (s. XIV) [I,5.47].

The manuscripts of the following versions were also examined, but did not contain any of these names: Greek-Latin translation (F 4); Hermann of Carinthia (F 8); "Mélanges" (F 10); Thirteenth century commentary (F 12.2); Reworkings (F 12.5 and 12.6; 12.7 does not contain the Books here involved); al-Nayrīzī (Latin transl., F 14; also not contained in the Arabic MS, Leiden or. 399, registered in F 6.1); Roger Bacon (F 18); Nicole Oresme (F 19).

In most of the texts, Arabic and Latin, the names are simply written in the margin, near the diagrams accompanying the theorems. Only in a few texts are they incorporated in the current text (Arabic: Paris 2500; al-Ṭūsī, MSS and edition of 1594; Latin: Albertus Magnus [F 17], Vienna, Domin. 80/45; Adelard III, Ven. Marc. f.a. 332 [only for I,5]).

I,5

Arabic: hādhā 'l-shakl yulaqqabu bi'l-ma'mūnī ("this theorem is named 'the Ma'mūnian'"): Rabat 1101; Paris 2500; al-Ṭūsī, MSS and 1594 / *al-shakl al-ma'mūnī* ("the Ma'mūnian theorem"): Oxf. Thurston 11 / *al-ma'mūnī* ("the Ma'mūnian"): Uppsala / *hādhā 'l-shakl yusammā 'l-nāfī* ("this theorem is called 'the One driving into exile'"): Cambr. Add. 1075.

Abd el-Hamid Sabra – Abd el-Hamid Lotfi, Cairo 1977 [on the Arabic title page: 1976]. Three manuscripts of the Isḥāq–Thābit version listed in F 6.2 were not available: Kabul, Kitābkhāne-i Wizārat-i Ma^{c}ārif [297 is not the shelf-mark of the MS]; Kastamonu 607; and Rampur, Raza Library ^{c}Arshī 200.

[6]This manuscript was not used in Busard's edition of Gerard's version, 1983.

Latin: Adelard I: *eleufuga* (Bruges 529); *elnefea id est fuga* (Oxf. Trin. 47); *& lifuga* (Paris 16201) / Adelard II: *elnefea id est fuga* (Vo)[7]; *elnefeca*, corr. in *elnefea* (Ea); *elefuga* (M, Pc, Pe); *ellefuga* (Be); *eleufuga* (Ma) / Adelard III: *hec dicitur elnefea id est fuga miserorum, dicitur etiam eleufuga id est fuga miserorum quia eius difficultate multi absterriti aufugerunt* (Venice BN Marc. f.a. 332) / Anon. comm. [F 12.3]: *hanc vocant elefugam ab ele quod est miser et fuga* (?) *quasi fugam* ...(?) *miserum* (?) ...(?) *in antiquis commentis etiam elnefea*[8] (Vat. Reg. lat. 1268) / Campanus: *elefuga id est fuga captiui* (?)[8] (Cambr. Add. 6866); *Figura miserorum*, and – in another hand – *aliter* (?) *figura* ...(?) *fuga*[8] (Erfurt Amplon. fol. 37); *Ellefuga* (Vienna 2465) / Albertus (Magnus?): *Scias etiam quod quidam antiquorum hanc figuram vocaverunt elefugam, ab 'ele' quod est miser et 'fugare' quia fugat miserum desidiosum qui in disciplinalibus non intendit* (Vienna, Domin. 80/45, and ed. Tummers II 36, 16-18).

From the cited forms it is obvious that the theorem (or its diagram) was first called, in Arabic, *al-ma'mūnī*, "the Ma'mūnian", after the caliph al-Ma'mūn (reigned 813-833) under whose patronage many scientific Greek texts were translated into Arabic, among them Euclid's *Elements.* The reason behind this naming is not obvious. The name was corrupted, in the Arabic script (cf. Cambr. Add. 1075), into *al-nāfī* (which is easily possible with the omission of the syllable *mū* in the middle of the word), which would be the participle of the verb *nafā, yanfī*, "to drive into exile". In this form the name was read by the Latin transmitter, most probably Adelard of Bath, who tried somehow – though not fully correctly – to interpret the meaning of *al-nāfī*. The best Latin rendering is *elnefea* [= *al-nāfī*] *id est fuga* (perhaps originally the reading was *fugans* which would completely correspond to the Arabic). The other Western forms are corruptions thereof. A wrong derivation of the corrupt form *elefuga* (alluding to Greek ἔλεος, "pity, compassion")

---

[7]The names from this version are quoted according to the recent edition: *Robert of Chester's (?) Redaction of Euclid's* Elements, *the so-called Adelard II Version*, ed. H.L.L. Busard and M. Folkerts, i-ii, Basel etc.: Birkhäuser Verlag, 1992 (Science Networks – Historical Studies Vols. 8-9); see the description of the manuscripts (vol. i, 33 ff.) and the apparatus (vol. ii). The symbols refer to the following 14 MSS: Be = Berlin lat. qu. 510, middle or 3rd quarter 13 c.; Ea = Erfurt Amplon. Q. 352, beg. 13 c.; F = Oxf. BL Auct. F.3.13, 2nd half 13 c.; J = Florence BN Conv. soppr. J.I.32 (San Marco 206), mid 13 c.; L = Leipzig, Univ. Lib. Rep. I,68c, 1st half or middle 13 c.; M = Munich Clm 14353, about 1300; Ma = *ib.* Clm 3523, middle 13 c.; N = Naples BN VIII.C.22, 2nd half 13 c.; O = Oxf. Corpus Christi Coll. 251, last third 13 c.; Oa = *ib.* BL Auct. F.5.28, 1st half 13 c.; Pc = Paris BN lat. 7374A, end 13 c.; Pe = *ib.* lat. 11245, 1st half 13 c.; T = Oxf. Trinity Coll. 47, 12 c., fol. 104^{v}-138^{r} [note: 139^{r}-180^{v} contain *Elem.*, version I, Books I-VIII,22]; Vo = Vat. Ottob. lat. 1862, ca. 1200. The names from the "mixed" MS Vf (= Venice BN Marciana f.a. 332, 2nd half 13 c.) are quoted under "Adelard III" [= F 12.1]. In another "mixed" MS, S (= London BL Sloane 285, middle 13 c.), version II starts on 14^{v} (VI,5 ff.); the earlier part is a still unidentified redaction of the *Elements*; in its proofs, III,7 and 8 are both directly called by their names: *Pes anseris quinque passiones habet* ...; *Cauda pauonis quinque passiones habet* ... (this section is quoted above as "S*").

[8]Not well readable on the film.

seems to have developed during the later middle ages, cf. above Albertus (Magnus?) and the Anonymous commentary [F 12.3]. Heath 1926: 416 cites the same etymology from Roger Bacon (*Opus Tertium*, ch. vi), but in his further considerations falls victim to the same erroneous associations as the medieval authors.

I,47 (a)

Greek: Heath 1926: 417 quotes the name θεώρημα τῆς νύμφης, "the theorem of the Bride", found by Tannery in a Paris manuscript of Georgius Pachymeres (1242-1310). To this should be added a scholion on I,47 in Euclid, *Elem.*, ed. Heiberg, V (Leipzig 1888), 217,6: ... τοῦτο τὸ τῆς νύμφης θεώρημα. As Heath demonstrates this name stems from an old Greek tradition.

Arabic: hādhā 'l-shakl yulaqqabu bi'l-ᶜarūs ("this theorem is named 'the Bride'"): Rabat 1101; Paris 2500; al-Ṭūsī, the MSS, but not 1594. This name was not taken over in the Latin versions.

I,47 (b)

Latin: The Latin texts mention for this theorem another name of doubtlessly Arabic origin which, however, has not been found so far in any of the Arabic sources. The name occurs in the following versions: Adelard I: *Ista figura vocatur dulcaron* (Paris 16021; both the diagram itself and the caption seem to stem from a later hand than the text) / Adelard II: *dulcarnon* (Be, F, J, Pe); *dulcarnen* (Ea); *dulcarnon vel nan* (O) / Campanus: *Dulcarnon id est bicornis* (Cambr. Add. 6866); *Dulcarnan* (Madrid BN 8989); *Dicta carmen* (Munich Clm 28209; the same hand as for the name of IV,10, but apparently different from the text itself); *nota quod ista figura nominata est Dulcarnun quod interpretatur immolatio bouis quia pitagoras cum peruenit ad demonstrationem huius figure immolauit bouem pro gaudio* (New York, Col. Univ. Plimpton 156); in Erfurt Amplon. fol. 37, there is a short note written beside the diagram, but it is illegible on the film / Albertus (Magnus?): beside the diagram: *Dulcarnon*, and in the current text: *Hec figura ab antiquis accepit nomen proprium quod arabice quidem dicitur dulcarnon, quod latine est cornuta. Quidam tamen ignari virtutis vocabuli dixerunt dulcarnon dictam quia inventa ea propter utilitatem sui gyometre dulciter canebant. Alii autem rudiores eos tunc dicebant dulces carnes comedisse et ideo sic vocatam. Sola autem prima ratio ab auctoritate habetur* (Vienna, Domin. 80/45, and ed. Tummers II 93, 1-6).

The Arabic name underlying the Latin *Dulcarnon* is *dhū 'l-qarnayn*, "the One with two horns"[9]. This was, basically, a well-known Arabic surname for

[9]Heath 1926: 418 erroneously thought of a combination of Persian *dū*, "two", and Arabic *qarn*, "horn".

Alexander the Great, also occurring in the Koran. In the *Elements*, however, it seems to aim at the two squares emanating from the upper left and right sides of the rectangular triangle, in the diagram accompanying the theorem. In the Arabic sources examined for this study we did not find this name, but it should have existed in some manuscripts; otherwise the word could not have been introduced into the Latin texts. The best Latin interpretation of the name is in Campanus, Cambr. Add. 6866: *bicornis*. The association with "horn(s)" is also echoed by the explanation given in Albertus (Magnus?), *cornuta*. A far echo of the "horns" (of an ox) may perhaps also be found in the explanation given in Plimpton 156 (one of the Campanus MSS), where the author (not necessarily Campanus himself, but some reader or so) remembers the tale that Pythagoras sacrificed an ox when he had found this theorem – but the association of the theorem with Pythagoras is in itself doubtful[10]. Albertus (Magnus?) cites two more Latin speculative etymologies and rejects them as not authorative: *dulcarnon* as being derived from *dulciter canebant* or from *dulces carnes* (*comedisse*)[11].

III,7

Latin: pes anseris. This name is found only in the Latin texts, in this form, without explanatory additions: Adelard II (Be, J, L, N, O, Oa, S*); Adelard III (Oxf. BL Digby 174; Venice BN Marc. f.a. 332); Campanus (Cambr. Add. 6866; Erfurt Amplon. fol. 37; New York, Col. Univ. Plimpton 156). The name was apparently inspired by the figure of the diagram accompanying the theorem. It strongly respires an Arabic aroma, though it could not be traced in the Arabic sources. In Arabic, it must have been something like *rijl al-baṭṭa*, or *rijl al-iwazza* ("the Goose's Foot")[12].

III,8

Latin: Adelard I: *θenep atoz. cauda pauonis* (Oxf. Trin. 47)[13] / Adelard II: *hec figura dicitur dhenep atoz id est cauda pauonis* (T); *scenepacoch cauda pauonis* (O); *Denebeltoz cauda pauonis* (L); *gezelde* (Ea); *cauda pauonis* (Be,

[10]See for details Heath 1921: 133f., 144f.

[11]In passing, it should be noted that in the wide-spread popular fables in the *Speculum stultorum* by Nigellus Wireker (12th cent.) there occurs a cow named *Bicornis* (cf. E. Sievers, in: *Zeitschrift für deutsches Altertum und deutsche Litteratur* **20**, 1876, 215f.– I owe the knowledge of this text to Dr. B.D. Haage, University of Mannheim). It may well be that this word, as a name for a cow or an ox, was also present in the minds of many scholarly readers of the *Elements*.

[12]Heath 1926: 99 and 418 mentions the names of III,7 and 8, but without detailed explanation.

[13]Burnett 1985: 479 quotes (from this MS) the names for III,8 and IV,10; he does not mention the name for I,5 in the same manuscript.

J, N, Oa, S*) / *cauda pauonis* further in Adelard III (Oxf. BL Digby 174; Venice BN Marc. f.a. 332); Campanus (Cambr. Add. 6866; Erfurt Amplon. fol. 37; Florence BN Magliab. XI 112; New York, Col. Univ. Plimpton 156).

Here, as in III,7, the name could not be traced in the Arabic sources. Again, it appears to be inspired by the figure of the diagram. However, in contrast to III,7, here the original Arabic form was preserved in the Latin texts, in transliteration; it can be identified as *dhanab al-ṭāwūs*, "the Peacock's Tail". The rendering of the Arabic *dh* as θ (in Adelard I) and *dh* (in Adelard II) points to Adelard as the translator of those Arabic names into Latin. In a list of Latin transliterations for certain Arabic consonants (in the margins of MS Bruges 529 [s. XIII], Adelard I), Δ is given for Arabic *dh* and θ is proposed for Arabic *th*[14]. The spellings θ and *dh* for our *dhanab* seem to be a loose reminiscence of that.

IV,10

Latin: Here, once more, the name was only found in Latin texts (including transliterated Arabic forms), but not in the Arabic sources. Adelard I: *hec est figura albeliz id est demonis* (Oxf. Trin. 47) / Gerard of Cremona: *Seclibiz* (Madrid Bibl. Univ. 102; uncertain whether in the same hand as the text or in a later hand) / Adelard II: *figura albeliz id est demonis* (T); *sesclalbliz* (Ea); *Segglebiz* (N); *equibisa vel elbeliz vel figura demonis* (J); *hec figura vocatur figura demonis id est scientis*[15] *propter eius difficultatem* (Pe); *Teckebyse* (S*, in a much younger hand) / Adelard III: *Equibise* (Oxf. BL Digby 174); *Seclabin* (Venice BN Marc. f.a. 332) / Campanus: *Figura demonis* (Munich Clm 28209; cf. above, *sub* I,47).

Some of the Latin forms contain the transliteration of the full Arabic name, *shakl Iblīs* ("the diagram/theorem of Iblīs [i.e. the Devil]")[16], while others only repeat the proper name of *Iblīs*. By simple distortion, or deliberately – in order to restore an assumed Arabic article *al* in front of the word –, they wrote *albeliz* etc. (instead of *abeliz* or the like; cf. the well-known *Albumasar* for Abū Maᶜshar; etc.). The reason for the naming is unknown, but apparently, this time, it was not inspired by the shape of the diagram. The argument brought forward by MS Pe (Adelard II) appears to be purely speculative.

*

[14]Cf. Kunitzsch 1991: 8, note 8.

[15]The question mark here added in ed. Busard–Folkerts is unnecessary. The medieval Western understanding of *demon* as *sciens* was discussed by Beaujouan 1981: 318.

[16]Burnett 1985: 479 writes *al-iblīs*; but *Iblīs* is used, in Arabic, without the article *al-* (and is inflected as a diptote). The word as such is assumed to derive from Greek διάβολος; see *Encyclopaedia of Islam*, new ed., iii (Leiden 1971), 668f., article "Iblīs". For *al-shakl*, MS Bruges 529 (Adelard I) presents, in the margin, the transliteration *elscekl*; cf. ed. Busard, 1983, p. 392, and Burnett 1985: 478.

Of these names one (I,47a) is of classical Greek origin and continued into the Arabic *Elements*, but it was not set forth in the Latin tradition. The others seem to have originated with the Arabs. As it seems, three of these (I,47b; III,7.8) were inspired by the shape of the diagrams accompanying the theorems, whereas for the other two (I,5; IV,10) the reason behind the naming remains unclear. In the Latin tradition they seem to have been rather popular, perhaps – as Roger Bacon argues (*apud* Heath 1926: 416) – as a pedagogical aid for memorizing at least some of the theorems. The introduction of the Arabic names into the Latin area is most probably connected with Adelard of Bath.

The transmission of the names in the Arabic tradition is less dense than in the Latin[17]. But also in the Latin it appears that practically no manuscript contains all the five names. Comparing all manuscripts, however, all the names are represented. It seems that the scribes proceeded at random, picking up one and omitting the other of the names. Altogether, from this state of the transmission we may conclude that what we now have in our hands is but a small portion from a much richer tradition, just a tip of an iceberg.

For the practical use of these names, outside the Euclidean tradition itself, a few specimens shall be mentioned.

MS Lüneburg, Ratsbibliothek Misc. D 4° 48 (presumably around 1200; cf. F 10: "Mélanges"), contains on fol. 1^r-7^v and 18^r-21^r fragments of an anonymous treatise on the construction of the astrolabe, *De artificio plane spere que dicitur walzacora,* Inc. *Omnia que de statu superioris machine* ...[18]. The author seems to have been a well-learned scholar, with a good library at his disposal. He cites Ptolemy's *Canones* and *Almagest* (the latter in the translation of Gerard of Cremona, Toledo, ca. 1150-1180); Albatenius; Alfraganus (in the Latin translation of John of Seville); and of Latin authors Vitruvius and Macrobius. Further, he knows and partly cites the astrolabe treatises of (Ps.-)Gerbertus and Herman the Lame. In addition, he makes ample use of Euclid's *Elements* for the mathematical demonstration of the construction of various circles etc. on the astrolabe. Here it is evident that he used the *Elements* in the version commonly called "Adelard II"[19]. In his mathematical argumentation he continuously refers to the *Elements* (much in the same way as, e.g., Jordanus de Nemore [1st half 13th c.?], "version 2"[20]), with those abbreviated terms common to thirteenth century mathematical texts: *per VI.*

[17]Of 15 MSS (including al-Nayrīzī, Leiden or. 399) of the Arabic *Elements* examined for our study only four contain some of the names: Cambr. add. 1075 (I,5); Oxf. BL Thurston 11 (I,5); Rabat 1101 (I,5.47a); Uppsala (I,5). Further, the abbreviation of Book I in Paris BN ar. 2500 and al-Ṭūsī's *Taḥrīr* (3 MSS seen) have I,5 and 47a.

[18]The text appears to be unknown until now; it is not mentioned in Thorndike–Kibre's *Incipits*.

[19]Cf. *inter alia* the words, or formulas, *elmuharifa* (2ra, -3); *theoreuma* (2rb, 5; cf. the *theoreumata* in the redaction of the *Elements*, 14ra ult.); *dicet aduersarius* ... (3vb, 7-8) and *Redarguatur itaque* ... (*ib.*, line 9).

[20]Cf. the ed. by R.B. Thomson: *Jordanus de Nemore and the Mathematics of Astrolabes: De plana spera,* Toronto 1978, pp. 87ff.

secundi (1vb, 4), etc. In some instances, he refers to Euclidean theorems by name. For I,5 we find: *per V. primi scilicet elnefeam* (5ra, -10), and for I,47: *per dulĉ* (1vb, 5) and *per dulcarnon* (2ra, -4). Since the three texts contained in the MS (astrolabe; *Elements*; planetary theory) show some relationship in the citation of sources and the mathematical style of argumentation, it would not seem improbable that they are all by the same (unknown) author.

Another author to be cited is Richard of Wallingford (d. 1336). In his *Albion* we find: ... *ex resolucione Dulkarnon* (North 1976, i: 250,13); and in the *Quadripartitum*: ... *per Dulcarnon, que est penultima primi* (*ib.*, i: 38, 39).

Third, we may mention Philippe Éléphant (middle 14th c.). From his *Mathematica*, Beaujouan cites *elefuga* (I,5), *cauda pavonis* (III,8) and *figura demonis sive intelligentis* (IV,10): Beaujouan 1957: 453, and *idem* 1981: 313[21].

Apart from these, there could be mentioned citations from Alexander Neckam (1157-1217), *De naturis rerum* II, 173: *dulcarnon* (in Latham 1989: 463), Roger Bacon: *elefuga* (in Heath 1926: 416) and Chaucer: *dulcarnon* (in Heath 1926: 416, 418; North 1976, i: 251 note 4).

A further development may be seen in Jean de Murs (first half 14th cent.) who calls certain figures formed in the diagrams of two propositions in his *De arte mensurandi*, *crista pauonis* ("the Peacock's Comb, or Crest")[22]. It seems probable that the choice of this name was provoked by the existing and well-known name *cauda pauonis* from Euclid's *Elements*.

Beyond the names of Greek and Arabic provenance discussed above, there existed several more names that seem to have originated in medieval Europe and which are outside the scope of this present study; see Heath 1926: 415f., 417f., 418, and Beaujouan 1975: 450, 453, 455f., and *idem* 1981: 312f., 318f.

---

[21] The citations from MS Oxf. Trinity Coll. 47 in Beaujouan 1981: 318 note 44 (based on excerpts made by Danielle Jacquart) are not fully exact. For I,5 read: 172^v (instead of 170^v); in III,8 read *dhenep atoz* (instead of *adhenepatoz*) – the wrong spelling also misled Beaujouan to explain the name from the Arabic plural *adhnāb* (instead of the singular *dhanab*) *al-ṭāwūs*; on fol. 167^v read *θeneb* (instead of *teneb*). For *elnefea*, Beaujouan's conjectured *al-nafy* (l'expulsion, le bannissement) comes quite near to the true *al-nāfī* (*"fugans"*) documented above.

[22] *De arte mensurandi*, cap. VI, prop. 23 and 24. I owe this information to H.L.L. Busard who is currently preparing an edition of that text.

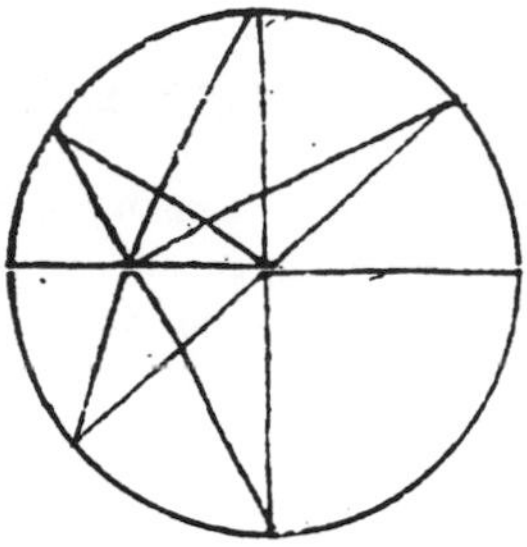

Diagram for III,7, adapted from MS Oxford, Trinity College 47, fol. 109v

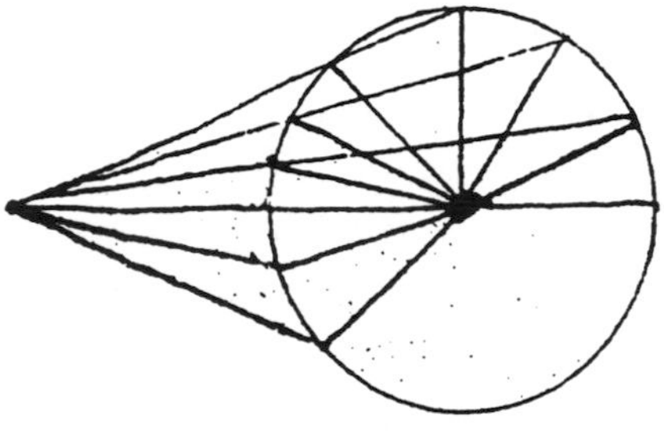

Diagram for III,8, adapted from MS Oxford, Trinity College 47, fol. 109v

Bibliography

(Editions of texts are cited in abbreviated form in the article; the full titles are given in Folkerts 1989)

Beaujouan 1975: G. Beaujouan, Réflexions sur les rapports entre théorie et pratique au Moyen Âge, in: *The Cultural Context of Medieval Learning*, ed. J.E. Murdoch and E.D. Sylla, Dordrecht/Boston, 437-484.

Beaujouan 1981: G. Beaujouan et P. Cattin, *Philippe Éléphant*, Paris (Extrait de l'Histoire littéraire de la France, t. xli, pp. 285-363).

Burnett 1985: C.S.F. Burnett, review of Busard's ed. of Adelard I, in: *Archives internationales d'histoire des sciences* **35**, 475-480.

Folkerts 1989: M. Folkerts, *Euclid in Medieval Europe*, Toronto (The Benjamin Catalogue for History of Science. Questio de rerum natura II).

Heath 1921: Th. Heath, *A History of Greek Mathematics*, I, Oxford.

Heath 1926: Th. Heath, *The Thirteen Books of Euclid's Elements*, transl., I (2nd. ed.), Cambridge.

Kunitzsch 1991: P. Kunitzsch, Letters in geometrical diagrams, Greek – Arabic – Latin, in: *Zeitschrift für Geschichte der Arabisch-Islamischen Wissenschaften* **7**, 1-20.

Latham 1989: J.D. Latham, Arabic into Medieval Latin (3): Letters D-F, *M.L.D.*, in: *Journal of Semitic Studies* **34**, 459-469.

North 1976: *Richard of Wallingford, An edition of his writings* ..., by J.D. North, i-iii, Oxford.

Sezgin 1974: F. Sezgin, *Geschichte des arabischen Schrifttums*, V: *Mathematik, bis ca. 430 H.*, Leiden.

Note added in proof: A late testimony (from 1747) for the name of I,5 is in Muḥammad ibn ᶜAlī al-Tahānuwī, *Kashshāf iṣṭilāḥāt al-funūn*, Istanbul 1318 [1900-01], I, p. 863. The author adds the – doubtlessly unhistorical – remark that the figure was called *al-Ma'mūnī*, because the caliph al-Ma'mūn was so much delighted with it that he had it embroidered on the sleeves of some garments. For this and three more names of geometrical theorems the author refers as a source to the work *Ashkāl al-Ta'sīs* of Shams al-Dīn al-Samarqandī (end of 13th century) and a commentary to it by Shams al-Dīn al-Bukhārī. (I am grateful to Dr. S. Brentjes, Leipzig, for bringing this source to my attention.)

XXVIII

LETTERS IN GEOMETRICAL DIAGRAMS, GREEK - ARABIC - LATIN

The letters used in geometrical diagrams in ancient and medieval texts of mathematical content will pose no major problem for editors and students of such texts as long as the texts were originally composed in one language and the letters were consistently applied within the accompanying text.[1] The matter, however, becomes more complicated when more languages are involved. Many of the major classical texts have suffered extensive processes of transmission - from Greek into Arabic, partly through Syriac, and hence further into medieval Latin. It is therefore interesting to see how the diagram letters were treated in the successive translations and what rules can be derived from a comparative study of them for the edition and understanding of translated mathematical texts.[2] For this purpose, in the following paper a number of "big" diagrams, i. e. with many letters, have been selected from some of the major mathematical texts existing in all the three languages involved, Greek, Arabic and medieval Latin, in order to investigate the "methods" followed by the translators in the adaptation of the diagram letters and, if possi-

[1] For the conversion of letters used in Arabic texts into modern symbols, a standard system has been developed and proposed by H. Hermelink and E. S. Kennedy, simultaneously published in *Sudhoffs Archiv* 45 (1961), 85, in *Isis* 52 (1961), 417, and in *Journal of the American Oriental Society* 82 (1962), 204 (the last one reprinted in E. S. Kennedy *et al.*, *Studies in the Islamic Exact Sciences*, Beirut 1983, p. 745). An expanded version by E. S. Kennedy is published in the present volume of ZGAIW, see below p. 21.

[2] A recent attempt to treat this subject did not prove successful in all aspects; see the Appendix (pp. 48–62) in R. Herz-Fischler, Theorem 14,** of the First "Supplement" to the *Elements*, in *Archives Internationales d'Histoire des Sciences* 38 (1988), 3–66.

ble, to discover the system or systems of conversion employed by them.

Before embarking on the subject, one important distinction must be made. Apart from the diagram letters, classical Greek and Arabic mathematical texts use another system of letters which are applied to designate numbers (in Arabic, this is called the *abjad* system or *ḥisāb al-jum[m]al*). Though related, these two systems are not fully congruent. The "alphanumeric" system aims at rendering the complete series of numbers (as far as the letters in the various alphabets reach) and may include symbols no longer used in actual writing, as in Greek. On the other hand, in the marking of diagrams the letters are taken in the sequence of the normal alphabet.

I. *Letters used as numbers*

As is well known, the Greeks received their letters from Semitic models and also retained the alphabetic order of the letters in the North-West Semitic, Canaanaean sequence. The numerical equivalents of the letters are similarly used, according to this Canaanaean sequence, both in Greek and in Semitic idioms (such as Syriac and Hebrew) and were later also retained by the Arabs (who introduced a new sequence for their "normal" written alphabet, but continued to use the old Canaanaean sequence for their *abjad* system). The question will not be discussed here whether the use of letters as numbers in this sequence, including the decimal divisions for the tens and hundreds, was invented at some point by the Greeks and then taken over from them by Semitic and other Near Eastern peoples, as is now generally assumed, or whether the development again went in the other direction, from the Semites to the Greeks. In order to comply with the Canaanaean sequence of the letters and their numerical value, the Greeks retained three symbols that had gone out of use in pronunciation and normal writing: *digamma* (Ϝ or ϛ), *koppa* (Ϙ) and *sampi* (ϡ). On the other hand, the existence of only one *s* with the Greeks (against three in the Semitic, viz. *sāmekh*, *ṣādē* and *sīn/shīn*) led to a rupture between Greek and Semitic after the letter Π. The Arabs, whose alphabet comprises 28 letters, added the "new" or additional letters of their alphabet to the end of the old Canaanaean series.

A complete survey of the alphanumeric systems is given in TABLE 1. The seven columns of this table contain the following:

column 1: the serial number;
column 2: the number values indicated by letters;
column 3: the corresponding Greek letters;
column 4: the corresponding letters in older Semitic idioms (Syriac, Hebrew);
column 5: the corresponding Arabic letters (*abjad*) according to the system current in Eastern Arabic and Islamic countries;
column 6: the deviations from the former as current in Western Arabic (Maghrebi) countries, i. e. North-West Africa and Muslim Spain;
column 7: Latin equivalents for the Maghrebi *abjad* letters as found uniquely inscribed on the so-called "astrolabe carolingien".[3]

II. *Letters used in geometrical diagrams*

In contrast to the alphanumeric letter system, where in Greek three obsolete letters were retained, in marking geometrical diagrams the letters were used in the sequence of the normal Greek alphabet as employed in writing. That means that the Greek E was directly followed by Z, and Π by P, and the series ended at Ω. If still more letters were required, hereafter some of those obsolete letters and other symbols were employed. Again, I was sel-

[3] This, as it appears, was an astrolabe made in Muslim Spain and brought to the Christian region in the North-East of Spain, where the inscriptions were added, in Latin, by some Christian monk who anxiously imitated the inscriptions usually found on the contemporary Arabic instruments. Uniquely among all the known instruments and texts, the Arabic *abjad* numbers were not transformed into Roman numerals, but the Arabic letters were converted, unchanged, into corresponding Latin letters. This Latin operation may be dated to around A.D. 980. See M. DESTOMBES, Un astrolabe carolingien et l'origine de nos chiffres arabes, in *Archives Internationales d'Histoire des Sciences* 15 (1962), 3–45. The instrument is now kept in the Museum of the Institut du Monde Arabe, Paris.

dom used (i. e., after Θ there usually follows K), perhaps because the letter was too tiny and the danger of mistaking or overlooking it was feared. This system brought about some difficulty for the Arabs, who in imitating the Greek letters had to jump over their *w*, *y* and *ṣ*, which were "normal" letters in their alphabet. On the whole, however, the Greek alphabet followed the old Semitic sequence of the *abjad* series and allowed the Arabs to keep in line with Greek. Very different was the situation for the Latin translators confronted with Arabic. Not only is the sequence of letters in the Latin alphabet more different from the *abjad* sequence than the Greek alphabet, but Arabic also contains many consonants for which there are no adequate equivalents in Latin. The Latin translators therefore had to make greater efforts to render the Arabic letters consistently; and we shall soon see that various translators did not adopt the same Latin equivalents for several Arabic letters, but arrived at considerably different solutions.

A few remarks on the method of this research should be made. For our comparison the diagrams themselves were chosen, usually leaving aside the accompanying text. In the diagrams the location of each letter is unequivocally fixed at its proper point. The progression in the use of the single letters, according to the alphabetic sequence, can easily be followed in the construction of most diagrams. On the other hand, in the course of the text the letters are more liable to confusion, which would be difficult to trace and correct. This especially applies to Arabic, where similar letters, which are distinguished only by their diacritical dots, are extremely easily confused when the dots are omitted or wrongly placed, as, e. g., in س [*s*] and ش [*sh*]; ب [*b*], ت [*t*] and ث [*th*]; ف [*f*] and ق [*q*]; ص [*ṣ*] and ض [*ḍ*]; etc. Other confusions found in the manuscripts occur between ر [*r*] or ز [*z*] and ن [*n*], or between ن [*n*] and ق [*q*], etc. The correct reading of these letters can be established much better in the diagrams, with the Greek also at hand, than in continuous text. Once the equivalents employed by the Latin translators for the single Arabic letters are firmly established, it also becomes possible to recognize incorrect readings of the Arabic made by these translators; their false readings belong to the same categories of confusing cognate Arabic letters as mentioned above.

To our surprise, it was found that the Arabic lettering of the big diagrams in Euclid's *Elements* is totally different from the Greek. This is a new point in the intricate history of transmission of the *Elements*. We have therefore decided to treat the *Elements* in a separate section.

A. *Diagrams fully congruent in Greek, Arabic and Latin*

1. Ptolemy, *Almagest*, III,4 [in Greek: 22 letters, A–X, including I; in Ar. and Lat. I is omitted]; X,7 [in Greek: 19 letters, A–Υ, omitting I; in the three succeeding partial sections of the diagram two more letters are introduced: Φ and X].

Sources:[4] Greek: transl. MANITIUS I, p. 168 (III,4); II, p. 177 (X,7) with the three partial sections on pp. 184, 185 and 187;
Arabic: MS Leiden, or. 680 [transl. al-Ḥajjāj];
MS Tunis 07116 [transl. Isḥāq ibn Ḥunayn as emended by Thābit ibn Qurra];
Latin, transl. Gerard of Cremona: MS Paris, B.N. lat. 16200 (class 𝔄); MS Vat. lat. 2057 (class 𝔅).

2. Theodosius, *Sphaerica*, II,22 [in Greek: 27 letters, A–Ω, including I, adding ϛ, ϙ and ϡ; I is represented in Ar., but omitted in Lat.]; III,6 [in Greek: 24 letters, A–Ω, omitting I, adding ϛ]; III,7 [in Greek: 21 letters, A–Ψ, omitting I and X]; III,8 [in Greek: 23 letters, A–Ω, omitting I].
Greek: ed. HEIBERG, p. 101 (II,22); 135 (III,6); 139 (III,7); 143 (III,8);
Arabic (translator unknown):[5] MS Istanbul, Ahmet III, 3464;
MS Lahore, private library M. Nabi Khan;

[4] In order not to overcharge the notes, all the source texts used will be quoted in abbreviated form only. The editions of Greek texts here mentioned are all published in the Teubneriana (Leipzig). I wish to express my sincere thanks to Prof. M. Folkerts and Dr. R. Lorch, Munich, for their advice and help in selecting and preparing the source material for this study.

[5] The *Taḥrīr* of al-Ṭūsī was also examined in several manuscripts and the Hyderabad edition, but proved useless in this connection because al-Ṭūsī changed several letters so that his series differs from the consistent Greek-Arabic-Latin chain of transmission.

6

MS Paris, B.N. hebr. 1101 (Arabic in Hebrew characters);
MS Cambridge, Un. Lib. Add. 1220 (another Arabic version, probably by Qusṭā ibn Lūqā; Arabic in Hebrew characters);
Latin, transl. Gerard of Cremona: MS Paris, B.N. lat. 9335;
MS Vat. lat. 1548;
MS Vat. Ottobon. 2234.

From a comparison of these diagrams in the sources mentioned a system of correspondences for diagram letters can be established (individual mistakes remain unnoticed). It becomes evident that the Arabic closely follows the Greek, whereas the Latin translator, Gerard of Cremona, had to devise a "system" of his own for rendering the Arabic diagram letters. The results of this comparison are shown in TABLE 2.

A few inconsistencies here become visible with some of the last letters. In Arabic, the sequence of the *abjad* letters would require ض [*ḍ*] for Ω and ظ [*ẓ*] for ς; the manuscripts however have ظ [*ẓ*] in the place of Ω and ص [*ṣ*] in the place of ς. Two reasons may have influenced the use of these letters: on the one hand, the omission of ص [*ṣ*] in its proper place may have provoked a tendency to add this somewhere near the end; second, in spoken Arabic there is a frequent confusion between ض [*ḍ*] and ظ [*ẓ*]. This could have led to a switch between ض [*ḍ*] and ظ [*ẓ*], so that for Ω the manuscripts inserted ظ [*ẓ*], whereas the ض [*ḍ*] which then had to follow was changed into the ص [*ṣ*] that had been omitted in its proper place. For the 27th letter, marked in HEIBERG's edition of the *Sphaerica* by the symbol ↑, the Arabic inserted و [*w*] which had also been omitted in its proper place of the sequence. As for Gerard, he shows some inconsistencies in the last eight letters, from Υ to ↑. Of special interest is his rendering of the Arabic ص [*ṣ*] and ظ [*ẓ*]; here he seems to have invented two artificial symbols, *7* (which resembles the abbreviation for "*et*" in 12th century Latin manuscripts)[6] for ظ [*ẓ*], and ħ (which resembles the astronomical symbol for the planet Saturn of later times) for ص [*ṣ*]. The change between *x* and *h* for X (in

[6] The same explanation is also given by BUSARD in his edition of the *Elements*, version Gerard of Cremona, Introduction, p. XXIVb.

Almag. X,7) rests on his different reading of the Arabic: he put *x* where he had read خ [*kh*], and *h* where he had read ح [*ḥ*] (without the dot). Perhaps the use of *i* for ت [*t*] in *Almag.* X,7 is also due to a reading as ث [*th*], for which we find *u*, *y*, *l* and *i* variously.

B. *Diagrams congruent in Arabic and Latin only* (*Euclid's* Elements)

In the transmission of Euclid's *Elements* there can be observed several developments different from the situation in the texts described above. First, in all the "big" diagrams examined the distribution of letters in the Arabic is totally different from the Greek as edited by Heiberg. This means that the Arabic translators found the diagrams in their Greek source manuscripts in a form different from that found and rendered by Heiberg. It is unlikely that the different distribution of these letters was the work of the Arabs themselves, for it can be assumed that they followed their sources – in this matter as in others – faithfully. Second, there appear no differences in the diagram letters between the Arabic translation of al-Ḥajjāj (which exists, in Arabic, to the end of Book VI, and perhaps for Books XI–XIII in MS Copenhagen) and the translation of Isḥāq ibn Ḥunayn as emended by Thābit ibn Qurra. Third, a basic difference in the choice of letters is to be recognized between Books I–XIII in the translations of al-Ḥajjāj and Isḥāq-Thābit (which generally omit the Arabic letters *w* and *y*) and Books XIV–XV, which were translated by Qusṭā ibn Lūqā (and which always use *w* and *y*). Fourth, in the Latin we have three different translations, by Gerard of Cremona, Adelard of Bath (generally called "Adelard I") and Hermann of Carinthia. Each of the three translators followed his own system of converting the Arabic diagram letters into Latin, as will be seen below in Table 3; Adelard and Hermann seem to incline towards a phonetic representation of certain Arabic letters, whereas Gerard follows a purely formal system.

For our purpose a number of "big" diagrams with numerous letters were examined. From these a unified standard list of correspondences for the letters was derived (individual scribal er-

8

rors remain again unnoticed). Here also some inconsistencies occur in all the sources in some of the late letters in the series. Gerard's Latin equivalents mostly conform with those in his other translations shown in TABLE 2. Also Hermann's system is consistent with what he used in his translation of Ptolemy's *Planisphaerium*.[7] For Adelard, one manuscript, Bruges no. 529, in three marginal notes (fols. 9^v at the bottom, 12^r on top, 42^r to the outer right) gives explicit Latin equivalents for eleven Arabic letters (three of them are mentioned twice) which, however, are not all carefully rendered like that in the manuscripts in the various diagrams.[8]

Diagrams of the following theorems in the *Elements* were examined for our purpose (the numbering follows HEIBERG's edition of the Greek text; the page numbers in HEIBERG's edition, where each diagram is found, are also mentioned):

II,8 (HEIB. I, p. 139): 19 letters, Α–Υ, omitting Ι.

VI,28 (II, p. 162): 21 letters, Α–Χ, omitting Ι; of these 16 only appear in Ar. and Lat.

VI,29 (II, p. 169): 18 letters, Α–Π, omitting Ι, plus ΦΧΨ. Ar. and Lat. have only 15 letters, omitting *h*; after *s* the three letters ʿ, *f* and *ṣ* are omitted, there follow directly *q*, *r* and *sh*.

[7] I compared the *Planisphaerium* (the Greek text of which is lost) in the Arabic MS Istanbul, Aya Sofya 2671, and in HEIBERG's edition of Hermann's Latin version (Teubneriana, Leipzig 1907). Since the diagrams in the Arabic MS were not drawn (the space for them was left blank), the Arabic letters had to be collected from the text itself. I found up to 17 letters, ا ['] until ر [*r*], with the omission of و [*w*], ي [*y*] and ص [*ṣ*]. Hermann's equivalents are the same as in his translation of the *Elements* shown here in TABLE 3 with the exception of ق [*q*], which appears as *q* in ch. 16 (as in the *Elements*), but as *o* in ch. 17 and 18.

[8] The following equivalents are given:

ح = *H*	ع = *i*	ث = θ
ط = *t*	ق = *c*	خ = *q*
ي = *j*	ش = *s*	ذ = Δ
س = *ſ*	ت = *t* (or *T*?)	

(BUSARD, Introd. to his edition of Gerard's version, p. X, note 11 has read ا, *alif*, instead of ع). Note the difference between *ſ* for س [*s*] and *s* for ش [*sh*] which, unfortunately, has not been followed by BUSARD in his edition of Adelard I, where he simply gives *s* for both letters.

X,94 (III, p. 293): 22 letters, A–X, omitting I. Ar. has 19 or 20 letters, Lat. only 19.

XI,25 (IV, p. 75): 24 letters, A–Ω, including I. Ar. and Lat. also 24 letters.

XI,31a (IV, p. 95): 28 letters, A–Ω, including I, but omitting N, and adding ϛ, ϙ, ϡ, õ [which expresses the symbol for zero], ͵α [designating 1,000 in the system of number letters]. Ar. and Lat. 23 to 25 letters.

XI,31b [= XI,32 in Ar. and Lat.] (IV, p. 97): 24 letters, A–Ω, including I. Ar. and Lat. also 24 letters, *alif–ṣ*, with *l* occurring twice.

XIV (V, p. 11): 13 letters, A–Ξ, omitting I. Ar. 13 letters, including *w* and *y*, but omitting *ḥ*.

XVa (V, p. 41 below): 10 letters, A–Λ, omitting I. Ar. 10 letters, *alif–l*, including *w*, omitting *y* and *k*.

XVb (V, p. 43): 18 letters, A–T, omitting I. Ar. 18 letters, *alif–q*, including *w* and *y*, omitting *ṣ*.

XVc (V, p. 45): 14 letters, A–O, omitting I. Ar. 14 letters, including *w* and *y*.

XVd (V, p. 47): 14 letters, A–O, omitting I. Ar. 13 letters, *alif–ʿ*, including *w* and *y*, but omitting *m*, *n*, *s*.

These items were inspected in the following editions and manuscripts:[9]

Greek: ed. HEIBERG;

Arabic:[10] MS Teheran, Malik 3586 (very early, dated A. H. 343 = A. D. 954/5!);
MS Copenhagen 81;

al-Nayrīzī (Anaritius), for VI,28–29:
Arab.: ed. JUNGE – RAEDER – THOMSON;
Latin (Gerard of Cremona): ed. CURTZE;

Gerard of Cremona: ed. BUSARD, Leiden 1983;
MS Bruges no. 521;

[9] The material concerning the *Elements* is listed in full detail in: M. FOLKERTS, *Euclid in Medieval Europe*, The Benjamin Catalogue for History of Science, Winnipeg, Canada, 1989 (Questio de rerum natura, II).

[10] As with Theodosius (cf. note 5, above), al-Ṭūsī's *Taḥrīr* of the *Elements* was not included in the present comparison because of the change of several critical letters in some diagrams. The MS consulted was Munich no. 848.

MS Vat. Rossiano 579;
MS Oxford, Bodl. Digby 174;
Adelard of Bath ("Adelard I"): ed. BUSARD, Toronto 1983;
MS Oxford, Trinity Coll. 47;
MS Bruges no. 529;
MS Oxford, Bodl. D'Orville 70;
Hermann of Carinthia: ed. BUSARD (I–VI: Leiden 1968; VII–XII, Amsterdam 1977);
MS Paris, B. N. lat. 16646.

The diagrams in all these versions are mostly basically identical in shape though shifts (right to left, upside down and even internal twistings) sometimes occur. The Arabic is quite consistent, but scribal errors of the sorts mentioned above (p. 4) are sometimes found. Echoes of such wrong readings are variously found in all the Latin versions. Of these, as usual Gerard's is the most reliable (although it had been found that the version of the *Elements* circulating under his name does not represent his original text but shows signs of later redaction[11]). The consistency in his rendering of the diagram letters allows the identification of occasional misreadings of some of the Arabic letters (cf. the specimen of VI,29 below in TABLE 4). Adelard's system of rendering the diagram letters has not always been faithfully expressed in BUSARD's edition. Here a special problem was posed by the difficulty to distinguish in the manuscripts between *H* (which stands for Arabic *ḥ*) and *n* (the shape of which, written as a capital, mostly looks nearly identical to *H*). A comparison of the Arabic would have helped to avoid confusion in the Latin edition.[12] The two Arabic letters س [*s*] and ش [*sh*] were carefully distinguished in the Adelard manuscripts as *ʃ* and *s*, but were both given identically as *s* in the edition. Some of the manuscripts' θ's (for Arabic ث [*th*]) were given as *e* (XI,25 = p. 317; XI,31a = p. 322) or *d* (XI,31b [= 32] = p. 323) in the edition. A special letter, perhaps intending a capital *T*, for Arabic ت [*t*], appears in the edition as *i*

[11] Cf. P. KUNITZSCH, Findings in some texts of Euclid's Elements (Mediaeval transmission, Arabo-Latin), in: *Mathemata, Festschrift für Helmuth Gericke* (Boethius, Bd. XII), Stuttgart 1985, pp. 115–128, esp. p. 119f.

[12] Of this sort the following cases may be mentioned: BUSARD *h* (correct *n*): VI,29 (p. 191); BUSARD *n* (correct *H*): XI,25 (p. 317); XI,31a (p. 322); XI,31b [32] (p. 323); XIII,1 (p. 354); XV (p. 388).

(XI,31a = p. 322) or *r* (XI,31b [= 32] = p. 323). This procedure resulted in using some letters twice in the same diagram. Special interest is provoked by MS Trinity College 47, which is dated to the 12th century. Here Arabic ق [*q*] is twice rendered as *Q* (VI,28 and 29; but *c* in II,8); further, in VI,29 in the place of *r* there was first written *z*, which then was struck out and *r* written above it, in the same hand. This could be interpreted as being an element from the working period itself where the translator first mistook the Arabic letter as ز = *z* and afterwards realized that it should be ر = *r* (*z* exists elsewhere in the diagram, but was omitted in this manuscript). Likewise *Q* could have been the translator's first proposal to render the Arabic ق [*q*] which sometimes alternated with, and in later manuscripts was completely replaced by, *c*. The extraordinary closeness of this manuscript to the translation process was already stated earlier[13] and finds new support in these details. Of Hermann's version only one complete manuscript exists (Paris, B.N. lat. 16646). In it diagrams are only given in the former part of the text, and they are all left without letters. The letters used in this version therefore had to be collected from the text of the theorems. Here also the system is consistent enough to recognize occasional misreadings of some Arabic letters. In X,94 Hermann uses a simplified model with only 12 letters (instead of 19 in the other versions). In XI,25 Hermann's procedure is also different and cannot be directly compared to the other versions. In XI,31a only 13 letters correspond to those in the Arabic manuscripts here used, 11 letters remain uncertain and cannot be assigned to the corresponding Arabic letters. The results of this comparative study of the letters used in the *Elements* are shown in TABLE 3.[14, 15]

[13] M. CLAGETT; cf. BUSARD, Introd. to his edition of Gerard's version, p. X, where however he expresses his opinion that MS Bruges 529 "goes back to an earlier text than that of the Trinity MS." The Arabic letters cited in the margins of MS Bruges (cf. note 8, above) - apparently in the same hand as the main text - were visibly copied by a scribe who did not know Arabic himself and who distorted their forms heavily.

[14] Since there is no congruence here between Greek and Arabic, as stated above, Greek letters are not included in this table.

[15] In Syriac, there exist the anonymous fragments from I,1–23 and 37–40, edited by G. FURLANI, in *Zeitschrift für Semitistik* 3 (1924), 27–52; 212–235. In the diagrams of this version, all of which belong to the smaller

The version generally called "Adelard II" was not included in this survey.[16] In harmony with the assumption that this is but a compiled, revised and modified version, also the diagrams are here treated differently from the other, "original" versions. Mostly, if at all, in the diagrams one letter only is assigned to a full diagram, or a surface etc., and no individual points are marked by letters as in the other texts. An additional sign of the "Latinization" of this version is found in the sequence of letters which runs *a*, *b*, *c*, *d* etc., using *c* where the translated versions had *g* (for the Arabic ج [*j*]). Often "*a*" is here called *alif*, taking over this full name of the letter from some of the versions translated from the Arabic, whereas the other letters are spelled in their Latin form as "*be*", "*ce*", "*de*" etc. One exception is MS Trinity College 47 which, in our specimens II,8; VI,28 and 29, gives the same full lettering of the diagrams as in version "Adelard I" contained in the same manuscript (with a few minor deviations).

In order to demonstrate the complicate situation in the transmission of the *Elements*, in TABLE 4 a complete survey of the situation in VI,29 is given. The table shows the importance of the Arabic for the study of the Latin versions. It is practically not possible to establish the Latin texts and to ascertain their correct readings without permanent comparison of the Arabic. The effect of misreading Arabic letters is well recognized in the Latin versions. Since there is no correspondence between Greek and Arabic, the Greek letters are not included. For the identification of the points in the four parts of the diagram, first a rough sketch is given with numbers added for identification. In TABLE 4 the letters are set against the respective numbers.

sort, up to ten letters could be registered which follow in this sequence: ʾ [*ālaph*], *b*, *g*, *d*, *h*, *z*, *ḥ*, *ṭ*, *k*, *l*; *w* and *y* are omitted. The series thus equals the Arabic series in the versions of al-Ḥajjāj and Isḥāq-Thābit as shown in our TABLE 3. This would corroborate FURLANI's assumption that the Syriac paraphrase was derived from an Arabic source rather than from a Greek one.

[16] For comparison, the following MSS of "Adelard II" were inspected: Copenhagen, Kgl. S. 277 fol.; Leipzig, Univ. Lib. Rep. I, 68c; London, Brit. Lib. Add. 34018; dto., Sloane 285; Munich, Clm 13021; Oxford, Trinity College 47; Paris, B.N. lat. 16647; Prague, Univ. Lib. III.H.19; Vatican, Reg. lat. 1137.

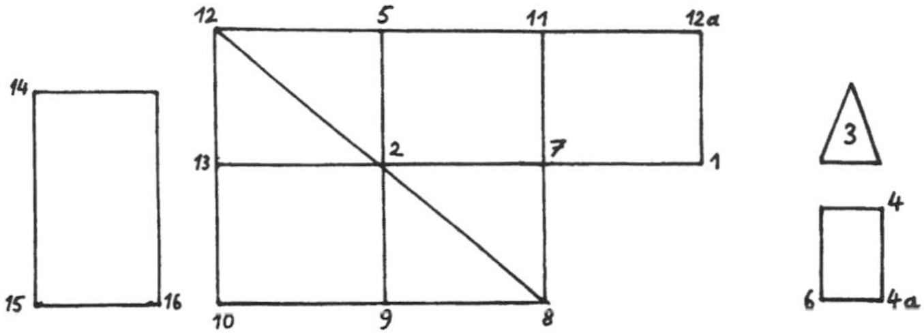

Figure of diagram in *Elements* VI,29

C. *Diagrams in a Western Arabic* (*Maghrebi*) *text*

As a specimen for a text written independently in Arabic, we discuss as an additional item the *Kitāb al-hay'a*, also *Iṣlāḥ al-majasṭī*, in nine Books, by the Spanish-Arabic astronomer Jābir ibn Aflaḥ (died perhaps around A.D. 1145). The work was also translated into Latin by Gerard of Cremona where the author became well-known as *Geber*. Living in the Western Arabic area, Jābir naturally applied the diagram letters according to the Maghrebi series (cf. below, TABLE 1).

For our comparison we used a diagram from Book II (on p. 27 of the 1534 Latin edition; 17 letters) and one from Book VIII (on p. 133 of the 1534 edition; 16 letters). Jābir did not translate from the Greek and therefore was free to use the letters for the diagrams according to his own will. We find that he omits *w* and *y*, like several Eastern Arabic authors and translators, but also *n*. The place after the (omitted) *n* is taken by ص [*ṣ*], س [*s*] following after *r*, according to Maghrebi usage.

Gerard keeps to his system for the conversion of the letters as found below in TABLES 2, 3 and 4, except for ص [*ṣ*], which he here renders as *p* (Book II) or *o* (Book VIII).

For our comparison the following sources were used:

Arabic: MSS Escorial 910 and 930; Berlin, AHLWARDT 5653;

Latin: MSS Madrid, Bib. Nac. 10006; Milan, Ambros. G 70 sup.; Vatican, Ottobon. 2234; Vat. Lat. 2059. In the printed edition of Nuremberg, 1534, the letters were added in hand-cut shape and sometimes cannot easily be recognized.

The results of our comparison are shown in TABLE 5 (individual scribal errors remain unnoticed).

TABLE 1

serial number	number value	Greek	Semitic	Arabic (East)	West Arabic (Maghrebi)	astrolabe carolingien
1.	1	Α	ʾ	ا [ʾ]		A
2.	2	Β	*b*	ب [*b*]		b
3.	3	Γ	*g*	ج [*j*, *ǧ*]		C
4.	4	Δ	*d*	د [*d*]		D
5.	5	Ε	*h*	ه [*h*]		E
6.	6	ϛ	*w*	و [*w*]		V
7.	7	Ζ	*z*	ز [*z*]		z
8.	8	Η	*ḥ*	ح [*ḥ*]		h
9.	9	Θ	*ṭ*	ط [*ṭ*]		T
10.	10	Ι	*y*	ى [*y*]		I
11.	20	Κ	*k*	ك [*k*]		K
12.	30	Λ	*l*	ل [*l*]		L
13.	40	Μ	*m*	م [*m*]		M
14.	50	Ν	*n*	ن [*n*]		N
15.	60	Ξ	*s* [*sāmekh*]	س [*s*]	ص [*ṣ*]	O
16.	70	Ο	ʿ	ع [ʿ]		G
17.	80	Π	*p*	ف [*f*]		F
18.	90	ϙ	*ṣ*	ص [*ṣ*]	ض [*ḍ*]	∂
19.	100	Ρ	*q*	ق [*q*]		
20.	200	Σ	*r*	ر [*r*]		
21.	300	Τ	*sh* [*shīn*]	ش [*sh*, *š*]	س [*s*]	
22.	400	Υ	*t*	ت [*t*]		
23.	500	Φ		ث [*th*, *ṯ*]		
24.	600	Χ		خ [*kh*, *ḫ*]		
25.	700	Ψ		ذ [*dh*, *ḏ*]		
26.	800	Ω		ض [*ḍ*]	ظ [*ẓ*]	
27.	900	ϡ		ظ [*ẓ*]	غ [*gh*, *ġ*]	
28.	1000			غ [*gh*, *ġ*]	ش [*sh*, *š*]	

Table 2

serial number	Greek	Arabic	Latin
1.	Α	ا [ʾ]	*a*
2.	Β	ب [*b*]	*b*
3.	Γ	ج [*j*]	*g*
4.	Δ	د [*d*]	*d*
5.	Ε	ه [*h*]	*e*
6.	Ζ	ز [*z*]	*z*
7.	Η	ح [*ḥ*]	*h*
8.	Θ	ط [*ṭ*]	*t*
9.	Ι	ے [*y*]	–
10.	Κ	ك [*k*]	*k*
11.	Λ	ل [*l*]	*l*
12.	Μ	م [*m*]	*m*
13.	Ν	ن [*n*]	*n*
14.	Ξ	س [*s*]	*s*
15.	Ο	ع [ʿ]	*q*
16.	Π	ف [*f*]	*f*
17.	Ρ	ق [*q*]	*c*
18.	Σ	ر [*r*]	*r*
19.	Τ	ش [*sh*]	*o*
20.	Υ	ت [*t*]	*p* / *Alm.* X, 7: *i*
21.	Φ	ث [*th*]	*Alm.* II, 4: *u* *Alm.* X, 7, MS Par.: *l*, MS Vat.: *i* *Sph.*: *y*
22.	Χ	خ [*kh*]	*x* / *Alm.* X, 7 (p. 184, 185): *h*
23.	Ψ	ذ [*dh*]	*Sph.* II, 22: *the* (in the text also *thel*, *theld*) *Sph.* III, 6, 7, 8: *i*
24.	Ω	*Sph.* II, 22: ظ [*ẓ*] *Sph.* III, 6, 8: ص [*ṣ*]	7 ħ
25.	ς	*Sph.* II, 22: ص [*ṣ*] *Sph.* III, 6: --	ħ 7
26.	ϙ	غ [*gh*]	*ge*
27.	ϡ	و [*w*]	*u*

TABLE 3

serial number	Arabic	Gerard of Cremona	Adelard of Bath	Hermann of Carinthia
1.	ا [ʾ]	*a*		
2.	ب [*b*]	*b*		
3.	ج [*j*]	*g*		
4.	د [*d*]	*d*		
5.	ه [*h*]	*e*	*h*	*e*
6.	و [*w*] (Qusṭā)	*u*	*u*	–
7.	ز [*z*]	*ž*		
8.	ح [*ḥ*]	*h*	*H*	*h*
9.	ط [*ṭ*]	*t*		
10.	ے [*y*] (Qusṭā)	*i* [XIV: *b*]	XIV: *f* [pro *i*]	–
11.	ک [*k*]	*k*	XIV: MS illeg., ed.: *r*	
12.	ل [*l*]	*l*		
13.	م [*m*]	*m*		
14.	ن [*n*]	*n*		
15.	س [*s*]	*s* [VI, 29: *x*, pro ش ?]	MSS: *ſ*, ed.: *s*	*c*
16.	ع [ʿ]	*q*	*i*, *j*	*y*
17.	ف [*f*]	*f*		

18.	ص [*ṣ*]	XI, 25: *i* [read as ض?] XI, 31a,b: MSS: *7*, ed.: *ç*	XI, 25: MS: *j*, ed.: *y* XI, 31a: MS: –, ed.: *y* XI, 31b: MS, ed.: *y*	XI, 31a: *u* XI, 31b: *o*
19.	ق [*q*]	*c*	*c* [VI, 28, 29, Trin.Coll.: *Q*]	*q*
20.	ر [*r*]	*r*		
21.	ش [*sh*]	*o* II, 8: *x* [cf. VI, 29]	MSS, ed.: *s*	*s*
22.	ت [*t*]	*p*	II, 8; XI, 25: *t* XI, 31a: MS: *T* (?), ed.: *i* XI, 31b: MS: *T* (?), ed.: *r*	XI, 31b: *x*
23.	ث [*th*]	II, 8: *y* XI, 25, 31b: *u* XI, 31a: *i*	MSS: θ, ed.: θ, *e*, *d*	XI, 31b: *p*
24.	خ [*kh*]	*x*	XI, 25: MS: *J*, ed.: *j* XI, 31a,b: *q*	XI, 31b: *h*
25.	ذ [*dh*]	*y*	XI, 25: *D* XI, 31a: *t* XI, 31b: MS: *d*, ed.: *e*	XI, 31b: *v*
26.	ض [*ḍ*]	XI, 25: MSS: *7* [read as ص], ed.: *z* [pro *ç*]	XI, 25: *p*	–
27.	أ [XI, 31a: extra letter, *alif* with a thick dot on top]	*u*	*a*	–

Table 4

serial number	Teheran, Malik	Copen-hagen	al-Ṭūsī Mü. 848	Ger.Crem. ed. Busard	Bruges 521	Vat.Ross. 579	al-Nayrīzī ed. Arab.	Anaritius, Ger.Crem. ed. Curtze	Adelard I ed. Busard	Trinity Coll. 47	Bruges 529	Herm.Car. ed. Busard (= BN 16646)	Adelard II Trinity Coll. 47
1.	ا [']			*a*			ا [']	*a*	*a*			*a*	*a*
2.	ب [*b*]			*b*			ب [*b*]	*b*	*b*	*f*[12]	*b*	*b*	*f*[19]
3.	ج [*j*]			*g*			ج [*j*]	*g*	*g*			*g*	*g*
4.	د [*d*]			*d*			د [*d*]	*d*	*d*			*d*	*d*
4a.							ه [*h*]	*e*					
5.	–	–	ه [*h*]	–	–	–	–	–	–	–	–	–	–
6.	ز [*z*]			*z*			ز [*z*]	*z*	*z*	–	*z*	*z*	*d*[20]
7.	ح [*ḥ*]			*h*			ح [*ḥ*]	*h*	*H*	*H*	*n*[17]	*h*	*H*
8.	ط [*ṭ*]			*t*			ط [*ṭ*]	*t*	*t*			*t*	*t*
9.	ك [*k*]			*k*			ك [*k*]	*k*	*k*			*k*	*k*
10.	ل [*l*]			*l*			ل [*l*]	*l*	*l*			*l*	*l*
11.	م [*m*]			*m*			م [*m*]	*m*	*m*			*m*	*m*
12.	ن [*n*]			*e*[2]			ز [*z*][8]	*n*	*h*	*H*[13]	*n*	*n*	*H*
12a.							ن [*n*]	–					
13.	س [*s*]			*z*[3]	*x*[4]	*x*[4]	س [*s*]	*o*[10]	–	–	–	–	–
14.	ق [*q*][1]			*f*[5]			ف [*f*][9]	*c*	*c*	*Q*[14]	*c*	*q*	*Q*[21]
15.	ر [*r*]		–	*n*[6]			ن [*n*][9]	*r*	*r*	[15]	*r*	*r*	*r*
16.	ش [*sh*]			*s*[7]			س [*s*][9]	*s*[11]	*s*	*f*[16]	*f*[16]	*c*[18]	*f*[22]

[1] the Arabs omitted ع [ʿ], ف [*f*] and ص [*ṣ*] and assigned to this rectangular figure the next three letters *q*, *r* and *sh* in the *abjad* sequence.

[2] in Gerard's Arabic source the letter ه [*h*], which is missing from our Arabic copies (except al-Ṭūsī, who may have inserted it himself to fill the gap), was displaced from point 5 to point 12.

[3] I do not see why Busard has here used a second *z* (cf. point 6) for the *x* of the MSS.

[4] Gerard's *x* yields a reading in the Arabic of ش [*sh*] instead of س [*s*]; cf. Table 3, nos. 15 and 21.

[5] misreading in the Arabic: ف [*f*] for ق [*q*].

[6] misreading in the Arabic: ن [*n*] for ر [*r*].

[7] misreading in the Arabic: س [*s*] for ش [*sh*], which would normally be *o* with Gerard.

[8] miswriting (or misreading) in the Arabic: ز [*z*] for ن [*n*].

[9] cf. notes 5–7.

[10] Gerard read ش [*sh*] instead of س [*s*].

[11] here Gerard read س [*s*] instead of ش [*sh*].

[12] probably the translator's first reading of this letter was ف [*f*] instead of ب [*b*].

[13] in the MS this letter looks nearly identical to that of point 7; it is intended as a capital *N*; Busard has wrongly inserted *h* in this place.

[14] probably, Adelard's first rendering of Arabic ق [*q*].

[15] first written *z*, then struck out and written *r* above it (by the same hand).

[16] reading س [*s*] in the Arabic instead of ش [*sh*]; the latter would appear as *s* in the Adelard MSS.

[17] copying error (wrong reading of a capital *H* in the source MS).

[18] Hermann read س [*s*] instead of ش [*sh*].

[19] cf. note 12.

[20] instead of *z*, a second *d* (perhaps due to a corresponding misreading in the Arabic).

[21] cf. note 14.

[22] cf. note 16.

Nos. 4a and 12a seem to be two extra letters in al-Nayrīzī not found in our Arabic copies of the *Elements*.

20

TABLE 5

serial number	Arabic	Latin
1.	ا [ʾ]	*a*
2.	ب [*b*]	*b*
3.	ج [*j*]	*g*
4.	د [*d*]	*d*
5.	ه [*h*]	*e*
6.	ز [*z*]	*z*
7.	ح [*ḥ*]	*h*
8.	ط [*ṭ*]	*t*
9.	ك [*k*]	*k*
10.	ل [*l*]	*l*
11.	م [*m*]	*m*
12.	ص [*ṣ*]	II: *p* / VIII: *o*
13.	ع [ʿ]	*q*
14.	ف [*f*]	*f*
15.	ق [*q*]	*c*
16.	ر [*r*]	*r*
17.	س [*s*]	*s*

XXIX

The Transmission of Hindu-Arabic Numerals Reconsidered

For the last two hundred years the history of the so-called Hindu-Arabic numerals has been the object of endless discussions and theories, from Michel Chasles and Alexander von Humboldt to Richard Lemay in our times. But I shall not here review and discuss all those theories. Moreover I shall discuss several items connected with the problem and present documentary evidence that sheds light—or raises more questions—on the matter.

At the outset I confess that I believe the general tradition, which has it that the nine numerals used in decimal position and using zero for an empty position were received by the Arabs from India. All the oriental testimonies speak in favor of this line of transmission, beginning from Severus Sēbōkht in 662[1] through the Arabic-Islamic arithmeticians themselves and to Muslim historians and other writers. I do not touch here the problem whether the Indian system itself was influenced, or instigated, by earlier Greek material; at least, this seems improbable in view of what we know about Greek number notation.

The time of the first Arabic contact with the Hindu numerical system cannot safely be fixed. For Sēbōkht (who is known to have translated portions of Aristotle's *Organon* from Persian) Fuat Sezgin[2] assumes possible Persian mediation. The same may hold for the Arabs, in the eighth century. Another possibility is the Indian embassy to the caliph's court in the early 770s, which supposedly brought along an Indian astronomical work, which was soon translated into Arabic. Such Indian astronomical handbooks usually contain chapters on calculation[3] (for the practical use of the parameters contained in the accompanying astronomical tables), which may have conveyed to the Arabs the Indian system. In the following there developed a genre of Arabic writings on Hindu reckoning (*fī l-ḥisāb al-hindī*, in Latin *de numero Indorum*), which propagated the new system and the operations to be made with it. The oldest known text of this kind is the book of al-Khwārizmī (about 820, i.e., around fifty years or more after the first contact), whose Arabic text seems to be lost, but which can very well be reconstructed from the surviving Latin adaptations of a Latin translation made in Spain in the twelfth century. Similar texts by al-Uqlīdisī (written in 952/3), Kūshyār ibn Labbān (2nd half of the 10th century) and ʿAbd al-Qāhir al-Baghdādī (died 1037) have survived and have been edited.[4] All these writings follow the same pattern: they start with a description of the nine Hindu numerals (called *aḥruf*, plural of *ḥarf*; Latin *litterae*), of their forms (of which it is often said that some of them may be written differently), and of zero. Then follow the chapters on the various operations. Beside these many more writings of the same kind were produced,[5] and in later centuries this tradition was amply continued, both in the Arabic East and West. All these writings trace the system back to the Indians.

The knowledge of the new system of notation and calculation spread beyond the circles of the professional mathematicians. The historian al-Yaʿqūbī describes it in his *Tārīkh* (written 889)—he also mentions zero, *ṣifr*, as a small circle (*dāʾira ṣaghīra*).[6] This was repeated, in short form, by al-Masʿūdī in his *Murūj*.[7] In the following century the encyclopaedist Muḥammad ibn Aḥmad al-Khwārizmī gave a description of it in his *Mafātīḥ al-ʿulūm* (around 980); also he knows the signs for zero (*aṣfār*, plural) in the form of small circles (*dawāʾir ṣighār*).[8] That the meaning of *ṣifr* is really "empty, void" has been nicely proved by August Fischer,[9] who presents a number of verses from old Arabic poetry, where the word occurs in this sense. It may thus be regarded as beyond doubt that *ṣifr*, in arithmetic, indeed renders the Indian *śūnya*, indicating a decimal place void of any of the nine numerals. Exceptional is the case of the *Fihrist* of Ibn al-Nadīm (around 987, that is, contemporaneous with the encyclopaedist al-Khwārizmī). This otherwise well-informed author apparently did not recognize the true character of the nine signs as numerals; he treats them as if they were letters of the Indian alphabet.[10] He juxtaposes the nine signs to the nine first letters of the Arabic *abjad* series and says that, if one dot is placed under each of the nine signs, this corresponds to the following (*abjad*) letters *yāʾ* to *ṣād*, and with two dots underneath to the remaining (*abjad*) letters *qāf* to *ẓāʾ* (with some defect in the manuscript transmission). This sounds as if he understood the nine signs and their amplification with the dots as letters of the Indian alphabet. Even a Koranic scholar, Abū ʿAmr ʿUthmān al-Dānī (in Muslim Spain, died 1053), knows the zero, *ṣifr*, and compares it to the common Arabic orthographic element *sukūn*.[11] (For all these authors it must be kept in mind that the manuscripts in which we have received their texts date from more recent times and therefore may not reproduce the forms of the figures in the original shape once known and written down by the authors.)

Of some interest in this connection are, further, two quotations recorded by Charles Pellat: the polymath al-Jāḥiẓ (died 868) in his *Kitāb al-muʿallimīn* advised schoolmasters to teach finger reckoning (*ḥisāb al-ʿaqd*) instead of *ḥisāb al-hind*, a method needing "neither spoken word nor writing"; and the historian and literate Muḥammad ibn Yaḥyā al-Ṣūlī (died 946) wrote in his

Adab al-kuttāb: "The scribes in the administration refrain, however, from using these [Indian] numerals because they require the use of materials [writing-tablets or paper?] and they think that a system which calls for no materials and which a man can use without any instrument apart from one of his limbs is more appropriate in ensuring secrecy and more in keeping with their dignity; this system is computation with the joints (*ʿaqd* or *ʿuqad*) and tips of the fingers (*banān*), to which they restrict themselves."[12]

The oldest specimens of written numerals in the Arabic East known to me are the year number 260 Hijra (873/4) in an Egyptian papyrus and the numerals in MS Paris, BNF ar. 2457, written by the mathematician and astronomer al-Sijzī in Shīrāz between 969 and 972. The number in the papyrus (figure 1.1)[13] may indicate the year, but this is not absolutely certain.[14] For an example of the numerals in the Sijzī manuscript, see figure 1.2. It is to be noted that here "2" appears in three different forms, one form as common and used in the Arabic East until today, another form resembling the "2" in some Latin manuscripts of the 12th century, and a form apparently simplified from the latter; also "3" appears in two different forms, one form as common in the East and used in that shape until today, and another form again resembling the "3" in some Latin manuscripts of the 12th century.

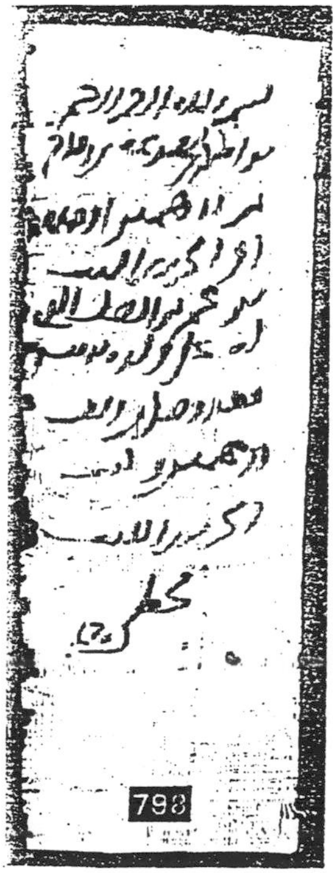

Figure 1.1
Papyrus PERF 789.
Reproduced from Grohmann, Pl. LXV, 12

Figure 1.2

MS Paris, B. N. ar 2457, fol 85v. Copied by al-Sijzī, Shīrāz, 969–972

This leads to the question of the shape of the nine numerals. Still after the year 1000 al-Bīrūnī reports that the numerals used in India had a variety of shapes and that the Arabs chose among them what appeared to them most useful.[15] And al-Nasawī (early eleventh century) in his *al-Muqniʿ fī l-ḥisāb al-hindī* writes at the beginning, when describing the forms of the nine signs, "Les personnes qui se sont occupées de la science du calcul n'ont pas été d'accord sur une partie des formes de ces neuf signes; mais la plupart d'entre elles sont convenues de les former comme il suit"[16] (then follow the common Eastern Arabic forms of the numerals).

Among the early arithmetical writings that are edited al-Baghdādī mentions that for 2, 3, and 8 the Iraqis would use different forms.[17] This seems to be corroborated by the situation in the Sijzī manuscript. Further, the Latin adaptation of al-Khwārizmī's book says that 5, 6, 7, and 8 may be written differently. If this sentence belongs to al-Khwārizmī's original text, that would be astonishing. Rather one would be inclined to assume that this is a later addition made either by Spanish-Muslim redactors of the Arabic text or by the Latin translator or one of the adapters of the Latin translation, because it is in these four signs (or rather, in three of them) that the Western Arabic numerals differ from the Eastern Arabic ones.[18]

Another point of interest connected with Hindu reckoning and the use of the nine symbols is: how these were used and in what form the operations were made. Here the problem of the calculation board is addressed. It was especially Solomon Gandz who studied this problem in great detail and who arrived at the result that the Arabs knew the abacus and that the term *ghubār* commonly used in Western Arabic writings on arithmetic renders the Latin *abacus*.[19] As evidence for his theory he also cites from Ibn al-Nadīm's *Fihrist* several Eastern Arabic book titles such as *Kitāb al-ḥisāb al-hindī bi-l-takht* (to which is sometimes added *wa-bi-l-mīl*), "Book on Hindu Reckoning with the Board (and the Stylus)." I cannot follow Gandz in his argumentation. It is clear, on the one side, that all the aforementioned eastern texts on arithmetic, from al-Khwārizmī through al-Baghdādī, mention the *takht* (in Latin: *tabula*) and that on it numbers were written and—in the course of the operations—were erased (*maḥw*, Latin: *delere*). It seems that this board was covered with dust (*ghubār*, *turāb*) and that marks were made on it with a stylus (*mīl*). But can this sort of board, the *takht* (later also *lawḥ*, Latin *tabula*), be compared with the abacus known and used in Christian Spain in the late tenth to the twelfth centuries? In my opinion, definitely not. The abacus was a board on which a system of vertical lines defined the decimal places and on which calculations were made by placing counters in the columns required, counters that were inscribed with *caracteres*, that is, the nine numerals (in the Western Arabic style) indicating the number value. The action of *maḥw*, *delere*, erasing, cannot be connected with the technique of handling the counters. On the other side, the use of the *takht* is unequivocally connected with writing down (and in case of need, erasing) the numerals; the *takht* had no decimal divisions like the abacus, it was a board (covered with fine dust) on which numbers could be freely put down (Ibn al-Yāsamīn speaks of *naqasha*) and eventually erased (*maḥw*, *delere*). Thus it appears that the Arabic *takht* and the operations on it are quite different from the Latin *abacus*. Apart from the theoretical descriptions in the arithmetical texts we have an example where an astronomer describes the use of the *takht* in practice: al-Sijzī mentions, in his treatise *Fī kayfīyat ṣanʿat jamīʿ al-asṭurlābāt*, how values are to be collected from a table and to be added, or subtracted, on the *takht*.[20] Furthermore it is worth mentioning that al-Uqlīdisī adds to his arithmetical work a Book IV on calculating *bi-ghayr takht wa-lā maḥw bal bi-dawāt wa-qirṭās*, "without board and erasing, but with ink and paper," a technique, he adds, that nobody else in Baghdad in his time was versant with. All this shows that the *takht*, the dust board of the Arabs, was really used in practice—though for myself I have some difficulty to imagine what it looked like—and that it was basically different from the Latin *abacus*.

Let me add here that the Eastern Arabic forms of the numerals also penetrated the European East, in Byzantium. Woepcke has printed facsimiles of the Arabic numerals appearing in four manuscripts of Maximus Planudes' treatise on Hindu reckoning, *Psephophoria kat' Indous*.[21]

So far, at least for the Arabic East, matters appear to be reasonably clear. But now we have to turn to the Arabic West, that is, North Africa and Muslim Spain. Here we are confronted with two major questions, for only one of which I think an answer is possible, whereas the second cannot safely be answered for lack of documentary evidence.

Question number one concerns the notion of *ghubār*. This term, meaning "dust" (in reminiscence of the dust board), is understood by most of the modern authorities as the current designation for the Western Arabic forms of the numerals; they usually call them "*ghubār* numerals."

It is indeed true that the term *ghubār*—as far as I can see—does not appear in book titles on Hindu reckoning or applied to the Hindu-Arabic numerals in the arithmetical texts of the early period in the Arabic East. On the contrary, in the Arabic West we find book titles like *ḥisāb al-ghubār* (on Hindu reckoning) and terms like *ḥurūf al-ghubār* or *qalam al-ghubār* for the numerals used in the Hindu reckoning system. The oldest occurrence so far noticed of the term is in a commentary on the *Sefer Yeṣira* by the Jewish scholar Abū Sahl Dunas ibn Tamīm. He was active in Kairouan and wrote his works in Arabic. This commentary was written in 955/6. In it Dunas says the following: "Les Indiens ont imaginé neuf signes pour marquer les unités. J'ai parlé suffisamment de cela dans un livre que j'ai composé sur le calcul indien connu sous le nom de *ḥisāb al-ghubār*, c'est-à dire calcul du *gobar* ou calcul de poussière."[22]

The next work to be cited in this connection is the *Talqīḥ al-afkār fī ʿamal rasm al-ghubār* by the North African mathematician Ibn al-Yāsamīn (died about 1204). Two pages from this text were published in facsimile in 1973;[23] on page 8 of the manuscript (= page 232 in the publication) the author presents the nine signs (*ashkāl*) of the numerals which are called *ashkāl al-ghubār*, "dust figures"; at first they are written in their Western Arabic form, then the author goes on: *wa-qad takūnu ayḍan hākadhā* [here follow the Eastern Arabic forms] *wa-lākinna l-nās ʿindanā ʿalā l-waḍʿ al-awwal*, "they may also look like this . . . , but people in our [area] follow the first type." (It should be noted that the manuscript here reproduced—Rabat K 222—is in Eastern *naskhī* and of a later date.) Another testimony is found in Ṣāʿid al-Andalusī's *Ṭabaqāt al-umam* (written about 1068 in Spain). In praising Indian achievements in the sciences this author writes: *wa-mimmā waṣala ilaynā min ʿulūmihim fī l-ʿadad ḥisāb al-ghubār alladhī bassaṭahu Abū Jaʿfar Muḥammad ibn Mūsā al-Khwārizmī* etc.,[24] "And among what has come down to us of their sciences of numbers is the *ḥisāb al-ghubār* [dust reckoning] which . . . al-Khwārizmī has described at length. It is the shortest [form of] calculation . . . , etc." This paragraph was later reproduced by Ibn al-Qifṭī in his *Tārīkh al-ḥukamāʾ* (probably written in the 1230s), but here the most interesting words of Ṣāʿid's text were shortened; in Ibn al-Qifṭī it merely reads: *wa-mimmā waṣala ilaynā min ʿulūmihim ḥisāb al-ʿadad alladhī* . . . , "And among what has come down to us of their sciences is the *ḥisāb al-ʿadad* [calculation of numbers] which al-Khwārizmī . . . etc."[25]

From these testimonies it is clear that in the Arabic West since the middle of the tenth century the system of Hindu reckoning as such was called "dust reckoning," *ḥisāb al-ghubār*—certainly in reminiscence of what the eastern arithmetical texts mentioned about the use of the *takht*, the dust board. It will then further be clear that the terms *ḥurūf al-ghubār* or *qalam al-ghubār* (dust letters or symbols) for the nine signs of the numerals used in this system of calculation basically described the written numerals as such, without specification of their Eastern or Western Arabic forms. This is corroborated by some known texts that put the *ḥurūf al-ghubār*, written numerals, in opposition to the numbers used in other reckoning systems that had no written symbols, such as finger reckoning and mental reckoning. In favor of this interpretation may be quoted some of the texts first produced by Woepcke. One supporting element here is what Woepcke derives from the *Kashf al-asrār* [or: *al-astār*] *ʿan ʿilm* [or: *waḍʿ*] *al-ghubār* of al-Qalaṣādī (in Muslim Spain, died 1486).[26] Further, in Woepcke's translation of a treatise by Muḥammad Sibṭ al-Māridīnī (*muwaqqit* in Cairo, died 1527), where the author cites words from the *Kashf al-ḥaqā'iq fī ḥisāb al-daraj wa-l-daqā'iq* of the Cairene astronomer Shihāb al-Dīn Ibn al-Majdī (died 1447), we read (of Ibn al-Majdī), "Cependant (Chehab Eddîn), . . . , s'est étendu dans l'ouvrage cité sur l'exposition de la méthode des (mathématiciens des temps) antérieurs, en fait du *maftoûh* et du *gobâr*."[27] Here the two systems, *ḥisāb maftūḥ* (mental reckoning) and *ḥisāb al-ghubār* (Hindu reckoning, with written numerals), are clearly set apart. In another paper Woepcke gave the translation of a treatise *Introduction au calcul gobârî et hawâï* (without mentioning an author or the shelf-mark of the manuscript) where, again, the "calcul *gobârî* " (Hindu reckoning, with written numerals) is opposed to the "calcul *hawâï* " (mental reckoning, i.e., without the use of written symbols).[28]

From these testimonies it can be derived that the written numerals in the Hindu reckoning system were called *al-ḥurūf al-tis'a* (the nine letters, or *litterae*) or, in *Mafātīḥ al-'ulūm*, *al-ṣuwar al-tis'* (the nine figures) or *ashkāl al-ghubār* (dust figures, in Ibn al-Yāsamīn) and *ḥurūf* or *qalam al-ghubār* (dust letters) by other Western Arabic authors. The designation thus refers to the written numerals as such, as opposed to numbers in other reckoning systems that did not use written symbols. I should think that, therefore, it is no longer justified for us to call the Western Arabic forms of the Hindu-Arabic numerals "ghubār numerals." Rather we should speak of the Eastern and the Western Arabic forms of the nine numerals.

The second, most difficult, question in connection with the Arabic West concerns the forms of the written numerals in that area, their origin and their relationship with the "Arabic numerals" that came to be used in Latin Europe.

Here one might ask why the Arabic West developed forms of the numerals different from those in the East. It is hard to imagine a reason for this development, especially when we assume—in conformity with our understanding of the birth and growth of the sciences in the Maghrib and al-Andalus in general—that the Hindu reckoning system came to the West like so many texts and so much knowledge from the Arabic East. About the mathematician and astronomer Maslama—in Spain, died 1007/1008—for example we learn from Ṣā'id al-Andalusī[29] that he studied the *Almagest*, that he wrote an abbreviation of al-Battānī's *Zīj* and that he revised al-Khwārizmī's *Zīj* (this work has survived in a Latin translation by Adelard of Bath and has been edited); he also knew the Arabic version of Ptolemy's *Planisphaerium* and wrote notes and additions to it that survive in Arabic and in several Latin translations.[30] Thus he, or his disciples, will certainly also have known al-Khwārizmī's *Arithmetic* and, together with it, the Eastern Arabic forms of the numerals. Not quite a century later Ṣāʿid al-Andalusī knows of al-Khwārizmī's *Arithmetic* under the title *ḥisāb al-ghubār*, as we have just heard.

Certainly, in this connection one has to consider that also some more elements of basic Arabic erudition took a development in the West different from that in the Arabic East: first, the script as such—we think of the so-called Maghrebi ductus in which, beyond the general difference in style, the letters *fāʾ* and *qāf* have their points added differently; second, the sequence of the letters in the ordinary alphabet; and, third, the sequence of the letters in the *abjad* series where the West deviates from the old Semitic sequence that was retained in the East and assigns to several letters different number values.[31] As far as I can see, linguists have also not brought forward plausible arguments for these differences.

That the Eastern Arabic numerals were also known in al-Andalus is demonstrated by several Latin manuscripts that clearly show the Eastern forms, for example, MSS Dresden C 80 (2nd half 15th century), fols. 156v–157r; Berlin, fol. 307 (end of 12th century), fols. 6, 9, 10, and 28; Oxford, Bodleian Library, Selden sup. 26; Vatican, Palat. lat. 1393; and Munich, Clm 18927, fol. 1r, where the Eastern figures are called *indice figure*, whereas the Western forms are labeled *toletane figure*;[32] the zeros are here called *cifre*.

However that may be, the evidence for the Western Arabic numerals in Latin sources begins in 976; a manuscript—the "Codex Vigilanus"—written in that year and containing Isidor's *Etymologiae* has an inserted addition on the genius of the Indians and their nine numerals, which are also written down in the Arabic way, that is, proceeding from right to left, in Western Arabic forms.[33] The same was repeated in another Isidor manuscript, the "Codex Emilianus," written in 992.[34] Hereafter follow, in Latin, the "apices," the numeral notations on abacus counters, which render similar forms of the numerals.[35] Here, the Western Arabic forms are still drawn in a very rough and clumsy manner. A third impulse came in the twelfth century with the translation of al-Khwārizmī's *Arithmetic*; from now on the forms of the numerals become smoother and more elegant.[36]

Unfortunately, the documentary evidence on the side of Western Arabic numerals is extremely poor. So far, the oldest specimen of Western Arabic numerals that became known to me occurs in an anonymous treatise on automatic water-wheels and similar devices in MS Florence, Or. 152, fols. 82r and 86r (the latter number also appears on fol. 81v). Two other texts in this section of the manuscript are dated to 1265 and 1266, respectively (figures 1.3a–b).[37] Here we have the symbols for 1, 2, 3, 4, 5, 8, and 9. The figures for 2 and 3 look like the corresponding Eastern Arabic forms and are not turned by 90° as in other, more recent, Maghrebi documents. The meaning of these numerals

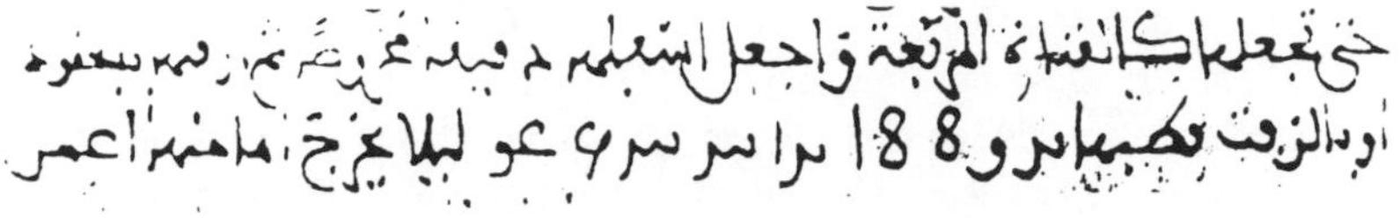

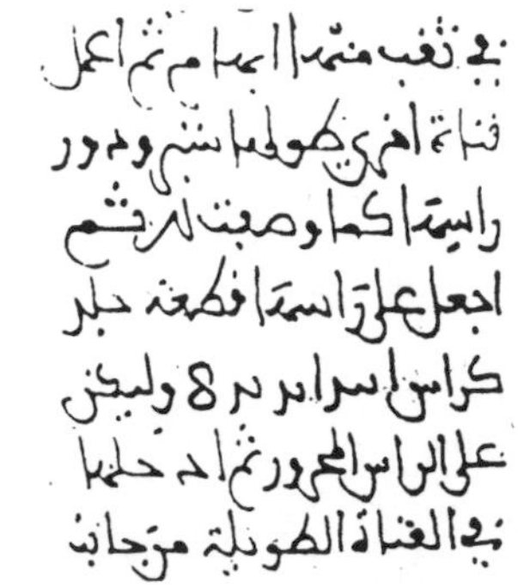

Figure 1.3a
MS Florence, Or. 152, fol. 82r
(dated 1265–1266)

Figure 1.3b
MS Florence, Or. 152, fol. 86r
(dated 1265–1266)

in the present context remains unexplained to me. The numerals in two other Maghrebi manuscripts that fell into my hands (figures 1.4–1.5)[38] resemble the forms found in the specimens reproduced in facsimile by Labarta—Barceló from Arabic documents in Aragon and Valencia from the 15th and 16th centuries.[39]

While specimens of Western Arabic numerals from the early period—the tenth to thirteenth centuries—are still not available, we know at least that Hindu reckoning (called *ḥisāb al-ghubār*) was known in the West from the tenth century onward: Dunas ibn Tamīm, 955/6; al-Dānī, before 1053; Ṣāʿid al-Andalusī, 1068; Ibn al-Yāsamīn, 2nd half of the 12th century. It must be regarded as natural that, together with the reckoning system, also the nine numerals became known in the Arabic West. It therefore seems out of place to adopt other theories for the origin of the Western Arabic numerals. From among the various deviant theories I here mention only two. One theory, also repeated by Woepcke,[40] maintains that the Arabs in the West received their numerals from the Europeans in Spain, who in turn had received them from Alexandria through the "Neopythagoreans" and Boethius; to Alexandria they had come from India. Since Folkerts's edition of and research on the Pseudo-Boethius[41] we now know that the texts running under his name and carrying Arabic numerals date from the eleventh century. Thus the assumed way of transmission from Alexandria to Spain is impossible and this theory can no longer be taken as serious. Recently, Richard Lemay had brought forward another theory.[42] He proposes that, in the series of the Western Arabic numerals, the 5,

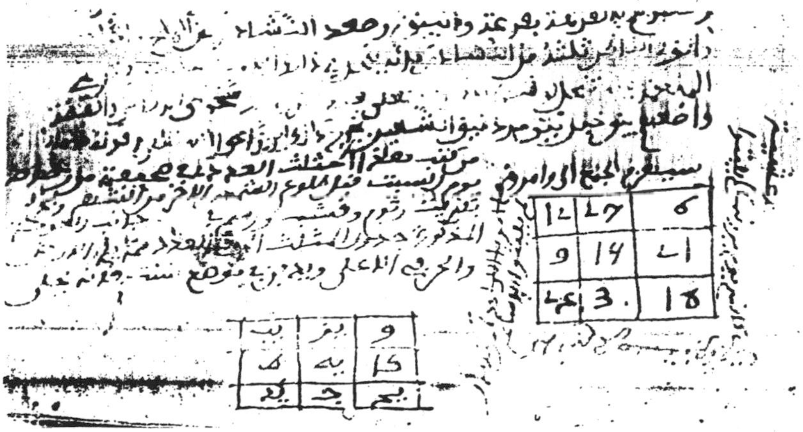

Figure 1.4
Rabat, al-Khizāna al-ʿĀmma, MS 321, p. 45 (after 1284)

6, and 8 are derived from Latin models, 5 as rendering the Visigothic form of the Roman v, 6 as a ligature of vi in the same style, and 8 as the *o* of *octo* with the final *o* placed above. This might appear acceptable for the Arabic numerals used in Latin texts. But since the Western Arabic numerals are of the same shape, that would mean that the Western Arabs broke up their series of nine numerals and replaced their 5, 6, and 8 by forms taken from European sources. This seems highly improbable. The Western Arabs received their numerals from the East as a closed, complete, system of nine signs, and it would only appear natural that they continued to use it in this complete form, not breaking the series up and replacing single elements by foreign letters.

When one compares the Eastern and the Western Arabic forms of the numerals, one finds that they are not completely different. The Western forms of 1, 2, 3, 4, 5, and 9 can be recognized as being related to, or derived from, the corresponding Eastern forms. Major difficulty arises with 6, 7, and 8. It may not be accidental that the oldest existing Latin re-working made from the translation of al-Khwārizmī's *Arithmetic* mentions just these three figures (plus 5) as being differently written.[43] As I have already said earlier, this notice can hardly stem from al-Khwārizmī himself; rather it may have been added by a Spanish-Arabic redactor of al-Khwārizmī's text. He would have been best equipped to recognize this difference. The Latin translator, or Latin adapters, would less probably have been able to notice the difference between the Eastern and Western Arabic forms of these four numerals. We cannot explain why, and how, the three Western figures were formed, especially since we have no

14

Figure 1.5
MS Ait Ayache, p. 192 (after 1344)

written specimens of Western Arabic numerals before the thirteenth century. For further research into the matter, therefore, the discovery of older, or old, documents remains a most urgent desideratum.

Lastly, I want to mention a curious piece of evidence. Somebody in the Arabic West once found out that the Western Arabic forms of the nine numerals resemble certain letters in the Maghrebi script and he organized their description in a poem of three memorial verses (in the metre *kāmil*). The poem is reported by the Spanish-Arabic mathematician al-Qalaṣādī (died 1486) in a commentary on the *Talkhīṣ fī ʿamal al-ḥisāb* of Ibn al-Bannāʾ (died 1321 or 1324) and, afterwards, by Ḥusayn ibn Muḥammad al-Maḥallī al-Shāfiʿī (died 1756, an Eastern Arabic author) in a commentary on an arithmetical work of al-Sakhāwī (died after 1592, also an Eastern author). The two *loci* are cited by Woepcke.[44] The text of the poem is as follows:

alifun wa-yāʾun thumma ḥijjun [*wa-*] *baʿdahu* ⋆
ʿawwun wa-baʿda l-ʿawwi ʿaynun tursamu
hāʾun wa-baʿda l-hāʾi shaklun ẓāhirun ⋆
yabdū ka-l-khuṭṭāfi idhā huwa yursamu
ṣifrāni thāminuhā wa-alifun baynah[*um*]*ā* ⋆
wa-l-wāwu tāsiʿuhā bi-dhālika yukhtamu

That is, 1 is compared to an *alif*, 2 to a final *yāʾ* (but to *ḥāʾ* in al-Maḥallī; both comparisons are possible), 3 to the combination *ḥāʾ-jīm*, 4 to the combination *ʿayn-wāw*, 5 to *ʿayn*, 6 to (an isolated) *hāʾ*, 7 to a *khuṭṭāf* (i.e., an iron hook), 8 to two zeros above each other and linked by a stroke, and 9 to a *wāw*. These memorial verses may be much older than al-Qalaṣādī's time. They seem to have become a topic since they are cited even by an Eastern Arabic author. Perhaps one can conclude from this standardized description that the written forms of the Western Arabic numerals were less variable than the Eastern ones.

To sum up, we can register that the history of the transmission of the Hindu numerals and Hindu reckoning to the Arabs in the East appears to be clear. For the Arabic West it is known that all the cultural and scientific achievements of the East were transferred there. In the stream of this cultural movement the knowledge of Hindu reckoning and the nine numerals must also have passed there. The oldest known testimony for the acquaintance with the Hindu system is documented for 955/6 in Kairouan. So far no written evidence of Western Arabic numerals for the tenth to the thirteenth centuries have been found; documents are only known from the thirteenth century on. But these numerals must have existed earlier since the first evidence in Latin sources—which took up these numerals from the Arabs in Spain—dates from 976. The most important task for further research would therefore be to find older Western Arabic material for the knowledge and use of the Hindu numerals in that region.

Appendix

An inspection of microfilms of the manuscripts of Leonardo of Pisa's *Liber abaci* (AD 1202) shows that a group of older manuscripts has numerals similar in shape to those in the New York MS of al-Khwārizmī's *Arithmetic* as visible in the facsimiles of its recent edition (Folkerts 1997): MSS Florence, BN, Conv. Sopp. C.1.2616 (beg. 14c.? Here the series of the nine symbols, at the beginning of the text, looks different, more "modern"; but in the text itself and in the diagrams and tables etc., they are of the Khwārizmī-MS N-type. This manuscript was used by Boncompagni for his edition, 1857–1862); Siena, Bibl. Publ. Comm., L.IV.20 (2nd half 13c.); Florence, Magliabecchi XI, 21.

On the other hand, more recent, "modern(ized)," forms of the numerals are used in MSS Florence, BN II.III.25 (16c.); Vat. Palat. 1343 (end 13c.?); Milan, I. 72 (15c.?). It thus appears evident that the numerals in the Leonardo manuscripts follow the forms current in the known Latin arithmetical texts. Contrary to what is sometimes assumed, they do not show the intrusion of new Arabic influence resulting from Leonardo's oriental travels and his personal contacts with trade centers in the Arab world.

Postscript

For the Maghribi manuscript Ait Ayache, Ḥamzawīya 80, quoted in this article, it is now established that it was copied shortly after AD 1600; see the detailed description by Ahmad Alkuwaifi and Monica Rius, "Descripción del Ms. 80 de Al-Zāwiya al-Ḥamzawīya," *Al-Qanṭara* 19 (1998), 445–463. Therefore the manuscript can no longer serve as a testimony to early forms of Western Arabic numerals.

Notes

1. See Nau.

2. Sezgin V, 211.

3. See al-Bīrūnī, *India*, ch. 14, *apud* Woepcke 1863, 475f. (note 1), *sub* 13°, 19° and 24° (= repr. II, 407f.).

4. Al-Uqlīdisī: Saidan 1973 and 1978; Kūshyār: Levey-Petruck; al-Baghdādī: Saidan 1985.

5. About fifteen such titles up to the middle of the eleventh century are quoted by Sezgin, V.

6. Al-Ya'qūbī I, 93; cf. Köbert 1975, 111.

7. Al-Mas'ūdī I, p. 85 (§152).

8. Al-Khwārizmī, 193–195.

9. Fischer, 783–793.

10. *Fihrist*, I, 18f.; cf. Köbert 1978.

11. Fischer, 792; Köbert 1975, 111.

12. Pellat, 466b.

13. Grohmann, 453f., no. 12, and Plate LXV, 12.

14. Prof. W. Diem, Cologne, who has studied and edited such papyri for many years, informs me (in a letter dated 6 August 1996) that the understanding of the symbols as a year number is not free from doubt, because an expression like *fī sanat* ("in the year . . ."), which is usually added to such datings, is here missing. Furthermore, he confirmed that a second dating of that type in another papyrus, understood by Karabaček, 13 (no. 8), as the Hindu numerals 275 (888/9), is not formed by Hindu numerals, but rather by (cursive) Greek numeral letters. This document, therefore, must no longer be regarded as the second oldest occurrence of Hindu-Arabic numerals in an Arabic document.

15. See the quotation by Woepcke 1863, 275f. (= repr. II, 358f.).

16. Translated by Woepcke 1863, 496 (= repr. II, 428).

17. Saidan 1985, 33.

18. Cf. on this also Woepcke 1863, 482f. (= repr. II, 414f.).

19. Gandz 1927 and 1931.

20. It is in §2 of the treatise. I owe this information to Richard Lorch. Dr. Lorch is preparing an edition of al-Sijzī's text.

21. Woepcke 1859, 27, note *** (= repr. II, 191).

22. First cited by Joseph Reinaud in an Addition to his "Mémoire sur l'Inde," 565, from one of the four Hebrew translations that were made from Dunas's original Arabic text, which itself has survived only in part.

23. Ibn al-Yāsamīn, 232f.; a German translation was given by Köbert 1975, 109–111.

24. Ṣā'id al-Andalusī, 58.

25. Ibn al-Qifṭī, 266, ult.—267,3.

26. Woepcke 1854, 359, *sub* 3° (= repr. I, 456).

27. Woepcke 1859, 67 (= repr. II, 231).

28. Woepcke 1865–66, 365 (= repr. II, 541).

29. Ṣā'id al-Andalusī, 169.

30. Edited by Kunitzsch-Lorch.

31. Cf. *Grundriss*, 176ff., 181f., 182f.

32. For Selden and Pal. lat., cf. the table in Allard, 252; for Clm 18927, cf. Lemay 1977, figure 1a.

33. See the reproduction in van der Waerden-Folkerts, 54.

34. Reproduced also in van der Waerden-Folkerts, 55.

35. For reproductions, see, *inter alios*, van der Waerden-Folkerts, 58; Tropfke, 67; Folkerts 1970, plates 1–21.

36. See the photographs in Folkerts 1997, plate 1. etc., from the newly found and so far oldest known manuscript of a re-working of the Latin translation of al-Khwārizmī's *Arithmetic*.

37. I owe the knowledge of this manuscript to the kind help of Dr. S. Brentjes, Berlin, which is gratefully acknowledged. A detailed description of the manuscript was given by Sabra 1977.

38. Rabat, al-Khizāna al-'Āmma, MS 321, p. 45. The preceding text, ending on p. 44, is dated in the colophon to 683/1284. P. 45 was left blank by the original writer; a later hand added in the upper part an alchemical prescription and at the bottom a magic square with directions for its use. I am grateful to Prof. R. Degen, Munich, for bringing this page to my attention, and to Prof. B. A. Alaoui, Fes, and M. A. Essaouri, Rabat, for procuring copies of the relevant pages from the manuscript.—Morocco, Ait Ayache, MS Ḥamzawīya 80. On p. 201 of the manuscript, in an excerpt from the *Zīj* of Ibn 'Azzūz al-Qusanṭīnī, there is a calculated example for July–August 1344 (cf. Kunitzsch 1994, p. 161; 1997, p. 180).

39. It should be added that in the table of *ghubār* numerals given by Souissi, 468, the numerals in the first two lines (said to date from the 10th century and ca. 950, respectively) are not (Arabic) *ghubār* numerals, but rather Indian numerals (cf. Sánchez Pérez, the table on p. 76, lines 8–9). It should also be noted that the date given by Sánchez Pérez, 121, table 1, for the specimen in line 9 ("Año 1020") is the Hijra year (= AD 1611/12); the author there mentioned, Ibn al-Qāḍī, died in Fes 1025/1616. Similarly, the specimen in line 12, ibid., from MS Escorial 1952, must belong to the 11th century Hijra; the manuscript contains a commentary by Abu 'l-'Abbās ibn Ṣafwān on the summary of Mālik ibn Anas' *al-Muwaṭṭa'* of Abu 'l-Qāsim al-Qurashī.

40. Woepcke 1863, 239 (= repr. II, 322).

41. Folkerts 1968, 1970.

42. Lemay 1977 and 1982.

43. Folkerts 1997, 28: MS N, lines 34–38.

44. Woepcke 1863, 60f. and 64f. (= repr. II, 297f. and 301f.).

BIBLIOGRAPHY

Allard, A. 1992. *Al-Khwārizmī, Le calcul indien*. Paris and Namur.

Fischer, A. 1903. "Zur Berichtigung einer Etymologie K. Vollers'." *Zeitschrift der Deutschen Morgenländischen Gesellschaft* 57, 783–793.

Folkerts, M. 1968. "Das Problem der pseudo-boethischen Geometrie." *Sudhoffs Archiv* 52, 152–161.

Folkerts, M. 1970. *"Boethius" Geometrie. Ein mathematisches Lehrbuch des Mittelalters*. Wiesbaden.

Folkerts, M. 1997. *Die älteste lateinische Schrift über das indische Rechnen nach al-Ḫwārizmī*. Munich.

Gandz, S. 1927. "Did the Arabs know the abacus?" *American Mathematical Monthly* 34, 308–316.

Gandz, S. 1931. "The Origin of the Ghubār Numerals or The Arabian Abacus and the Articuli." *Isis* 16, 393–424.

Grohmann, A. 1935. "Texte zur Wirtschaftsgeschichte Ägyptens in arabischer Zeit." *Archiv Orientální* 7, 437–472.

Grundriss. 1982. *Grundriss der arabischen Philologie*, ed. W. Fischer, I: *Sprachwissenschaft*. Wiesbaden.

Ibn al-Nadīm. 1871–1872. *Kitāb al-fihrist*, ed. G. Flügel, I–II. Leipzig.

Ibn al-Qifṭī. 1903. *Ta'rīḫ al-ḥukamā'*, ed. A. Müller and J. Lippert. Leipzig.

Ibn al-Yāsamīn, Abū Fāris. 1973. "Dalīl jadīd 'alā 'urūbat al-arqām al-musta'mala fī l-maghrib al-'arabī." *Al-Lisān al-'Arabī* 10, 231–233.

Karabaček, J. 1897. "Aegyptische Urkunden aus den königlichen Museen zu Berlin." *Wiener Zeitschrift für die Kunde des Morgenlandes* 11, 1–21.

al-Khwārizmī, Muḥammad ibn Aḥmad. 1895. *Liber Mafâtîh al-olûm*, ed. G. van Vloten. Leiden.

Köbert, R. 1975. "Zum Prinzip der *ġurāb*-Zahlen [sic pro *ġubār*] und damit unseres Zahlensystems." *Orientalia* 44, 108–112.

Köbert, R. 1978. "Ein Kuriosum in Ibn an-Nadīm's berühmtem *Fihrist*." *Orientalia* 47, 112f.

Kunitzsch, P. 1994/1997. "'Abd al-Malik ibn Ḥabīb's *Book on the Stars*." *Zeitschrift für Geschichte der Arabisch-Islamischen Wissenschaften* 9, 161–194; 11, 179–188.

Kunitzsch, P., and Lorch, R. 1994. *Maslama's Notes on Ptolemy's* Planisphaerium *and Related Texts*. Munich.

Labarta, A, and Barceló, C. 1988. *Números y cifras en los documentos arábigohispanos*. Córdoba.

Lemay, R. 1977. "The Hispanic Origin of Our Present Numeral Forms," *Viator* 8, 435–462 (figures 1a–11).

Lemay, R. 1982. "Arabic Numerals." *Dictionary of the Middle Ages*, ed. J. R. Strayer, vol. 1. New York, 382–398.

Levey, M., and Petruck, M. 1965. *Kūshyār ibn Labbān, Principles of Hindu Reckoning.* Madison and Milwaukee.

al-Masʿūdī. 1965ff. *Murūj al-dhahab*, ed. C. Pellat, Iff. Beirut.

Nau, F. 1910. "Notes d'astronomie syrienne." III: "La plus ancienne mentionne orientale des chiffres indiens." *Journal asiatique*, sér. 10, 16, 225–227.

Pellat, C. 1979. "Ḥisāb al-ʿAqd." *Encyclopaedia of Islam*, new ed., III. Leiden, 466–468.

Reinaud, J. 1855. "Addition au Mémoire sur l'Inde." *Mémoires de l'Institut Impérial de France, Académie des Inscriptions et Belles-Lettres* 18, 565f.

Sabra, A. I. 1977. "A Note on Codex Biblioteca Medicea-Laurenziana Or. 152." *Journal for the History of Arabic Science* 1, 276–283.

Sabra, A. I. 1979. "ʿIlm al-Ḥisāb." *Encyclopaedia of Islam*, new ed., III. Leiden, 1138–1141.

Ṣāʿid al-Andalusī. 1985. *Ṭabaqāt al-umam*, ed. Ḥ. Bū-ʿAlwān. Beirut.

Saidan, A. S. 1973. *al-Uqlīdisī, al-Fuṣūl fī l-ḥisāb al-hindī.* Amman.

Saidan, A. S. 1978. *The Arithmetic of al-Uqlīdisī.* Dordrecht and Boston.

Saidan, A. S. 1985. *ʿAbd al-Qāhir ibn Ṭāhir al-Baghdādī, al-Takmila fī l-ḥisāb.* Kuwait.

Sánchez Pérez, J. A. 1949. *La aritmetica en Roma, en India y en Arabia.* Madrid and Granada.

Sezgin, F. 1974. *Geschichte des arabischen Schrifttums*, V: *Mathematik, bis ca. 430 H.* Leiden.

Souissi, M. 1979. "Ḥisāb al-Ghubār." *Encyclopaedia of Islam*, new ed., III. Leiden, 468f.

Tropfke, J. 1980. *Geschichte der Elementarmathematik*, vol. 1, 4th ed. Berlin and New York.

van der Waerden, B. L., and Folkerts, M. 1976. *Written Numbers.* The Open University Press, Walton Hall, Milton Keynes, GB.

Woepcke, F. 1854. "Recherches sur l'histoire des sciences mathématiques chez les orientaux. . . ." *Journal asiatique*, 5e série, 4, 348–384 (= repr. I, 445–481).

Woepcke, F. 1859, *Sur l'introduction de l'arithmétique indienne en Occident. . . .* Rome (= repr. II, 166–236).

Woepcke, F. 1863. "Mémoire sur la propagation des chiffres indiens." *Journal asiatique*, 6e série, 1, 27–79; 234–290; 442–529 (= repr. II, 264–461).

Woepcke, F. 1865–1866. "Introduction au calcul gobârî et hawâï." *Atti dell'Accademia de'Nuovi Lincei* 19, 365–383 (= repr. II, 541–559).

Woepcke, F. 1986 (repr.). *Études sur les mathématiques arabo-islamiques, Nachdruck von Schriften aus den Jahren 1842–1874*, ed. F. Sezgin, I–II, Frankfurt am Main.

al-Ya'qūbī. 1883. *Historiae*, ed. M. Th. Houtsma, I–II. Leiden.

Index of Names

Arabic names are transliterated here according to the system current in English, even for the articles in French and German; the article *al-*, or abbreviated *'l-*, is ignored; *ibn*, in genealogical application, is abbreviated as "b.". The Index contains names of historical interest; names of modern authors, editors etc. are not registered.

Index of Manuscripts

Paris, Bibliothèque nationale de France
ar. 2457: XXIX 5f.
ar. 2482: V 31
ar. 2483: V 31
ar. 2485: III 204f., 207, 209
ar. 2493: XIII 154
ar. 2500: XXVII 205f., 208, 211
ar. 4821: VI 84; VII 151f.
ar. 5098: XIII 153
hebr. 1100: V 31; VII 148f.
hebr. 1101: XXVIII 6
lat. 7374A: XXVII 207
lat. 7412: VIII 97–99; XV 197; XVI 114–120; XVII 395f.; XVIII 656f., 664–667, 669; XIX 58, 62, 68f.; XX 244f., 247–250; XXI 181f., 185
lat. 8663: XV 199
lat. 9335: I 76; XXVIII 6
lat. 11245: XXVII 207
lat. 16200: XXVIII 5
lat. 16201: XXVII 206–208
lat. 16646: XXVIII 10, 12, 18
lat. 17868: XV 197; XX 248

Prague, University Library
Ms. III.H.19: XXVIII 12

Rabat, al-Khizāna al-ʿĀmma
Ms. 321: XXIX 13, 18
Ms. 1101: XXVII 206, 208, 211

Rampur, Raza Library
ʿArshī 200: XXVII 206

Salzburg, St. Peter
a.V.2: XXII 200

L. J. Schoenberg, Florida
LJS 268: V 32–37

Siena, Biblioteca Publica Comunale
L.IV.20: XXIX 16

St. Gallen, Stadtbibliothek
Vad. 412: XXV 20

Tehran, Fakhr al-Dīn Naṣīrī
Ms. 789: V 31f.; VII 148

Printed and bound by CPI Group (UK) Ltd, Croydon, CR0 4YY
21/10/2024
01777084-0011